T0275913

FROM KNOWLEDGE TO POWER

FROM KNOWLEDGE TO POWER

THE RISE OF THE SCIENCE EMPIRE IN FRANCE,

1860–1939

HARRY W. PAUL

Professor of History
University of Florida

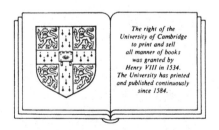

The right of the
University of Cambridge
to print and sell
all manner of books
was granted by
Henry VIII in 1534.
The University has printed
and published continuously
since 1584.

CAMBRIDGE UNIVERSITY PRESS

CAMBRIDGE

LONDON   NEW YORK   NEW ROCHELLE
MELBOURNE   SYDNEY

PUBLISHED BY THE PRESS SYNDICATE OF THE UNIVERSITY OF CAMBRIDGE
The Pitt Building, Trumpington Street, Cambridge, United Kingdom

CAMBRIDGE UNIVERSITY PRESS
The Edinburgh Building, Cambridge CB2 2RU, UK
40 West 20th Street, New York NY 10011–4211, USA
477 Williamstown Road, Port Melbourne, VIC 3207, Australia
Ruiz de Alarcón 13, 28014 Madrid, Spain
Dock House, The Waterfront, Cape Town 8001, South Africa

http://www.cambridge.org

First published 1985
First paperback edition 2002

A catalogue record for this book is available from the British Library

Library of Congress Cataloguing in Publication data
Paul, Harry W.
From knowledge to power.
Includes index.
1. Science – France – History.   I. Title.
Q127.F8P36   1985   509′.44   84-27508

ISBN 0 521 26504 5 hardback
ISBN 0 521 52524 1 paperback

Plato thought nature but a spume that plays
Upon a ghostly paradigm of things;
Solider Aristotle played the taws
Upon the bottom of a king of kings;
World-famous golden-thighed Pythagoras
Fingered upon a fiddle stick or strings
What a star sang and careless Muses heard:
Old clothes upon old sticks to scare a bird.
– W. B. Yeats, "Among School Children"

Prophéties! Prophéties! l'aigle encapuchonné
du Siècle s'aiguise à l'émeri des Caps.
– Saint-John Perse, *Amers*

CONTENTS

CONTENTS

ACKNOWLEDGMENTS

I want to thank the National Science Foundation (Program in the History and Philosophy of Science) for supporting my research during several hot summers and a damp fall and winter in Paris, Lille, Lyon, Angers, and in several places of enlightenment recommended by Gault and Millau. The American Philosophical Society also favored me with a grant. The University of Florida, especially the Division of Sponsored Research, has looked kindly on my answering the mating call of the research imperative.

I am grateful to Adrienne Turner for her great skill with floppy disks. Steve Boyett generously typed several drafts of the manuscript while not obsessed with science fiction and his charming *Ariel*. Louisa Dunlop prepared the index.

The following publications and publishers have granted permission to quote from copyrighted works.

Cambridge University Press, Cambridge, England: Harry W. Paul, "Apollo Courts the Vulcans: The Applied Science Institutes in Nineteenth-Century French Science Faculties," in Robert Fox and George Weisz, editors, *The Organization of Science and Technology in France, 1808–1914* (1980). (Tables 1–4 in Chapter 4 are from this source.)

Contemporary French Civilization: H. W. Paul and T. W. Shinn, "The Structure and State of Science in France," volume VI (1981–2).

Societas – A Review of Social History: volume 1, no. 4 (Autumn 1971).

Elsevier Scientific Publishing Company, Amsterdam: H. W. Paul, "The Role and Reception of the Monograph in Nineteenth-Century French Science," in A. J. Meadows, editor, *Development of Science Publishing in Europe* (1980).

Editions Gallimard, Paris: Saint-John Perse, *Oeuvres complètes* (© Editions Gallimard 1957).

INTRODUCTION
FROM DREAM TO DREAM IN LAYING
THE FOUNDATIONS

L'histoire est un "roman vrai."

– Paul Veyne

I N his *Histoire de la civilisation en Europe* (Sorbonne lectures, 1828–30), François Guizot showed without difficulty that French civilization was the truest representative of European culture. That this judgment was a truism for cultured men of the day was evident in John Stuart Mill's ready assent that only a clod could not know that "the history of civilization in France *is* that of civilization in Europe." What was the nature of this civilization that enthralled the European mind for at least two centuries? The superb German survey entitled *The Civilization of France* (1930) by Ernst Curtius makes it mainly a matter of literary culture: "Literature plays a far larger part in the cultural and national consciousness of France than it does in that of any other nation." Victor Hugo summed it up: "Literature is civilization." Curtius argued that "in France literature fulfills the function which among us [Germans] is divided between philosophy, science, poetry and music." So even Cartesianism, "the most important factor in the intellectual history of France," was distilled into the literary cliché of *clarté française*. France's intellectual hegemony in Europe was facilitated by the cultural universality of its language, characterized by "logic, lucidity, brilliance, naturalness, and taste." This leaves little room for the cult of Isis, widely touted elsewhere as the goddess to whom we must pay tribute for our modernity. Curtius makes only a few remarks concerning the ideological value of science in France as a source of liberation from "the authority of dogma, society and custom."[1]

Like most fascinating generalizations, that of Curtius on the role of science in French civilization has just enough truth in it to be quite misleading. In spite of the importance of the medieval Sorbonne as a center of scholastic criticism providing a point of origin for the scientific theories of the sixteenth and seventeenth centuries, by the time of the Renaissance the university was so moribund that the monarchy permitted humanists to lecture outside the university – this was the "founding" of the Collège de France. This was also the beginning of the classic French policy of creating new institutions rather than fighting the battle of reforming old ones. François Furet assumes that a law is in operation: "The history of French higher education obeys a law of

1

peripheral development."[2] It is true that the institutionalization of the Académie française (1635) established a literary connection with the state, but the sciences were not far behind with the founding of the Académie des sciences (1666). In the seventeenth century, Paris was an international scientific center, although many of the savants were foreign. Because the government was interested in good administration and concerned with defense, the eighteenth century was an especially creative period in the founding of higher scientific-technical-engineering schools: the Ecole des ponts et chaussées (1715), the Ecole royale du génie (1748), the Ecole des mines (1783), and the Ecole polytechnique (1794). From about 1750 to 1840 France enjoyed scientific preeminence in Europe. Later in the nineteenth century, the rise of German and British scientific research had the comparative effect of reducing France to *one* of the centers of science in the Western world.

France remained a leader in scientific teaching and research. Few countries had such a galaxy of comparable institutions: the Sorbonne, the Ecole normale supérieure, the Collège de France, the Muséum d'histoire naturelle, the Polytechnique, the Observatoire, and a whole gamut of technical institutions. The significant institutional creations of the late nineteenth century were the Ecole pratique des hautes études (1868), the Institut Pasteur (1888), the Ecole supérieure de physique et chimie industrielles (1883), the Ecole supérieure d'électricité (1894), the laboratories of marine biology, and a few good provincial faculties of science. By the time of the Great War, state-supported or subsidized institutions were doing a respectable job in scientific education and research. Research related to industry became increasingly important, especially in provincial faculties of science. Even after the devastation wrought by the two world wars, France was able to develop a new structure of scientific education and research; that it could do so was in some measure a result of the scientific and technological potential built up over the preceding century.

The French budget plan for 1983 included 52.2 billion francs for research and industry, an increase of 35.8 percent over 1982.[3] There was a clear priority given to scientific research, technological development, industrial restructuring, and the development of national energy sources. The establishment of a Ministry of Research and Industry by the Socialists shows their commitment to the belief that basic research is closely linked to industrial prowess. The Socialist decision to build the much-debated Museum of Science and Technology at La Villette, a temple for the new religion of science, shows both the government's desire to make the public aware of the transforming power of science and technology and Socialism's devotion to its historical heritage.[4] While recently praising Jack Lang's notorious verbal fandango in Mexico, François Gros harkened back to the seminal period of French scientism in the nineteenth century, a period when men knew the liberating virtues of science. Socialist action is the most recent of a series of intensive efforts and financial investments by government and industry since the Second Empire. Of course the big science that came into being after World War II takes us in many ways into a different world. By the 1930s big science may not have begun, but the epitaph of little science had already been written.

While pushing the growth of science many scientists innocently came to know sin, but it does not seem that the pursuit of an elusive power led to a weakening of the human lust for knowledge, science's primal pulse. Knowledge and power are really inseparable. The history of modern science makes it clear that knowledge can be achieved only through the medium of some type of power, by the close alliance of scientists with the state while forging or maintaining their own cognitive and disciplinary identities. In his *Science and Polity in France at the End of the Old Regime*, Charles Gillispie has intrigued us with the intersections between government and science.[5] In the nineteenth and twentieth centuries these intersections progressed to the point of interface or even intussusception. My account of this development in France certainly recognizes the importance of "the integration of science into history through the medium of events and institutions," but it does not claim to reject "configurations of ideas or culture." Perhaps this is my heretical homage to *The Edge of Objectivity*, still our best *Essay in the History of Scientific Ideas*.[6]

A major theme of this work is the astounding growth of university science in France from the Second Empire (1860s) to the period of incipient disintegration of the Third Republic (1930s). Outside the auto-titillating world of literary criticism, context is still viewed as of some importance in understanding the discourse. French scientific growth makes sense only within an elucidation of the distinctive nature of French economic, political, and social life. The growth of science was stimulated by a number of essential tensions: Parisian versus provincial priorities in scientific development, the eternal university problems of a division of resources between teaching and research, the hoary distinction between pure and applied science, science as objective icon versus science as ideological fetish, and prolific if sometimes fruitless quarrels between scientists over the direction of science, its methods, and the partition of the spoils of its victories. "Le sens de l'histoire appelle tout d'abord son contresens."[7]

It is now commonplace to argue that "conceptions of science are shown to be sustained by particular social and political contexts."[8] This seems a reasonable generalization about the early years of the Third Republic, when,

3

as George Weisz has shown with Benedictine fidelity to the evidence, politicians and power brokers took educational reform seriously, with a keen interest in science, assuming that reform "was indispensable socially and politically."[9] Unfortunately, this enlightening dogma must immediately be restricted to modest scope by the frank admission that little is known about the class basis of the Third Republic; class is not a very useful category of analysis in this case. Still, there are a certain number of tags and ideas that have become part of linguistic usage in the historical world, and little harm is done in using them with the realization that their explanatory power is limited, much like using an equation to "explain" gravity. In the analysis of local situations the precision of the data and the micro-nature of the problem sometimes justify a relaxation of the moral and critical scruples that are essential in the high-level generalizations necessary to achieve Mandarin status in theology and the social sciences.

A naïve epistemologist might be forgiven for believing that there is more certainty in economics than in social theory. It does seem that there is at least less confusion in economic retrodiction than in economic prediction, but even in retrodicting the historian is faced with an embarrassingly rich choice of historical models of growth. Among the better known of these clever concoctions are the Rostow model of "the airborne economy," the Gerschenkron model of "the deprived economy," and the Parsonian system with a modified version available in Aron's model, both of which are entrepreneurial or managerial hypotheses. "The battered and tired French economy of the nineteenth century" is a "test-bed for many of the historical models of economic growth." Although Clive Trebilcock leans more to a pessimistic view of French growth, accepting the steady growth theory of recent economic historians but limiting it with the slower pace of their honorable ancestors, he reluctantly concludes that "Spengler was wise to set aside the so-called objective factors limiting French growth (the limited demographic and capital resources of France), to replace the coal with soul explanations: French value patterns, French tastes, the heterogeneous nature of France's cultural heritage." Economics clearly comes under the empire of Derrida's deconstructionist axioms: "Everything can be given at least two equally cogent explanations; in the temporal process of thinking about anything, one explanation collapses into its contrary."[10] It will be obvious that this work leans more toward accepting the optimistic view of French development than the pessimistic view; the rosy view generally fits my account of scientific and technical development better than the dark view. The glaring deficiencies of French economic development are allowed their chiaroscuro function when justified. It may be, as Trebilcock notes with satisfaction, that "the revisionist tactics for the economic history of France

4

appear to be passing out of style, but "French entrepreneurial history remains a shooting war with learned articles serving as projectiles."[11] The historian has the luxury of waiting for the victory of one gang or another.

In spite of the argument that the Restoration was a period of great growth in industrial production, the clear dualism afflicting French technology before 1850, with the archaic far outweighing the advanced technologies, makes it dubious to posit a French industrial revolution for the period 1830–50. The general Continental practice of founding technical schools to overcome economic backwardness continued with the founding of the Ecole centrale des arts et manufactures in 1829 in order to promote "the creation of a new breed of men, intellectually armed with a knowledge of *la science industrielle* and morally fortified by the precepts and discipline of three years at their Parisian school."[12] From the 1830s on even the university faculties began to show signs of life. And it would not be long before they would begin to flatter the technical schools by trying to imitate them in forging as best they could some industrial connections. Technical schools may promote industry, but industry frequently looked like the source of scientific life and growth for the faculties. Later in the century some of the faculties gave birth to versions of the great Parisian technical schools, a development inconceivable without the existence of the very industry that these faculties so assiduously promoted.

Abolished in 1793, reestablished in 1808, the collection of independent faculties – letters, sciences, medicine, law – designated for linguistic convenience as French universities, showed little sign of dynamism before the Third Republic.[13] The ten original faculties of science, to which five more were added in 1854, were small groups of professors, frequently from lycées, chiefly concerned with reproducing themselves in sufficient numbers to fill a small number of secondary school and faculty jobs. Paris was an obvious exception. Some of the greatest contemporary scientists were associated with the founding of the Sorbonne faculty of sciences. Research was not given a key role in the university system, but it did exist and grew slowly, with an important spurt in the 1830s, until the reform movement of the 1860s institutionalized and legitimized research with the founding of the Ecole pratique des hautes études in 1868. Devoted to the research ethos, the Ecole pratique was a part of Duruy's excessively ambitious reforms aiming at the promotion of a revolution in higher education, then beginning to be the victim of unfavorable comparisons with foreign exemplars. True, the Ecole was just an administrative mechanism for the distribution of funds, but what could be more important, especially as a beginning? Good results were soon evident in all areas of science, with a new burst of creativity in physics and the maturing of a research cycle in marine biology. Along with the rise of the

University of Paris, one must note the corresponding decline of the Muséum and, to a lesser extent, the Polytechnique and the Collège de France as centers of research. After 1937, with the arrival of Leprince-Ringuet, the Polytechnique became important again in basic research in physics. The basis for the rapid expansion of the university in the last quarter of the nineteenth century was firmly established, but the faculties needed subsidized students, money for construction and equipment, and professors interested in research as well as in teaching and service. By 1900 the provision of these ingredients by the Third Republic had produced one of the world's great university systems, a system that was particularly strong in the sciences.

In his study of the political role of French universities, George Weisz has shown the failure of higher education to live up to the expectations of politicians who wanted universities to "play a major role in fostering social integration." The faculties of science were probably the most feeble contributors to the social mission of the universities, although, as Weisz admits, even "science could not remain totally isolated from the ideological battles of the Third Republic."[14] The Third Republic did not lack scientists willing to proclaim an alliance between democratic republicanism and science or to concoct the basis of an ideological marriage. The republic was not without powerful weapons in its quest for ideological support in the faculties: appointments to new chairs, the influence of the Conseil supérieur de l'instruction publique, and even the system of university inspection by Parisian professors. But it proved notoriously difficult, if not finally impossible, to squeeze much ideological sap out of science itself. Catholics and Marxists have found science equally refractory to their ideological wooing. Not only did science qua science show itself to be generally immune to the infection of political ideology, but the new science of biology was nearly as politically sterile as physics and chemistry, with all failing miserably to supply much ammunition for the republican *Kulturkampf*. No genuine fusion of ideology and science took place, even with the ostensibly promising opportunity of a meeting of biology and positivism.

Yet for a while about mid-nineteenth century it looked as if positivism and biology would form a fruitful alliance and French politics would benefit from its secular offspring. Certainly the scientifically sanitized positivism of Littré and his acolytes played a role in the Société de biologie, but even here the influence of Charles Robin must be weighed against the preponderance of Claude Bernard. That both positivism and French biology opposed evolution, especially the Darwinian variant, had little consequence in the forging of a common ideological outlook. The fact is that the ideas of Comte and even of the few medical scientists who embraced one type of positivism or another were unacceptable to most biologists and of little importance in the definition of research programs. As physiology emerged as the supreme biological science, at least in its own powerful opinion, the divergence between Bernard's nonphilosophy and positivism became the stumbling stone for the ideologues whining for a scientific *Weltanschauung* or a scientifically inspired *Kulturkampf.* Littré's positivism was popular with the first generation of republican politicians, who thought it to be the ideological gyro for university reform and national unity with social integration. Alas, by the 1870s it was plain that scientism of all sorts was a weak reed incapable of elicting much intellectual support and certainly not able to sustain a national consensus.[15]

Considerable attention has been focused on the fortunes of French physics and on the shortcomings of organic chemistry in the nineteenth century. Little attention has been directed toward biology, except to lament the failure of French biologists to ingest Darwinism, a failure probably without much significance for research because of the generally contested status of Darwinism as a paradigm and as a research program. About 1850 the Muséum and Collège de France, whose reputation in biology was much admired in England, were much stronger scientific institutions than the faculty of sciences in Paris. But by the turn of the century the faculty was much better in biology than the Muséum and, considering the wide spectrum of biological research, more active than the Collège de France. The Collège rarely deviated from the respectability of physiological research, still an important biological area but much less so than in the heyday of Claude Bernard. The state-stimulated growth of secondary education, with its necessary fillip to faculties, and the key research role played by doctoral students go a long way toward explaining the new research dominance of the faculty of sciences in Paris.

The development of marine biology was a general Euro-American phenomenon in which France played a major research role. Here was a classic case of the interaction of national, local, university, and scientific interests, as was also the case in the growth of university institutes of applied research. And whatever we may think about the old-fashioned great-man theory of history, whose version in the history of science is now under attack by the epistemological disease of ethnomethodology, great individuals, defined as such by others as well as subtly by themselves, were conspicuous in creating this area of research.

In spite of striking success in marine biology, the general state of French biology early in the twentieth century was quite unsatisfactory for many of the biologists, especially when they compared the conceptual advances in Germany and England with the seeming stagnation in France. Even if one makes due allowance for the constitutional inclination of the French to find that foreigners are doing it better, a century's impressive growth in biological research and teaching ended in intellectual confusion and pedagogical smugness and stodginess. The success of the research program in marine biology did not seem transferrable to other areas of biological research. The question of what direction French biology should take led to a division of leading researchers into two enemy camps with no immediate benefit to research or teaching. Beginning as a bitter academic monk's quarrel, a squabble between two of France's leading biologists soon enveloped the whole field of research, with serious questions being raised on the orientation of the major research areas and on France's ability to keep up with the work being done in other countries. One group even argued that the whole direction of research, largely derived from marine biology, was hopelessly old-fashioned. By the 1920s it was inevitable that the structure of higher education be blamed for a striking scientific failure that had been snatched from the jaws of success.

One of the most distinguishing features of the provincial faculties of science was the establishment in the late nineteenth century of institutes and programs of applied science; indeed, without these ventures in technology many of the faculties would have remained local stagnant fens with no tributaries into the mainstream of science. Victor Duruy was full of Baconian optimism in hoping that the universal exhibition of 1867 would show businessmen that the "wealth of industry flows, like a river from its source, out of the chemist's laboratory and the physicist's and naturalist's study."[16] Yet the theorem worked well enough to convince governments and a few industrialists to push technology into a position of dominance in higher scientific education. After the stimulus of defeat in 1870, French private interests, in collaboration with the Ministry of Commerce, promoted the establishment of commercial and industrial schools in many of the major cities. It is not unreasonable to make a positive correlation between the growth of the French economy in at least metallurgy, chemicals, and electricity, especially in Paris, Lyon, Nancy, and Toulouse, and the rise of the university institutes and programs of applied science, which were churning out about 500 industrial degrees annually by 1913.[17] With the 550 higher-level graduates of the *grandes écoles* and the 600 lower-level products of the *écoles d'arts et métiers*, France was among the few countries in the world endowed with substantial human resources over a broad spectrum of the engineering and technical needs of a great power.[18]

It is curious that two of the areas in which French science has had considerable success, biology and applied science in the universities, were experiencing difficulty early in the century. Nothing fails like success. Applied science became excessively vocational. Scientific research and

8

industry remained in a state of hostile virginity instead of entering into a fruitful intercourse. But these deficiencies should not blind us to the "significant methodological and ideological interactions" of nineteenth-century science and technology.[19] This is part of what Kuhn calls the second scientific revolution of the first half of the nineteenth century, although the interactions seem more evident in the second half of the century. The French case also raises considerable doubt as to the universal necessity of maintaining the separate institutional arrangement of "the universities for *Wissenschaft* and the Technische Hochschulen for industry and the crafts" in order to have first-rate science and technology. Across the channel were the English sisters of the French provincial universities. The civic universities of Birmingham, Liverpool, and Manchester, which owed much to their "intimate connections and interrelations with industry," were "leading centres of scientific advance" in almost all industrial areas.[20] A mutually beneficial parasitism seems to have prevailed between science and technology since the eighteenth century.

Agricultural research and related teaching were paltry affairs in the universities when compared with the wealth of industrially related programs. No diploma mills for agricultural studies existed in the faculties. To some extent, this was the result of the adequacy of education provided by the Ministry of Agriculture, vigilant in jealously guarding its prerogatives from imperialistic ventures of the Ministry of Education. But in ferreting out the connections between the faculties of science and agriculture one must look beyond the few formal institutes and programs producing the lowly *brevets* and certificates, for there did exist close relations between science and agriculture at theoretical and practical levels. The late nineteenth-century spin-offs from pursuing agricultural work in small faculties could be important, especially in gaining local support. Isidore Pierre at Caen could wax ecstatic: "L'étude des applications des sciences à l'agriculture a pris, dans les universités françaises, un essor dont on ne saurait trop se féliciter."[21] Many professors recognized this and preached incessantly about the services science could render to agriculture. No one was a more insistent and powerful advocate of the virtues of commercialization and of research than Louis Grandeau, first in Nancy and then in Paris. Scientists in previously peripheral subjects like botany found in the late nineteenth century that their research was of economic value or could be made to appear so with a little effort. So plant physiology and biology attracted support for laboratories – essential status symbols as well as vital instruments for the advance of science – and various kinds of agricultural institutes and stations. The regional importance of agricultural science is clear in cases like that of Bordeaux, where industries based on wine and pine drew heavily upon chemistry to found the Station oenologique and the Institut du pin, both offspring of the University. The structure of a national system of agricultural research was clearly visible in France by 1914.

In the short run the high-quality research network was more important for the scientific elite than for the peasantry. The peasant's republic did little for his agricultural education. The input of science into agricultural production was limited. Certainly French agriculture was a less competitive industry on the world scene than it should have been, but this probably had little to do with science. By the end of the century science had become firmly attached to the end of the agricultural chain: Consumption gave rise to thorny questions concerning quality and safety. Answers were mainly the job of the lower ranks of the scientific army, while the higher ranks often tried to keep a certain distance from the dangerous quarrels arising out of public health. The interest of a serious scientist in consumer issues is often, as a critic said of one of Barthes's intellectual flirtations, like that of a man interested in women but who can appreciate them only by an X-ray machine. The professionalization of the public health movement was dependent on the development of scientific analytical techniques and the production of personnel by the state's faculties of science.

In higher education the French state enjoyed a monopoly until 1875, when Catholics were given the right to establish faculties, and retained easy dominance after the establishment of limited freedom for private interests. In many ways the five Catholic universities came to occupy a role in France's educational system out of proportion to their educational significance, although that was considerable. A prominent role was given to the sciences in the Catholic universities, with serious attention paid to the wishes of local industrial and agricultural elites, who had to supply most of the funds for the new institutions. But Catholics, who also had access to the excellent state system, were unwilling to build the monuments of glory envisaged by the hierarchy. The new universities were also hindered by political and legal shackles as a result of coming into maturity during the rise of radical anticlericalism. Only for a while did it look as if Catholicism might be as stimulating for French capitalism as Confucianism has been for Japanese industrial success.[22] While Catholic and anticlerical were caught in a sort of Barthesian clash of "organized networks of obsessions," the nation suffered, although there was inevitably some national benefit derived from the competition, however unequal, between Catholic and state institutions. Some dormant state faculties finally came alive when confronted by the threat of a Catholic faculty, and others were spurred on to greater efforts than would ever have been thought possible before 1875. The impact of Catholic higher education was far less than Catholic elementary and

10

secondary education, but it is a largely forgotten chapter of French history that deserves a modest rehabilitation. "Et salut, témoins purs de l'âme en ce combat/Pour l'affranchissement de la lourde nature!"[23]

The pattern of professional scientific communication definitively established in the nineteenth century was clearly different from that in the humanities and most social sciences. Scientists came to publish the results of their research "in journals read only by the profession. Books are exclusively textbooks, compendia, popularizations, or philosophical reflections, and writing them is a somewhat suspect, because non-professional, activity."[24] Probably truer of the twentieth than of the nineteenth century, this statement nevertheless identifies a distinguishing characteristic of scientific publication that had its origins in the seventeenth century but became a striking characteristic only in the nineteenth century. Research and publication also became "unquestionably key elements in any successful academic career."[25] Doctoral dissertations, more original and unfortunately longer, became more important in the degree-granting process and in job getting.[26] The increasing loss of scientific importance for books did not mean that there were fewer published: the need for textbooks in the expanding educational system, the growing lay demand for popularizations, and the publication of the classics of science produced a great increase in books and monographs. The French developed a Gallic textbook model widely regarded, especially by the French, as superior to the pedagogical efforts of their neighbors. The publication of *Les Atomes* (1913) by Jean Perrin showed that they were not always wrong in their assumption of French superiority in continuing the Enlightenment tradition of *haute vulgarisation*. William Ramsey was conventional enough in praising Perrin's style as a product of French sensitivity that was unattainable by people of the North. J. D. Bernal was struck by another virtue of Perrin's book: The brilliant experimentalist turned publicist showed a combination of the imaginative experimental methodology with refined German mathematics, presented in its full complexity and abstraction without a sacrifice of reality and comprehensibility.[27] Perrin's scientific reputation was firmly established before his work of popularization, but the book gave him a more general fame and enhanced his reputation even within the scientific community, where one specialist often has to understand another's specialty through a popularization. The scientist's creative work is another matter.

Kuhn states that "the first half of the nineteenth century . . . witnessed a vast increase in the scale of the scientific enterprise, major changes in patterns of scientific organization, and a total reconstruction of scientific education." The processes accelerated in the second half of the century, especially with the growth of scientific education and research in the university system.

11

Scientific societies with their own journals proliferated as old disciplines became "professions with their own institutional forms," subsidized by both governments and private interests. New fields, like physics and biology – previously "clusters of research fields" – jelled.[28] The transformation of science did not necessarily mean that all old priests were replaced by new presbyters, as was clear in the survival of natural history. Modern science of high quality came to be increasingly associated with the existence of a university faculty; this was nearly axiomatic in provincial France, where the faculty of sciences was inevitably the only research and teaching institution in the region, and even in Paris the faculty of sciences accelerated the professionalization of science. In the new tower of Babel, specialist spoke only to specialist, the generalist's color was abandoned for mathematical nuance, but the tower stood firm in the advancement of knowledge.

> Pas la couleur, rien que la nuance!
> Oh! la nuance seule fiance
> Le rêve au rêve.[29]

New laboratories and buildings, more professors and *maîtres de conférences* attended by even more laboratory technicians, an increase in the quantity and quality of research, expansion of scientific education at all levels, and publication of monographs and journals: all highly desirable, even a necessity for a powerful program in science, and all expensive, virtually unattainable without substantial and steady funding. In the last century of French scientific growth, there is a possible paradox: For most of the period, there seems to have been low-level funding by a parsimonious state and less by private interests; yet France ended up as one of the world's few significant scientific and technological powers endowed with a powerful military technology. The simple escape from the paradox is to take the short-range view that the French science empire is of recent vintage, less than fifty years old. Certainly there is a great difference in scale between all aspects of the funding of science in the 1930s and in the 1980s. There are also important differences in the funding mechanisms of the two periods. Perhaps, in the end, the best explanation is that of sold souls. As Freeman Dyson said about the pact between Oppenheimer and Groves, the new position of science is the result of a Faustian pact between scientists and the military state.

Historians are notoriously reluctant to admit the total novelty of any major intellectual or institutional change. So one can dimly detect the origin of the science empire in the late nineteenth century and certainly clearly see it in the first three decades of the twentieth century, with a mini-Faustian pact coming into existence during the First World War, chiefly as a result of the modest mobilization of scientists and as a result of the founding of new

12

organizations for funding certain types of scientific research of direct interest to the state.

In the midst of the supposed funding famine there was one area of French science that enjoyed the privilege of a perpetual financial feast. French science had too much money for prizes, for science done rather than to be done, rewarding achievement rather than encouraging research. We assume that people will do more and better science if they are paid for doing it rather than rewarded for having done it. After the founding of the Caisse des recherches scientifiques in 1901 France had an institution for funding individual research, resembling more a modern system of research grants than the Ecole pratique des hautes études, which funded individuals but not specific research programs. Once a professor was on the fiscal pipeline of the Ecole pratique, generally only retirement or death ended the subsidy. The Ecole had been a good beginning for stimulating research, but with the growth in the number of researchers it soon became inadequate. Unfortunately the legislation establishing the Caisse limited funding to biological research, chiefly medical research on diseases, especially TB and cancer, and problems of public health, water pollution in particular. Afflicted by legal limitations and paltry funds, the Caisse still established precedents in the type of funding and in procedures that fed into the different structure of funding arising in the late 1930s.

In spite of local successes in promoting many areas of scientific research, some major areas like organic chemistry and optics remained weak. It was clear that only heavy state support could strengthen these areas significantly. But that type of support came only to big physics and aviation, and there were serious problems even there. As a result of the First World War there was a great leap forward in the state's funding of science – a certain practical type of science, it is true – with the establishment of the Direction (later Office) des inventions, presided over by J.-L. Breton. This was the largest funding mechanism in the history of French science until the founding of the Centre national de la recherche scientifique (CNRS) in 1939. For the first time there was an active, systematic policy promoting the marriage of science and technology, with cooperation between scientists and engineers; connections were forged between scientific and technological research and governmental departments, with some friendly and mildly profitable relations established with industry and the military. The Office was killed in 1938 by the Popular Front government with the active support of dissident scientists. But its death was the necessary prologue to the birth of the CNRS. By the 1930s a new generation of scientists, led by Jean Perrin, an old scientist turned young Turk, led a campaign to get "big bucks" for science. A mechanism controlled by scientists themselves would distribute these funds

13

to scientists doing the right kind of science, especially the new physics. The vices of the old scientists when practiced by the new generation were transformed into virtues. Between 1935 and 1939 three new funding and administrative organizations were created in France. The CNRS, replacing the others, was to rule over basic and applied research in France.[30] The big science phase of the CNRS, at least in American terms, would begin only with its new charter in 1959. The idea of a national funding mechanism for science had triumphed in the 1930s; a generation later the dreams of its promoters became the reality of some survivors of the 1930s and of a later group for whom the new structure and generous state support of science were the normal conditions of scientific life.

1

FROM SECOND EMPIRE
TO THIRD REPUBLIC
THE GREAT SCIENTIFIC MUTATION

Le premier vice de notre système c'est que chez nous la recherche
scientifique est subordonnée à la mission d'enseignement.
– Maurice Barrès, *Pour la haute intelligence française* (1925)

The bourgeois context

I N spite of the "rise of the bourgeoisie" – a universal constant in Western
history whose explanatory power is exceeded only by the formula "It
began with the Greeks" – France was a peasant country in 1870. In
1814 the population of France was 29,000,000 rising to 36,000,000 in
1871, more than Great Britain's but nearly 5,000,000 less than Germany's.
Between 1846 and 1859, and again the period 1890–1946, the reproduc-
tive rate fell below 1.00. Although there was a rise in the rate between 1860
and 1889, the French population stood at only 39,500,000 in 1914, less
than Britain's and way below Germany's 67,800,000. No comparison of
economic or educational systems can ignore the direct implications of
different population bases. According to the 1891 census, 46 percent of the
population was living off agriculture, 25 percent off industry, 13.5 percent
off commerce and transportation, and 6.6 percent off the liberal professions.[1] The second half of the nineteenth century saw a decline in the
percentage of the active population in agriculture and domestic service and a
corresponding growth in other sectors, but the structure of the active
population was not radically modified.

Yet enough profound change took place to justify the statement that a
new France emerged between 1870 and 1914. Industrial output tripled,
national income perhaps doubled, and real wages in industry increased 50
percent. French investment abroad increased 600 percent. Economic growth
was unevenly distributed throughout the century: From 1840 to 1860 and
from 1905 to 1914 were more rapid growth periods than 1860 to 1883,
and the period from 1883 to 1896 was one of stagnation, except in
metallurgy, chemicals, and electricity. Material change was slower in France
than in England and Germany. In 1870 France was second in the world in
terms of industrial production and trade, but it had fallen to fourth position
in 1914, in spite of maintaining a favorable comparative per-capita
production and rate of growth. The harsh judgment by Landes cannot be

15

reversed.[2] "The nineteenth century saw a unified Germany rise to Continental hegemony on the strength of the Ruhr and Silesia; while France, slower to industrialize, was never again to enjoy the pre-eminence to which the *levée en masse* and the genius of Napoleon had raised her on the eve of the economic revolution."[3]

Traditional analysis tells us that about 1880 the segment of the bourgeoisie identified as *haute* lost political power to the *moyenne* segment, which dominated parliament and the cabinet after 1879 and kept its hold on the higher civil service posts until the rise of the *petite* bourgeoisie to political power after 1900. By 1914 the *nouvelles couches sociales* were well into a trajectory begun before 1870; a large proportion of the Radical deputies came from the *petite* bourgeoisie. There was a "downward shift in the locus of political power."[4] But the social structure remained relatively stable. The arrival of the *nouvelles couches* was a slow process; even in 1881, 50 percent of parliament came from the upper bourgeoisie.[5] Although between one-quarter and one-third of the active population was in the "industrial sector," and the sector was often pioneering, French industry was not expansionist. Career opportunities were supposedly better in commerce than in industry and engineering.[6] But it was a society in which science and technology assumed a far more important role than ever before.

The change downward in the social basis of political power was accompanied by an unprecedented expansion of the educational system and a steady growth of the liberal professions. Between 1876 and 1912 the number of medical doctors grew from about 11,000 to over 20,000. In 1831 the number of students in all secondary education was 60,432; by 1865 it was 140,253; by 1898 a slower growth had raised the number to 163,452. A 100-percent increase occurred between 1840 and 1880, although even by 1921 education was still only available to less than 5 percent of children. Financial aid was given to less than 8 percent of students. Education, especially at the secondary level, continued to be the preserve of the bourgeoisie, beyond the reach of most of the nation.[7] Higher education, especially in the *grandes écoles*, was mostly an affair for the "more affluent social groups," although 19 percent of the science students at the Ecole normale in the period 1868–79 came from the industrial and agricultural working classes.[8]

Social change and politics were closely connected. Noting the political dimension of the influence of the medical profession, Theodore Zeldin argues that "its rise to power in the state is one of the striking features of the [nineteenth] century." There were 72 doctor deputies in 1898. Teachers made up over 6 percent of the deputies, doctors and pharmacists made up 11 percent, while lawyers continued to dominate the assembly with about one-third of the representatives.[9] From the 1840s on, many teachers were pushed into republicanism by a harsh competitive system and political repression. An overcrowded legal profession also provided "a principal ingredient of the intellectual proletariat of most towns."[10] Universal male suffrage enhanced the political importance of rhetoric, but the lawyers kept the republic conservative through "effecting compromises" in an administrative role. John Weiss has pointed out, however, that under the Third Republic few of the science teachers he studied were directly involved in politics; they had no ideological devotion to the scientistic cause but rather a practical view of politics as useful for personal advancement. The "increasingly 'democratic' profile" of science teachers from 1846 to 1881 was encouraged by the creation of an average of fifteen new posts annually. Bourgeois to the core, they were obsessed with their careers and social status, and presumably they loved the music of Meyerbeer, Halévy, Gounod, and Chopin – "the music of the bourgeoisie" – and could not understand Bizet.[11]

Jean Lhomme has picked 1880 as the date for the end of the period of the monopoly of political power by the *grande bourgeoisie*.[12] If one accepts Lhomme's argument that the anticlericalism, antimilitarism, and socialism infecting the University made it suspect to the *grande bourgeoisie*, then one can argue that the passing of this group from political dominance and the political ascent of lower segments of the bourgeoisie, from which most University personnel originated, meant that the rulers and the University shared a common system of values more than ever before. The shift to secular republicanism in education paralleled the shift in the electorate and the government. Republicans tried to use the University as a creative source of ideology, purging or silencing dissenters with the help of its allies within.

The University

During the reign of Napoleon I, French higher education became formally a function of the state. The University, a term including lycées, collèges (less complete lycées), and faculties, may have lost its independence, but its bondage to the state was handsomely rewarded by a monopoly that permitted private education only under strict University regulation. A real monopoly was jealously kept in higher education until 1875. Each *chef-lieu d'académie* had its faculties of letters and of sciences. If the University labored under a quasi-military discipline, it also suffered in the poverty and atmosphere characteristic of a poor convent. Each faculty got a subsidy from five to ten thousand francs and borrowed from the neighboring lycée and communal collège their rhetoric and mathematics teachers; it had two or three other posts. Latin and mathematics were the foundation of lycée teaching; only with the Revolution had the sciences been admitted to a status comparable to that of classical languages. There were also twelve faculties of law and five faculties of medicine. By 1861 reorganization and expansion meant that France had sixteen faculties of letters, sixteen faculties of science, nine faculties of law, three faculties of medicine, three higher schools of pharmacy, twenty-two preparatory schools of medicine and pharmacy, and five schools preparatory to higher education in letters and in science.[13]

The decree of March 8, 1808, organizing the University instituted in each faculty three different degrees: baccalaureate, licence, and doctorate, all degrees attached to the faculty by the specific language of the diploma. On February 16, 1810, a statute, which was the work of the Conseil de l'Université, established a distinction between three types of licence and stated the precise regulations for the doctorate. A formal division between the traditional three branches of science did not appear in 1822. There was no act instituting three distinct licences or doctorates, although reference was made to such a division in the regulations on examinations issued in June 1848. By the decree of January 22, 1896, there was established one *licence ès sciences* and many certificates; three *certificats d'études supérieures* would constitute a licence. The Sorbonne had seventeen varieties of certified scientific cuisine. The new system permitted more flexibility in granting dispensations for people getting the doctorate; dispensations were rarely given before 1896 because the supplicant would have to be dispensed from the whole licence program – hardly recommended for anyone following the arduous route to the state doctorate. The new regime also made it possible for foreign and even French students to take the new University doctorate.[14]

More concerned with primary and secondary than with higher education, the July Monarchy did try after 1835 to revive the comatose higher creature by encouraging the growth of faculties in a few cities. In 1837 Salvandy (minister of education and *grand maître de l'Université*) asked Thenard to inform him what was going on in the faculties – a short assignment for one of the world's greatest chemists. The minister wanted to know how to get the faculties to produce useful results. Baron Thenard was not long in producing a masterly report lamenting the ghastly shortage of human and material resources that rendered the University powerless. Even the faculty of sciences in Paris, with only fourteen chairs, did not have the resources necessary to fulfill its mission of training competent lycée teachers.

Both the minister and Thenard would have liked all faculties to be as good as that in Paris. It was difficult enough even to create decent provincial faculties because of their small size and the paucity of professors waiting for nonexistent jobs. Montpellier and Lyon had seven science chairs each, Dijon and Caen had four each, and Grenoble had only three. The best that could be hoped for was that the government would adopt the Montpellier–Lyon model, ensuring that all the faculties would have the minimum number of basic professorships: physics, chemistry, zoology and animal physiology, botany and plant physiology, and mineralogy and geology. Five of the provincial faculties had only one professor to teach mineralogy, geology, botany and plant physiology, and zoology and animal physiology, subjects that made up six courses at the Sorbonne. This teaching load produced incomplete and superficial instruction with little if any research.

This was a serious deficiency, Thenard argued, because "the honor of the faculties requires professors to go beyond what is known." Public attention, the budding root of public support, would be attracted by the "excellence of doctrines and the importance of discoveries," and students would be attracted by courses utilizing the works of the professors. The choice of professors should not depend solely on brilliant examination results and successful theses but on "more irrefragable proofs of their scientific capacity," indeed, perhaps on a reputation based on original work. Another generation would fulfill the aims stated in Thenard's report. Low salaries and a shortage of chairs did not encourage bright young men to choose the obscure career of a provincial professor of science. Ah! if only France would imitate "neighbouring countries where the students pay the professors." A final illusion: "Les sciences mènent souvent à la fortune."[15]

Thenard's report led to Salvandy's decision to increase the number of provincial faculty chairs. And faculties of science and of letters would be created in cities already having significant educational facilities, as in the cases of Lyon and Bordeaux. Towns having a number of studious young people in schools of law and medicine would also qualify for the creation of faculties of science and of letters. It was at this point, in 1838, that a faculty of sciences was established in Bordeaux, although with only six chairs, including only one for botany, mineralogy, and geology. The fact that the new faculty had only two chairs in natural science, while Toulouse had three, gave the Bordelais something to complain about. The important thing is that a faculty had been established, part of a slowly developing movement in higher education that would over the next half-century result in the creation of an extensive network of faculties providing local populations with services and opportunities in science, technology, and agriculture that did not exist even in serious fantasies at the beginning of the nineteenth century.

It is difficult to contest, although necessary to modify in some respects, Prost's designation of the years 1800–80 as "the long stagnation" of the

University, following the "decadence" of the Enlightenment. The chief function of the few professors in the faculties of arts and of sciences was to give the examinations for the baccalaureate, a degree granted after a faculty examination on subject matter learned in the lycée. During the Restoration this hybrid degree was established as the point of entry into the liberal professions through public education.[16] Not much qualitative difference separated the baccalaureate from the licence. Early in the Third Republic, however, a group of professors from the sciences, especially in Lille, would denounce giving these examinations as a waste of the professor's time, which could be better devoted to research. Being paid for the examinations, the professors did not revolt.

While at Lille, Pasteur proclaimed that "the university professor must devote himself completely to teaching and scientific progress." But between 1854–5 and 1857–8 the number of baccalaureate candidates examined by the science faculty at Lille increased from 153 to 196; by 1874–5 the number was up to 375. The *bac* examination was a bloody affair in nineteenth-century France. In the examinations for 1862, when the success rate was up to over 55 percent from 45 percent the previous year, most weaknesses were in literature, chemistry, and natural history. In 1875, with 375 candidates for the sciences baccalaureate, 323 of whom wanted the *baccalauréat complet*, the success rate was 40 percent for the *complet* and 38 percent for the *restreint*. The *baccalauréat restreint*, having less mathematics and more science, was intended for medical students.[17]

Throughout the Second Empire there were few candidates at Lille for the licence, the teaching degree between the baccalaureate and the doctorate; the increase in licence students had to wait for the introduction of scholarships and the expansion of secondary education by the Third Republic. In 1859 only one out of four passed the examinations for the license in mathematics and two out of four in the physical sciences. Possibly part of the weakness of candidates arose from the suppression of the *baccalauréat ès sciences mathématiques* in 1852. Although the baccalaureate was excellent for those who wanted a liberal or an industrial career, it was insufficient for a teaching career because it barely reached the level of the class of special mathematics in the lycée. In 1862, when only one out of the four candidates passed, Girardin called for an intermediate degree between the baccalaureate and the licence in mathematics, because one of the major troubles with the licence was that it covered too much material. Since in the physical sciences candidates could not do the practical tests, there was a need to open up faculty laboratories to the lycée students for experiments. In 1877 there were eleven candidates at Lille for the three licences in mathematics and physical and natural sciences, and seven succeeded. Reform of the licence and the need for lycée personnel were directly related to the number of candidates turned out by the system.[18]

Not having a group of professional students, the physicist, if he wanted to have a role different from that of the lycée teacher, had to appeal to a general educated audience and especially had to show that his science was of practical use. Lallemand did this at Rennes, which even by the 1890s had no significant role for a scientist of the caliber of Pierre Duhem, who was also infuriated by the poverty of the library. Lallemand's idea of a course in physics (1855) was that it should include a detailed treatment of the most essential of the different parts of the science. It should be theoretical but use only elementary mathematics. Practical applications could have only a secondary emphasis, useful for exciting the interest of the audience and making it understand the degree of importance of the theoretical truths and experimental proofs whose methodical exposition made up the body of each lesson. But Lallemand argued that some extensive industrial developments of considerable importance required full lectures: The steam engine, the electric telegraph, and photography deserved detailed coverage, not merely to awaken the curiosity of the student but also to furnish him exact knowledge that he could use. Lallemand was here conforming to the model of industrial science, which probably existed in its purest form at Lille but was strongly represented in other provincial faculties.[19]

Toward the end of the Second Empire the practical applications of science were also being taught in night courses at Rennes. In 1869 the professors of chemistry, physics, and mechanics were giving courses. Massieu's course on the principles of mechanics and their application to machines filled the amphitheater with sixty persons, chiefly the heads of workshops, railroad workers, and arsenal and factory workers. The rector was delighted with the success of these courses, for they showed that the supposedly cold Bretons could become enthusiastic about useful and serious subjects taught with intelligence and talent.[20] The regionalization of the University took root in the Second Empire.

Having practically no genuine research students, professors in France's institutions of higher education turned to the general public for an audience. The Isocratic tradition was brilliantly upheld by Guizot, Villemain, and Cousin at the Sorbonne. Unfortunately, as Michelet and Renan discovered at the Collège de France in the Second Empire, the government could decide that the professor was teaching dangerous doctrines and suspend his courses. The philosopher Ollé-Laprune was also a victim at the Ecole normale during the early years of the Third Republic. But because salaries were generally paid, the chief damage was to the professor's ego. Law, medicine, and pharmacy, under strict state regulation, kept to the narrow path of a professional training whose parameters were set by examinations rather than by subject matter and research.[21] Perhaps he was inhibited by the inherent dullness of his typical subject matter, but the scientist rarely ran into such problems. Apart from the stellar example of Cauchy – tax-farmer Lavoisier must be left aside – French scientists have a remarkable record of being faithful servants of the state and have had little difficulty in changing allegiances to régimes; of course, the nature of French politics in the nineteenth century with its giddy change of régimes might have encouraged such prudence. The ideological commitment of a substantial nucleus of the scientific community to the Third Republic is therefore more interesting than the support for other régimes. Perhaps it was partly a matter of hostile reaction to the excessive demands for loyalty made by the Second Empire.

In its reorganization of the University, the Second Empire showed an unhealthy interest in the "establishment" in science, but its vices were anchored in the ideology of modernity. The government was thinking about the relationship between modernity and science at least as early as 1852. In a speech in April 1852 Hippolyte Fortoul, minister of education and religions, declared that "the new University will tie its life to that of modern societies through a fuller organization of the teaching of science, source of the wealth and political supremacy of nations."[22] Such an important segment of the University had to be supervised by a hierarchy of scientists who were known for their support of the regime. The key functions of the *inspection générale* were given to Dumas, Le Verrier, Flourens, Regnault, and Cournot. The government checked out its key scientific personnel very carefully before giving them the reins of power. Flourens's frank rallying to the government was welcomed in view of his great influence in both letters and sciences. Dumas was at the top of science. In his inspection tours he was singled out as eminently fitted to discover and encourage gifted but forgotten personnel. The mathematical physicist Blanchet of the Ecole normale was to have a great future, due chiefly to his fine research work. Adolphe Brongniart's name carried great authority in science and his alliance with the Dumas family gave the government the surest of guarantees. Le Verrier, a member of the Senate, was a hero because of his boldness in instigating reforms in public education; his devotion to the government needed little comment. Cordier of the Muséum was an *ancien pair de France*. Chasles was viewed as safe enough to have only his qualifications enumerated. Commissions investigated and reported on the observatory, the Bureau des longitudes, which were reorganized, and the Muséum d'histoire naturelle. Scientists were on the commissions; the commission for the observatory and the Bureau was chaired by Fortoul with Biot as vice president, and it included Maréchal Vaillant, Admiral Baudin, Dumas, Le Verrier, and Bivet (professor of astronomy at the Collège de France).[23] The New Atlantis would be built by a scientific estate with unquestioned loyalty to the new empire.

As minister of education, Fortoul tried to establish a well-coordinated, smoothly functioning system. Fortoul gave "science a new prominence in the curriculum of lycées." The triumvirate of Fortoul, Dumas, and Le Verrier reorganized secondary education by creating the system of bifurcation, introducing a baccalaureate in science parallel to the one in letters. Duruy abandoned the system in 1864. Good education had to be planned and closely supervised. No faculty should be independent of another; all higher education had an interlocking intellectual structure: The schools of medicine and pharmacy depended on the faculties of science for chemistry and natural history, at least in Fortoul's plan of the 1850s. Fortoul also included the applied sciences in his grand scheme. The nature of the careers opened up by higher education encouraged the unity of the system. "Puisque la force des choses et la nature même des carrières dont l'enseignement supérieur ouvre l'accès amènent de continuels rapprochements, d'une part entre les différentes branches des sciences, de l'autre entre les sciences et les lettres, il importe que le haut enseignement, dans son ensemble et dans ses détails, forme un tout proportionné et harmonieux, et que sous la variété infinie des aperçus se cache une unité puissante et féconde." Courses of the faculties should be available to the greatest possible number of people. Thus it was important for Fortoul to know the methods used in teaching and the *esprit* characterizing the courses; no doubt the latter was more important for philosophy and history than for chemistry and natural history. It was Fortoul's hope to give to higher education the precision it had lacked before. The advance of science could be best assured by following administrative rules, which would keep it on the right path, avoiding fruitless or whimsical digressions.[24] No *Homo ludens* would be tolerated by this Napoleon of educational administration.

Attempts were made to overcome the limitations imposed by the rigid structure of formal examination-centered courses in the faculties by the introduction of *cours libres*. In 1867 Duruy, a far less Draconian minister than Fortoul, proudly noted this significant development during the Second Empire. Thirty doctors were giving specialized courses in the faculty of medicine. About ten other professors had just been authorized to give courses in the sciences and in letters at the Sorbonne. These courses were all regular courses in higher education, very different from the 900 public lectures given all over France during 1867. Duruy made a point of emphasizing that the emperor and the country favored the development of private education in all forms and at all levels, just as they supported public education. Later observers and supporters of the *cours libres* would compare it in some respects to the German system of *Privatdozenten*.[25]

The official obsession with regulation eventually ensured the legitimation of the *cours libres*, which worried the educational bureaucrats because they were given by people who did not have an official function in the University. This problem was first solved for the Paris faculty of medicine and the former faculty of theology at Strasbourg. In 1882 it was recognized by the Ministry of Education that any regulation would have only the aim of promoting the progress of learning and science. Thus it was a simple matter when a person holding a doctorate in science or letters applied for permission to teach a course. But the ministry was aware that there could be cases where distinguished scientists did not have doctorates. The ministry showed its wisdom and liberality in deciding that if the opinion of the scientific community, stated by its most qualified members, was that a refusal to permit such persons to teach would deprive the faculties of useful teachers, then they should be given permission to teach a *cours libre*. Indeed, the ministry recognized that the French educational system had been too restrictive, especially in letters, and should have a greater variety of courses introduced without delay.[26]

In the 1880s and 1890s administrative decrees established the role of adjunct professors. A decree of December 28, 1885, declared that "the title of adjunct professor may be given by decree, on the proposition of the faculty council, following the opinion of the permanent section of the Conseil supérieur de l'instruction publique, to *chargés de cours* and to *maîtres de conférences* who have the degree of doctor and are distinguished by their service." Another decree of July 31, 1894, established that the number of adjunct professors in the faculties of science could be equal to one-third of the number of professorships. This was an attempt to reward personnel who deserved better than low positions in the system but who could not hope to get one of the limited number of chairs. It also made it possible to give them wider scope in teaching new courses and in doing research in the system. The government was genuinely concerned about the generally low level of the teaching of science in the provinces.[27]

It was the hope of Fortoul that the government could ensure the quality of the teaching of science in the provinces by insisting that appointments go to the best-qualified personnel. But an attempt to assign personnel on the basis of the best man for the job was soon foiled by the old-boy network and the passion of many scientists for the capital. In 1855 the botanist Duchartre and the zoologist Gratiolet ran the risk of incurring the permanent displeasure of the ministry because both refused provincial jobs. Fortoul wanted to establish the right of the ministry to determine who would get the jobs in Paris. His ideal was to reward merit and provincial service, thus bypassing the petty arrangements of the Parisian coteries. It was made clear that the ministry would not forget a refusal to accept a provincial post. Nicklès and Berthelot were designated as two distinguished scientists who should give evidence of their professorial abilities in Besançon and Strasbourg. Berthelot refused. Brongniart, Le Verrier, Dumas, and Thenard were key consultants for the minister in the shifting of personnel. Scientists were to be given an important share of the positions of rectors of academies. Thenard informed the minister that Masson, Abria, and Malagotti were three specialists in the physical sciences who deserved rectorats because of their writings, their discoveries, and their service; Thenard added that Dumas would probably also approve of this list. Le Verrier was consulted on the mathematician Vieille and Brongniart on the natural scientist Schimper, both potential rectors.[28] With some notable exceptions, most French scientists spent some time in the provinces. Perhaps one can conclude that in the Second Empire service in the provinces became a built-in feature of the scientific establishment. The eventual result was the development of a respectable and sometimes solid structure of scientific research and teaching in the provincial faculties of Lyon, Nancy, Lille, Bordeaux, and in some other towns for some areas of science by the last decades of the nineteenth century.

Scrutiny: the inspection générale and the state of science

One of the most distinctive features of French education was the system of *inspecteurs généraux*, which took on its definite form in the Napoleonic state. From the beginning, top scientists like Coulomb and Cuvier played important roles in this system of assuring state control of the educational system.[29] Titles varied from *inspecteurs généraux des études* (An XIV) to *inspecteurs généraux de l'Université* (1848) and *inspecteurs généraux de l'instruction publique* (1851). The number of inspectors reached thirty in the First Empire but was ruthlessly cut to twelve in 1815; raised to seventeen in 1824; the number was also cut back to twelve in 1830. In 1840 the *écoles de pharmacie* were attached to the University and after 1841 twenty *écoles préparatoires de médecine et de pharmacie* were constituted on new bases; so by 1845 credits were reestablished for an *inspecteur général des études médicales* – there had previously been two for medicine. The prestige of the posts in science was maintained during the Second Empire by Cournot, Isidore Geoffroy Saint-Hilaire, and then by Dumas, Le Verrier, and Brongniart.

Although the composition of the inspectorate might change with regimes – several clergy were added in the 1820s – its basic functions did not. This was made clear in 1851 under the Second Republic by Charles Giraud, minister of public education and religions, in a circular to the rectors of the academies encouraging them to lend their support to the inspectors. The inspector should take cognizance of all the details of administration. The minister noted that the presence of the inspector would increase the power of the rector, whose moral authority would be damaged if he attempted to conceal the real state of affairs or hide documents from the inspector. Being the direct representative of the central authority, the inspector had the same right to truth as the minister himself.[30]

According to the law of March 15, 1850, the *inspecteurs d'académie* were to be chosen from former inspectors, faculty professors, provisors and censors of lycées, principals of *collèges*, heads and teachers of higher classes of private secondary establishments, *agrégés* of faculties and lycées, and primary-school inspectors possessing a licence or having ten years' experience. For higher education the *inspecteurs généraux* included the same group plus former *inspecteurs généraux, inspecteurs supérieurs de l'instruction publique*, rectors and *inspecteurs d'académie*, as well as members of the Institut de France. In 1871 the government of national defense decreed the reorganization of inspectors into three categories for primary, secondary, and higher education, as in the decree of March 9, 1852, and gave them the old title of *inspecteurs généraux de l'instruction publique*. The results of the inspections were also to be published in the *Journal officiel de la République*.

In the finance law of December 29, 1876, the minimum salary of an inspector was fixed at 10,000 francs, the same as professors at the Collège de France and the Muséum d'histoire naturelle. Profiting from the *cumul* process, some scientists did very well indeed. Balard made 7,500 francs annually at the Collège de France (before the general raise to 10,000 in 1876) and 12,000 francs as an *inspecteur général*. Le Verrier collected 30,000 francs annually in a triple capacity as inspector (12,500), professor in the faculty of sciences (12,500), and a member of the Bureau des longitudes (5,000), although he received no pay as the director of the observatory.[31] It was not unusual for a scientist to fill one of his jobs with a junior man and pay him half the salary budgeted for the chair, which was an ingeniously cheap system for employing young academics, although open to a certain amount of abuse.

The system for the approval of courses was simple. According to article 18 of a decree of August 22, 1854, the rector convoked the faculties either together or separately in order to deliberate on the individual programs of each course and to coordinate them. A ministerial circular of November 30,

1855, detailed the division of instruction in the faculties. The ministry sent out an annual circular asking for the program of courses for the coming year. Each faculty then met and the professors presented their courses. The dean of each faculty sent the programs to the rector, who sent them to the ministry with his considered opinion.[32] The minister sent the programs to the appropriate inspectors for their opinions. Each inspector was responsible for examining the courses in his area before they were examined in the Comité de l'inspection générale, under the minister himself during the Second Empire. Any remarks of the inspectors were sent to the faculties through the rectors of the academies. Although the rectors sent in the programs and the minutes of the meeting of the assembly of professors along with their covering report before July 31 (changed to April 15 in 1856), there were complaints that the comments of the inspectors did not always arrive in time to be of use for the administrators and often arrived after the beginning of courses.

Especially in Paris it was often difficult for the bureaucracy to get professors to give the necessary amount of detail on their courses. In 1861 the inspector of the Academy of Paris complained that only the programs for the courses in differential and integral calculus (Lefébure de Fourcy) and rational mechanics (Liouville) were detailed enough to show that they conformed to the needs of candidates for the licence. Having audited some of the other courses, the inspector knew that they did conform to the licence requirements, but that did not excuse the professors from carrying out a ministerial request for detailed programs. The inspector noted that there seemed to be a tendency on the part of some professors to escape from any type of regulation in order to preserve full liberty as long as possible. He hastened to add that not too much importance should be attached to this observation! The programs for 1861–2 were fully detailed. By 1865–6 they were back to skeletal form with references to the official programs for details.[33] Unless there were changes in the official program it did seem rather pointless to keep repeating the same thing.

The functioning of the system of inspectors in higher education, considered only in relation to its impact on the teaching and content of courses in science, can best be seen by taking a look at specific cases in the various academies. Sometimes the impact might be just a minor modification of a course. In 1865 the Comité de l'inspection générale noted that Clos had submitted for his course in botany at Toulouse a good program, sufficiently detailed, but that it would perhaps be better in the existing state of science to shorten the twelve lectures devoted to summing up organography, in order to devote a few more lectures to cryptograms because of their unusual structure. In his reply to the ministry and to inspector Brongniart, Clos recognized the correctness of the comment but pointed out that according to instructions sent out in 1855 there should be from twelve to fifteen introductory lectures. Naturally the progress of science would change the viewpoint of the *inspection générale* over a period of ten years. Clos suggested that the change should be implemented by means of a general observation inserted in a circular rather than by personal observation. Thus a professor would not be rewarded by a painful surprise for following the rules strictly. This incident shows the professor in the bureaucratic system being prudent to the extent of following the rules more strictly than the inspector; it also shows the value of the inspector as an agent of change in a system whose extensive guidelines were a threat to innovation, even when scientific progress justified it. The system of inspection could also be an effective weapon against professorial inertia as well as a means of maintaining quality control in the teaching of science. Of course, the *inspection générale* had the potential for exerting undesirable influence, especially if the inspector was an important scientist not in favor of certain new trends in his discipline, but this was rare and certainly did not work over long periods of time.[34]

It was also in 1865 that the Comité de l'inspection générale voiced its fear that Leymerie, the professor of mineralogy and geology, might have planned a course in mineralogy that was too advanced and theoretical for the students at Toulouse. But it was left up to Leymerie to decide if there was really a problem and what its solution would be. He did not agree with the committee. The program had been the same for twenty-five years except for the improvements based on experience. No doubt such a serious course could not be followed by a large audience, but it did satisfy young men who wanted good solid instruction, especially the candidates for the *licence ès sciences physiques* and even the candidates for the *licence ès sciences naturelles*. Leymerie admitted that he was not much interested in people who came to courses in order to pass time. In any case, the number of serious students had grown every year since he had pioneered in establishing "la belle science de Haüy" in the region. Leymerie had eliminated from his course elements of chemistry and optics better given in other courses. This increased its popularity, partly due to his gift for clear exposition, which was also clearly seen in the very favorable reception given his two-volume work on mineralogy. After a spirited defense of his course, Leymerie made some tactical concessions: He would abridge certain developments in crystallography and give a few more lectures describing mineral species. Brongniart concluded that it was not necessary to ask Leymerie for another program, although it would be useful to call his attention to some modifications he himself recognized as possible. Both sides could retreat with honor.

The affair had been precipitated by a letter of the *proviseur* of the Lycée impérial de Toulouse to the inspector of the Académie de Toulouse complaining about the situation in natural history. There was a lack of agreement between the program of scientific studies in the lycée, which had eleven to twelve lectures for zoology, four to six for botany, and five to six for geology, and the program of the baccalaureate in the sciences, which had eighteen, twelve, and ten lectures respectively for the same subjects. The lycée pupils could not be properly prepared for the *baccalauréat ès sciences*. The demands of Leymerie in mineralogy when he was on the jury made the difficulty even greater. It was a situation requiring reorganization or diplomatic pressure on Leymerie; choosing the latter option could hardly do more than improve the situation slightly.[35]

The need to keep within the guidelines of the licence program had the dangerous potential of bringing institutional power to bear against desirable innovations and of keeping topics out of courses until officially approved. But there was a certain flexibility. The first part of the course in zoology and animal physiology was devoted to histology ("microscopic anatomy"). In 1865 the inspector noted that at Strasbourg there were perhaps too many lectures given in histology. This touched a raw nerve in the Strasbourg faculty of sciences. It had been taught there before it was introduced in Paris, in fact since 1838, when Lereboullet was named to the chair of zoology and introduced it into public courses in 1840. Being the indispensable basis of any solid physiology, histology was one of the most important parts of the course. Because histology was not taught in the medical faculty as a special subject, the course given in the science faculty was vital for medical students, who learned to use the microscope to examine tissues. The number of sessions had been reduced from twenty to fifteen; in view of important developments since 1840, it was impossible to reduce this minimal number.[36]

Among the inspectors Adolphe Brongniart seems to have been one of the most vigilant in indicating to delinquent professors the deficiencies of their programs and their failure to comply with ministerial instructions. In 1864 the program in geology and mineralogy of Lesquès at Marseille was brought under scrutiny in the Comité de l' inspection générale because he was not keeping within the specialty of his chair. Instead of a course in geology and mineralogy he was teaching paleontological zoology. For this subject to be given by Lesquès it would have to be treated from the viewpoint of stratigraphic geology based on paleontology. A geology course could not be a front for a zoology course. After an attempt to justify his position, Lesquès submitted a new program conforming to the nature of the chair he occupied.[37] Claude Jourdan of Lyon had his zoology course done at the printer and simply put in the dates. This type of standardization brought the less typical observation from the ministry that some variation in the course

29

could be achieved by a few modifications, such as giving more emphasis to molluscs, reptiles, and zoophytes, so important for paleontological geology. More typical of the bureaucratic reactions to programs were the requests that Fournet (Lyon) furnish the coverage of each of the six divisions of his geology course and that Faivre (Lyon) indicate the time he would give to each subject in his botany course as well as supply more detail than that given in his summary.[38] The ideal was to maintain a balance between the old and the new. In 1865 Rouville of Montpellier was congratulated for keeping up on the recent works in mineralogy but was also warned that concern for the latest novelty in science should not lead him to forget the older and basic parts.[39] The idea of science as a cumulative enterprise led to the policy neatly put in Pope's advice to be not the first by whom the new is tried, nor yet the last to cast the old aside.

Sometimes the demands of the system for general courses covering a wide range of topics clashed with the trend in science toward specialization. This clash could be interpreted, to some extent, as a clash of generations in the scientific community. The situation at Lille in 1864 illustrates the problems involved. On August 15 the ministry informed the rector that the examination of the program of Dareste's course in the natural sciences by the Comité de l'inspection générale showed that the topics treated were much too specialized. Although the topics had been the object of Dareste's research, it would deflect the faculties from their aim if specialties were admitted just because of the professor's research work, which was a reason advanced by Dareste to justify his course. A follow-up letter of Brongniart to Petit, head of the first division of the ministry, noted that Dareste's course needed modification. Another series of exchanges followed about Dareste's proposed lectures for the first semester of 1864. The rector asked the ministry if Dareste could devote the semester to the study of animal generation and embryogeny. Since the rector agreed with the ministry that the subject was too limited to inspire much interest in the young people taking natural history, mostly from the preparatory school of medicine and pharmacy, Dareste did not have much of a chance of getting his way. By November Brongniart was able to report to the ministry that Dareste had taken care of most of the objections to his course. Now, however, Brongniart agreed with Dareste's justification of embryology, which made up a big part of the course as a consequence of the serious treatment of generation.[40] It seems that in the face of professorial determination to change courses according to new developments in science, even sacred ministerial programs were not totally inflexible.

In the excessively integrated French system problems developed when one part of the system made an innovation not simultaneously introduced into

30

the other parts. For Berthelot (Collège de France), atomic theory was subtle, ingenious, and novel, but it dealt with numerical relations, not the bodies themselves. In 1875, with the creation of the chair of organic chemistry for Wurtz, the Sorbonne became an effective counterforce to Berthelot's doubts and opposition. Berthelot's resistance to atomic theory in chemistry did not mean that he kept it out of the system. Micé adopted it for his course in chemistry and toxiology at Bordeaux. The rector reported this to the ministry in 1875, noting the problem involved. Even the best-prepared students, medical students who had the *baccalauréat ès lettres* and the *baccalauréat ès sciences* (either *complet* or *restreint*), had some trouble in making the transition from *enseignement secondaire classique* to higher education. The situation was similar for the first-class pharmacy students. What was an advantage for students exclusively in the physical sciences was a disadvantage for those aiming at other careers. The real problem was for the second-class pharmacy students, who, having only a *certificat de grammaire*, were woefully ignorant. The rector declared that there was no relation between this category of students and Micé's program. There was also the general problem of reconciling the bad preparation of these students with the nature of higher education.[41] The problem pointed out the need for basic reform in the system itself.

In 1855 the Comité de l'inspection générale, to which the minister submitted all course programs, examined the science program as a whole and made some pertinent observations. The courses should be done in two years, as in Paris; this was sometimes difficult in one or two of the provincial faculties, which took three years for the licence program. The basis for the science programs was the examinations for the licence. Because it was not easy to cover all the material in such a short time, hard decisions had to be made on topics and the depth of treatment. In physics one-third of the course had to be devoted to the most indispensable and most general ideas, including the general properties of matter, the laws and applications of gravity and of heat. Emphasizing practical utility, the professor would give this part of the course in twenty to twenty-five lectures as detailed and analytic as possible. The remaining forty lectures did not permit a similar detailed treatment of other topics like light and electricity and magnetism; so a summary-type approach had to be used. The same procedure was employed for chemistry. In the twenty to twenty-five lectures the professor would treat the important subject of nonmetallic elements – e.g., chlorine, hydrogen, and oxygen – and in the forty lectures of the second semester he would give a comparison of metallic bodies and organic substances. Paris had the advantage of having the complementary courses of the Collège de France, the faculty of medicine, and the Muséum, but the programs of the faculty of sciences were the same as elsewhere in order to ensure uniformity in the system.[42]

Although the communication of personal research and opinion was still allowed and, indeed, recognized as one of the prerogatives of the University professor, it would have to be done in other lectures, not in the time allotted to the official programs. The committee also recognized that supplementary lectures were desirable for local industrial purposes: Lectures in chemical processes related to the sugar industry would be quite valuable at Lille. There was a need to announce practical courses that would give the University a role in national production. By pursuing this policy the faculties of science could become popular much more rapidly than would be the case if they limited themselves to official programs. In reality, the provincial faculties needed little encouragment in this direction.[43]

Even in Paris the Second Empire's University courses were nearly all organized according to the program of the *licence ès sciences* and brought into agreement with the division of studies at the Ecole normale supérieure. Two exceptions were the courses in higher mathematics – higher geometry taught by Chasles and the calculus of probabilities (including mathematical physics in the 1860s) taught by Lamé. Outside the licence program, the material of these courses was better suited for doctoral candidates. By 1870 four courses could be defined as having the speculative character of the courses offered by the Collège de France: the mathematical courses of Chasles, Hermite, and Briot and the mathematical astronomy of Puiseux. In the 1870s the general nature of the licence courses in physics and chemistry was complemented by the addition of specialized courses that could accommodate the advances made in both disciplines. By 1878, in physics, Desains dealt with heat, magnetism, electricity, electro-magnetism, and their chief applications, and Jamin covered acoustics and optics. In chemistry Sainte-Claire Deville dealt with the general laws of chemistry and the history of metalloids, Troost was concerned with metals as well as organic chemistry, and Wurtz gave special attention to the aromatic series. Physics and chemistry were moving toward providing the type of specialized, high-level courses long associated with mathematics. Such courses were vital for the doctoral program that had taken on new life in the laboratories funded through the Ecole pratique des hautes études. The production of licence students, however, remained the biggest educational business of the University.

Cardwell has lamented the fact that mid-nineteenth-century Oxford and Cambridge were really high-class lycées.[44] This situation was not unknown on the Continent, as was made clear in the report of A. Bertin on his physics course for 1855–6 at Strasbourg. In order for teaching in the faculties to be really higher education, the programs had to be those for the licence. But there were too few licence students to justify a special course for them. Bertin gave a third lecture each week in which he dealt with the more difficult parts of physics, especially modern optics. Unlike Leymerie, Bertin adjusted the level of the course to the average capacity of the auditors. Not having a suitable amphitheater, the faculty was unable to attract the class of people who like to give a certain part of their leisure to the study of the sciences. So physics in Strasbourg was taken by mostly young men who were candidates for the *baccalauréat ès sciences* and who could only follow elementary physics as taught in the lycée. In 1855–6 Gerhardt's course in chemistry was found to be too abstract by the inspector, who recommended that he follow the University program closer. In a report to the Directeur de l'enseignement supérieur, Berthelot found courses too close to lycée courses at Clermont, Grenoble (first part of chemistry course), Marseille (chemistry), and Nancy (chemistry). The Strasbourg situation was aggravated by the fact that many of the students were in the army and found continuity difficult in their studies due to changing garrison frequently.[45] The general problem, existing in all French science faculties, would not be solved until the state created a professional student class through scholarships.

Lack of facilities and of equipment was the eternal complaint of the scientific community. In 1856 the faculty and the dean at Rennes indicated the insufficiency of material for demonstrations. Zoology and geology needed models; physics, although classified as "assez riche," still needed new instruments; the professor of pure mathematics needed a sextant; the professor of applied mathematics needed everything. Nothing had been done by 1859. The letter of the rector to the ministry in that year lamented the fact that the faculty was so poor that it had to borrow the few instruments and apparatuses owned by the lycée. It was not surprising that results in the *bac* examinations in mechanics and cosmography were bad.[46] Another small faculty, that of Poitiers, had no complaints throughout most of the 1860s, but if the rector could report that "material conditions . . . are generally satisfactory," it was probably because not very much was going on there. By 1868 the dean reported that the area was developing a taste for the sciences in contrast to a nearly complete lack of interest before. Most of the students took courses in the winter. The special report of the dean in 1868 emphasized the poverty of the faculty: Complete collections were needed for modern natural history; physics desperately needed apparatuses and precision instruments; there was a need of space and facilities for astronomical observations and meteorological research; the library did not have the books needed for teaching and research; and the faculty, like that at Marseille, still had no *jardin des plantes*.[47] The situations at Rennes and

Poitiers were similar to that in most of the small faculties. It is not surprising. The interesting thing is that the faculties were complaining and getting the ministry to recognize the justice of their complaints, even if no immediate action was taken.

In the large provincial facilities dissatisfaction was equally great. In 1862 the rector of Nancy asked for two professors of mathematics and one of natural science. The lament was repeated in 1863. The candidates for the *licence ès sciences physiques* did not have a course in mineralogy and the candidates for the *license ès sciences naturelles* did not have a course in geology. In spite of the metallurgical and mining industries of the region, Nancy was one of only three faculties lacking a professor of geology and mineralogy. By 1864, however, the rector pointed out to the ministry that the facilities of the faculty were very good and had even been praised by Dumas. The income of the faculty was nearly 23,000 francs, making it third in status in the Empire. Of course, it still had only one professor of mathematics. When dealing with institutions one must understand that the satisfaction of existing needs does not mean an end to demands; it seems that the more needs are satisfied, the more demands increase, probably as a result of growth. By 1882 Nancy's needs were greater than ever before. In 1882 the dean prepared a long report on the organization of the teaching needs of the faculty, especially in the licence and *agrégation* programs. One of the most important requests was for a *third* professor of mathematics (astronomy); the minister was reminded that Marseille, Toulouse, Bordeaux, Lyon, Clermont-Ferrand, and Besançon already had their three chairs. This part of the program had been covered by a *maître de conférences* (Sauvage), and the rector added in his letter that the area would continue to be covered by a new *maître de conférences*. Deans' letters made requests that were often toned down by rectors. The dean also repeated a request for a chair of organic chemistry, an area now covered by Haller, a *maître de conférences*, in two lectures a week. Two new *maîtres de conférences* were requested in zoology and mineralogy. The chemistry and physics laboratories were insufficient for certain advanced students, and a laboratory assistant was needed.[48] The empire had learned that science is an expensive pursuit, and the republic soon found out that good science is financially insatiable. Contributions from municipal and departmental governments led to increased demands on the state. No sooner was the Lyon faculty installed in the Palais St. Pierre, at the expense of the municipal government, than it bombarded the ministry for the necessary apparatuses and collections to fill the building and to serve the students.[49] Within a generation the scientific estate had become a fiscal red queen who had to spend more and more in order not to fall behind in the "scientific race."

Republican reform

Growth of the University

In the last two decades of the nineteenth century, French higher education changed qualitatively and quantitatively to such a degree that it is justified to call it, as Prost does, a new arrival on the educational scene. Its basic structure was fixed for the twentieth century. The remarkable thing is that in the period of reform one finds "the rare conjunction of a great work to be done, strong personalities to undertake it, and a collective movement to sustain it."[50] A society existed for the study of higher education and published the excellent *Revue internationale de l'enseignement*. Associated with the movement were the great men of the University. Among them were Pasteur, Boutmy, G. Monod, Fustel de Coulanges, Durkheim, Bréal, Berthelot, Lavisse, Gréard, and Liard. Connections were maintained with members of parliament and the educational bureaucracy. A nucleus of genuine students was created in 1877 through 300 *bourses de licence* and in 1880 through 200 *bourses d'agrégation*. The annual number of degrees, including the baccalaureate, between 1831 and 1842 averaged 3,200, and between 1870 and 1880 it averaged 7,200. (The Ecole normale supérieure produced 30 graduates annually, the Ecole polytechnique 150, and the Ecole centrale 230.) In all areas, the trend was toward specialization, with different *agrégations*, licences, and certificates devised for each type of specialization. Professional competence in professorial specialties saw a spectacular improvement. The University as we know it was created in the late nineteenth century.

Late-nineteenth-century growth of the University was the result of new ideas on the role and function of higher education. As late as 1870 the character impressed on the faculties at their origin remained: the function of a special school limited to professional preparation or the role of preparation for examinations with some other teaching added. Outside Paris a faculty of sciences or letters was made up of five or six *maîtres*. Even in Paris there were no special courses for medieval history, contemporary history, comparative grammar, French medieval literature, organic chemistry, constitutional law, or for certain groups of diseases long recognized as special branches of medicine and surgery.[51]

What ideas explain the beginning of the University's drive to maturity in the 1880s? Part of the inspiration came from the hope that true universities, rather than collections of faculties, could be created in France. The foreign example, especially Germany, was important, but it was not the sole concern.[52] Prost insists that the idea of a university held by people like Liard and Lavisse was also based on three complementary ideas: efficiency of organization, which would permit only about six real universities in the large centers of population and learning, while the rest of the fifteen faculty groups would be far less important, especially in research; second, a wish to decentralize through associating the University with regional development, which could be done through scientific-technological-industrial connections, although the percentage of Paris students fell only from 52 percent of the total in 1888 to 42 percent in 1898 – obviously, this was an aspect of a broader French political and social issue; third, "the philosophical idea of science" as the source of unity of thought through its ability to give universal laws, which could only be achieved by grouping the sciences together in certain centers.[53] The reformers saw their inspiration also in the Revolution: "On a retrouvé la conception et la définition qu'en avait données, dès 1792, la Révolution française, laquelle avait rêvé, au sommet de l'enseignement nationale, de grandes écoles, où toutes les parties de la science eussent été groupées suivant leurs affinités naturelles."[54] But the independence and power of the faculties were never subordinated to any university structure on local levels; the laws of 1885 and of 1889 gave a legal structure and financial power to the faculties. The law of 1896 did not change this but recognized the old groups of faculties as universities, one for each academy (1880: Aix, Besançon, Bordeaux, Caen, Chambéry, Clermont, Dijon, Grenoble, Lille [some faculties transferred from Douai in 1887], Lyon, Montpellier, Nancy, Paris, Poitiers, Rennes, Toulouse, Alger). Yet the retention of the Napoleonic structure did not much matter, for the essential work of reform had been done, with higher education substantially extended and considerably improved. Contemporaries were rightly more impressed with the successes of the reformers than with their failures. Even reformers spoke of the danger of attempting to transplant to France English or German institutions that had developed over a long period of time. France had a different political situation and a different educational system characterized by centralization, highly valued state degrees, and public funding. Change – even if justified – would have to come slowly.[55]

By 1868 the prereform movement had been launched by Duruy. In the decade after 1876 all the faculties received 201 new chairs, 200 *cours complémentaires*, and 1,229 *maîtrises de conférences*. In the last quarter of the nineteenth century faculty scientists doubled to about 200 with a fourfold increase in support personnel. Weisz gives the following judicious estimates for increases: from 103 to 178 professors between 1865 and 1919, from 21 to 91 teachers ranked below professor between 1888 and 1919, and from 51 to 355 laboratory personnel between 1865 and 1908.[56] Whatever neo-Pythagorean guesses are used, the signal of the figures is steady, mod-

36

erate growth. Important works were begun: the expansion of the faculty of medicine in Paris and of the Muséum and the construction of the Ecole de pharmacie. Within the next decade, the reconstruction of the Sorbonne and the Ecole des langues orientales was begun, expansion of the Ecole de droit decided on, and the project for the growth of the Collège de France submitted to parliament. Construction was undertaken for the faculties of sciences at Bordeaux, Toulouse, Lyon, Nancy (Institut de chimie), Clermont, Rennes, and institutes for Lille and Montpellier; for medicine at Bordeaux, Lyon, and Montpellier; for letters at Bordeaux, Toulouse, and Lyon; and for law at Montpellier and Lyon. Regional-municipal participation was heavy: The cities supplied 50 million out of a total of approximately 115 million francs. The subsidy to higher education rose enormously, from 5,972,971 francs in 1870 to about 10 million in 1878; by 1889, it stood at 14,492,595 francs, of which the state paid under 10 million, for about 5 million was derived from fees and similar income. Shinn gives a 70-percent increase in the budget for science faculties between 1890 and 1898, with 13 million francs spent for renovation and construction; in Paris 9 million francs were spent for new buildings, including six new laboratories. In the provinces 75 percent of the 30 million spent was supplied by local governments; Lyon, Bordeaux, Nancy, and Lille account for 20 million of the total. Weisz thinks that a conservative estimate of the total subsidies and donations between 1898 and 1913 would be 35 million francs. Land and buildings were rarely included in the figures. Paris gobbled up 80 percent of the gifts.[57] In spite of a drop in annual scholarship funds from 720,000 to 484,000 francs in the period 1881 to 1898, the total number of students rose from 11,204 in 1876 to 42,037 in 1914; the number in the faculties of science rose from 293 to 7,330, from 2.6 percent to 17.4 percent of the total. By 1914 medicine had 8,533 and law 16,465 students in faculties and cafés, while letters entertained about 6,586 aspirants.[58] The Third Republican University had taken on its definite shape, which would not change substantially until after World War II.

Once the professors had been given their new facilities they discovered the elusive virtues of their old slums. In his history of the faculty of sciences at Bordeaux, Rayet recalled how in the cramped old quarters of the rue Montbazon professors lived their academic lives in familial intimacy. The old faculty was a moral being endowed with a common feeling; it was also a political animal with its representatives on the elected municipal council. In the new buildings the laboratories were vast compared with the old ones; each scientific areas was a kingdom with a royal chief, his helpers, and his subjects. Strangers to one another, the professors were masters in their own domain, each thinking only of his own work. The faculty had fissioned into a series of institutes that happened to be in the same building. Nor were the institutes reluctant to fight one another for privileges and favors from on high. "La solidarité ancienne est remplacée par l'individualisme moderne."[59] But even Rayet was willing to shoulder the burden of progress.

Ideology: the republic and science

No regime has ever been more committed to science as a source of ideology, the guiding beacon for life, than the Third Republic in the last quarter of the nineteenth century. Next to Berthelot, Pooh-Bah of the republic of science, Paul Bert was probably the most prolific fountainhead of scientifically inspired ideology. In a new comprehensive general scientific journal, *Revues scientifiques*, started in 1879, Bert, then as a deputy well into his infidelity toward physiology, gave a marvelously clear articulation of the role of science in the new political world, the place science deserved to occupy "in a truly democratic conception of our social organization." Politicians had to give science a greater role in their views, a role unknown till then but in harmony with its impressive daily growth. This meant a change in the social role of science, which had hitherto been given a status of luxury by monarchs, who condescended to protect scientists along with poets and artists. More recently, in an opposite error, the sciences were regarded as only a source of wealth, as a Rumpelstiltskin who could through fertile applications turn the dross of theory into riches for an entire nation. But for Bert the true role of science was above these two different appreciations. They were valid: It was the noblest of mental occupations for a man to plumb the mysteries of the universe; it was a worthy activity for science to harness natural forces to transform the material conditions of modern civilization. Above the solution of theoretical problems and the practical riches flowing from it, Bert worshiped science as the liberator of human thought and the potential regulator of society. The old forms of social organization based on faith and resignation, whose logical political form was royalty, had decomposed in cruel circumstances. What could reconstruct the edifice, if not science, which replaces belief by demonstration, resignation by the victorious fight, and which logically tends to democracy with its only possible expression, the republic. Much of this dangerous nonsense was rhetoric of the Revolution of 1789, which had been more serious about implementing some of these ideas than the Third Republic was. Long before the advent of Soviet Marxism, some Frenchmen came close to the idea that "the pursuit of science should be directed by the public authorities to serve the welfare of society," although the French thought that the authorities themselves would be scientists or their admirers.[60]

Bert envisaged science as a total guide to man's way of life. By its methods, its discoveries, and their applications, science was conquering the intellectual leadership of society; science still had to become society's moral guide. The University (educational system) had not played a large role in this transformation because it was the inheritor of Jesuit methods, which reduced science to a sterile enumeration of figures and facts. (Bert was an expert on the Society of Jesus: His erudite and titillating exposé of *La morale des Jésuites* was published in 1880.) The University itself had to be transformed. Another hurdle to the realization of the scientistic dream was the false idea of a separation between pure and applied sciences. A quotation from Pasteur could crush such an insidious idea (by basically avoiding the issue): "Il n'y a pas de sciences appliquées, il n'y a que des applications de la science." It was a good line anyway. Pasteur's *theoria* might have been closer to the ideology of the Jesuits than that of the republic, but his *praxis* made him acceptable – and perhaps needed at any cost – as a republican icon. Democracy, republic, science: a secular trinity guaranteeing the existence of modern man's *vita nuova*. "S'il est vrai que la République soit la mise en pratique de la démocratie et de la liberté, on peut dire que la science moderne est républicaine par ceci . . . qu'elle se libéralise et se démocratise de plus en plus."

Support for Bert came quickly from lycée teachers. Louis Crié saluted the transition from the empire to the republic as signifying the change from emphasis on the practical to emphasis on the theoretical, with greater support for natural science. Bert's natural-sciences textbook for the lycées along with his reforms would be important instruments of change. But Bert's reforms, when he was minister of education, were attacked by fellow scientists, especially for the errors in the program.[61]

In Bert's brutal epistemology to know was the opposite of to believe, as to prove was the opposite of to affirm. In the republican mythology taught by Bert, the great expansion of science had taken place since the sixteenth century, when human thought, breaking the chains of Catholicism, had begun the process of successively ridding Europe of the errors accumulated in official views and superstitions. Freedom of the mind was thus the tradition and the very essence of science. Science itself had a built-in process of error correction, for it too had accepted and established errors, once a necessary weapon in the discovery of truth. The secret of the purification of science was experimentation, the successor to a priori reasoning. Science had survived the slavery in which it was kept by the Church, the burdens of its own errors, and attacks by tyrannical heroes; it had evolved to be the business of the masses. Victory would be on the side of the big battalions. If France were to have a big scientific battalion, education would need reform.

A wave of secular reform was just around the corner; with it would come an increase in support for science. Spending for education rose from 51 million francs in 1876 (under 2 percent of the national budget) to 180 million in 1894 (more than 5 percent of the budget). Much of the increase went for science.

The icing on the ideological cake was provided by the social sciences when Durkheim allowed himself to be convinced by Louis Liard that he should give an annual compulsory course (1904–13) on the history of education in France for the students of the University of Paris taking letters and science *agrégations*.

> He made a systematic attempt to identify broad historical continuities, interpreting them as evidence of cultural traits or as answering fundamental social needs. Thus one can trace a number of recurring themes: the continuity of "formalism" . . . ; the continual reappearance of encyclopaedic ambitions . . . ; the gradual . . . growth of secularism and reason within education to an ultimate position of dominance; the extremely prolonged educational monopoly of studies relating to man, only lately matched by an equal concern for the rest of nature; the continuity of educational organization, reaching back to the Middle Ages.[62]

Durkheim had a key role in his generation's defining "afresh the nature, direction, and aims of education to assure such freedom and rationality as can be attained for a future generation,"[63] at least in terms of republican ideology. The sociologist's analysis of the history of French education was followed by "a number of prescriptive conclusions – pro-scientific and anti-literary, uncompromisingly secular and anti-classicist" – for secondary education. The teaching of science was "essential for the formation of the complete mind."[64]

Most republican intellectuals were of the same opinion. Marcelin Berthelot acquired a well-deserved notoriety in the front ranks of those demanding a more scientifically oriented education. Ideologues wished to replace the "old literary education" of the church with the "modern scientific education" of the republic. What actually happened is that the sciences did win a respectable and substantial place in education in the second half of the nineteenth century. If the University provided a continuity of structure, which evolved considerably, it also proved capable of allowing the moderate growth of some new disciplines, especially the sciences. The political and social structures in which this growth took place were considerably different from those in which the University had stagnated in the first half of the nineteenth century. Closer connections between politics and society and the University made that growth possible.

40

Conseil supérieur de l'instruction publique

As a result of the Jules Ferry's reforms in education, the Conseil supérieur de l'instruction publique was radically changed in composition and in function. The Section permanente was reestablished. The once powerful council (under a legion of names), dating from the days of Napoleon I, had been rendered powerless by Fortoul. The law of February 27, 1880, advertised itself as a national reform, not the creature of circumstance or the work of a party; it was the act of a government concerned with the rights of the state and jealous of its responsibilities, intent on restoring to the public domain an area of action in which the role of the state had been lessened in the preceding thirty years. This official rhetoric concealed the Radical plan to expand the secularization of education, thus pushing back Catholic gains, and to emphasize the role of the state in public education. It was a powerful stimulus to the great expansion of education in the late nineteenth century. On the grounds that social influence and moral interests should play a role in educational policy, previous governments had taken away part of the responsibility for public education from the Conseil supérieur by ensuring that the majority of the Conseil d'enseignement came from outside education. The result was that although private schools could be opened and given free rein in curricula, the state could not create a faculty or open a lycée, choose books, or discipline teachers without the consent of the Conseil supérieur, which was strongly influenced by private elements. The new Jacobins took a different view of the role of the Conseil supérieur: "C'est le grand comité de perfectionnement de l'enseignement national." All the noneducational representation was eliminated in a fit of Rousseauistic hatred of "partial societies." Part of the legislation of 1850 and 1873 was undone. By recognizing representatives of the state and of society, the law of 1850 negated the democratic and representative regime of the republic. Like the teaching church, the Etat enseignant must be its own master. "La société n'a pas d'autre organe reconnu . . . que l'ensemble des pouvoirs publics émanés directement de la volonté nationale, et cet ensemble s'appelle l'Etat."[65]

The governmental *projet de loi* had three chief features: the exclusion of religious members representing different social interests; the ensuring of representation by all the big establishments of public institutions and especially the three segments of the University; and the predominance of the elected part of the council. A Chamber amendment provided for representation by the communal colleges; the Senate secured representation for four members of the Institut. The Section permanente, made the center and base of the council, had fifteen members, of whom nine were named by decree. Paul Bert was president of the commission studying and elaborating the project finally embodied in the law of February 27, 1880.

The Ferry spirit was strongly represented in the Conseil supérieur by scientists like Bert and especially the eternal and powerful Berthelot. As minister of education, Berthelot echoed the Ferry line in his speech to the council in December 1886. He hailed the law of October 30, 1886, for changing the situation partly resulting from the defunct "reactionary law of 1850." He rejoiced in the fact that members of the teaching body had to belong to "la Société civile" – meaning elimination of religious personnel – and were inspired by a modern spirit and the republican faith. Primary teachers could show their love of progress through their presence in the departmental councils; and what was most important, female teachers would elect representatives to the councils, which showed the importance the republic placed on the education of women. (Perhaps they would become sufficiently emancipated from the clerical yoke to get the vote, but Berthelot did not say this.) It was all very satisfying for republicans to hear the voice of science prophesying the fulfillment of the dream of Condorcet: progress based on science in our time.

Berthelot also expressed the proper concern of a minister for the provincial universities, although as a young scientist he had boldly refused to follow the career pattern of many well-known scientists in passing a few years in a provincial faculty of science. In his speech of March 21, 1887, Berthelot interpreted the new organization of the faculties as tending to coordinate the faculties of each academy. This was a safe enough statement, worthy of the wooliest ministerial mind. But he indulged in the reform rhetoric of looking forward to the day when autonomous provincial universities would rival their foreign counterparts. He thought that excessive University centralization hindered the development of local intellectual and scientific life.[66] This is a dubious view, and it may well be that the lack of interest and of funding by both central and local authorities was mostly to blame. In any case, the situation was changing rapidly.

The scientific community was always represented on the Conseil supérieur by leading members of the scientific "establishment." During the Second Empire this meant that scientists like Dumas, Le Verrier, Faye, Milne Edwards, Claude Bernard, and Balard were in the Conseil impérial de l'instruction publique. During the Third Republic over a third of the Conseil could be scientists; in 1880 out of fifty-eight members, not including the minister of education as president, seventeen came from the scientific community. Berthelot was then vice president. Others who show up on the lists for the late nineteenth and early twentieth centuries are Wurtz, Chevreul, Bichat, Buquet, Darboux, Barrois, Mascart, Duclaux, Bert, Lacaze-Duthiers, Jules Violle, Perrier, Gaudry, and Jules Tannery, and Bouchard and Brouardel for medicine. After the Great War, it is evident that a new generation has appeared: Appell, d'Arsonval, Bochet, Carvalho, Duboscq, Gabellel, Guignard, Lacroix, Lambling, Picard, Lucien Poincaré, Roger, Wéry, and Widal; the components of the educational system represented remained generally the same as in 1880.[67]

The Conseil supérieur soon had an opportunity to show its support for university development outside Paris in the case of the plan to consolidate the faculties of Douai and Lille by shifting arts and law to Lille, already the home of the sciences. In the council discussion of March 23, 1887, Lacaze-Duthiers, who, with Pasteur, was one of the founders of the faculty of sciences at Lille, spoke of the vitality of the city and the resources available for the University. Because it was close to the frontier, many Belgian students came to study here and returned to universities like Liège as missionaries of "la science française." Some Belgian scholarship holders were sent to the faculty of sciences. The faculty of medicine had been founded at Lille in 1877 with 131 students and had grown to 281 students by 1887. Gréard throught it important for education to profit from the considerable resources of a big industrial center, as was the case with the German University in Leipzig. Fremy also supported the unification of the faculties, which was approved by the council and acted on by the ministry.[68] A collection of faculties looked more like a university if they were in the same town.

In 1883 Ferry, the minister of education, sent a circular to the faculties and the academic councils with questions concerning the creation of universities similar to those in other European countries. Generally, the faculties were in favor of the immediate creation of independent universities as a means of reducing isolation and professional egoism and of ending the loss inherent in lack of educational cohesion. Unity would produce a common spirit and encourage the development of existing points of contact in teaching and research. Ferry agreed with the faculties but was too conservative to push the scheme; opinion had to be prepared for the change, which would be more easily accepted if the new units were not called universities, implying an undesirable fission process of decentralization repugnant to the worshippers of the centralized state, whose chief weapon against internal enemies was the educational system. In France the word *university* meant the teaching state, the total of the three orders of public education. (The Université de France had disappeared *legally* on March 15, 1850.) The creation of universities might be interpreted as a rupture of the system of national education. Even Guizot had seen the virtue of Napoleon's idea of the unity of national education. Representatives of the municipal and general councils in the academic councils expressed the same fears as Ferry. Liard emphasized the dangers involved in transplanting foreign institutions to France. Convinced of the need for proceeding slowly, the government issued a decree (July 25, 1885) creating for each "university" a Conseil général des facultés, made up of elected representatives of the faculties. In this minor experiment on the grouping of faculties, the old prerogatives of the separate faculties were kept.[69] Since this was also the wish of the faculties themselves, there does not seem to have been much basis on which to create independent universities according to the German model. The "creation " of provincial universities in 1896, a much-publicized administrative regrouping, not without some fiscal and political importance, was a footnote added by Louis Liard (director of higher education) to the republican project.

Ecole pratique des hautes études: the triumph of the research imperative

In Germany, too, from the 1820s on, the universities were changing, though with a greater tension between teaching and research functions than existed in France.

Turner has placed the emergence of the Prussian university as a research center in the *Vormärz* period (1818–48). Universities had succeeded academies as scientific research centers by 1830, and by 1850 the typical career scientist was a university professor whose laboratory was part of the university. The professor's role was redefined to emphasize research and publication rather than teaching. What led to the triumph of this *Wissenschaftsideologie?* Turner's thesis emphasizes the role of state in setting the criteria for professorial appointments; it does not discard competition and decentralization, although in at least the areas of appointments and subsidies the trend was toward centralization. "The state's new, aggressive control of professorial appointments ... became the specific institutional mechanism through which the new scholarly values ... came so quickly to dominate the universities .... By stressing disciplinary criteria over the collegiate and pedagogical concerns of direct interest to the local, corporate university, the state gradually established the disciplinary criteria as accepted professorial norms."[70]

The funding of research through the Ecole pratique des hautes études came about partly because of a recognition by the government that the scientific hegemony of France was being contested by other countries. In his address to the Conseil impérial de l'instruction publique in December 1867, the minister of education, Victor Duruy, noted the threat to France from the sacrifices and efforts of her rivals. The deficiencies in higher education had to be eliminated. This meant development and renewal of the material structure of science in the faculties, a project requiring the spending of large sums of money, especially for salaries and laboratories. A start was made in 1867 with the establishment at the Sorbonne of "the only true physics laboratory in France." Duruy hoped that all branches of science in all institutions would soon be able to support the research of their best people by giving them laboratories, the best instruments, and the most intelligent assistants available.[71] The founding of the Ecole pratique des hautes études in the following year showed that the minister was not just indulging in political rhetoric. The research structure of French science, whatever its embryonic origins, had its birth in the Second Empire.

In his report to Louis Napoleon recommending the creation of the Ecole pratique des hautes études, Duruy emphasized the national importance of the plan.[72] There were really two connected projects: the creation of the Ecole pratique and the reorganization of laboratories into a system of teaching and research laboratories, with an increase in the number of research laboratories. The basic aim was stated in terms of international competition. French laboratories, the "arsenals of science," were rapidly becoming obsolete in comparison with the facilities bestowed upon science in the United States, Germany, England, and Russia. And new foreign schools, featuring famous scholars and scientists, were, in the drama of unceasing scientific progress, "a serious threat to one of our most legitimate ambitions." Establishments dating from past days of glory could not ensure French equality, let alone hegemony, in the new age of scholarship and science. New times, new needs.

Before the Ecole pratique des hautes études got under way, there were two other plans for "écoles pratiques." First there was a plan to create an Ecole pratique pour les naturalistes in the Muséum; this was relegated to the amiable bureaucratic category of "classer: projet inutile." Another more ambitious scheme called for an Ecole pratique des sciences for the important scientific establishments under the Ministry of Education. For graduates of the Ecole normale supérieure and for private students, this school would give a diploma after three years of study as well as shelter students working on doctorates. The existence of these plans shows than an active movement was about to improve French research capability.

Lest the emperor be alarmed by the possible cost of all the putative advantages to be derived from renewed scientific glory, Duruy pointed out the modest financial burden of the twin project. The costs of expanding the Sorbonne, the Muséum, and the Ecole de médecine would be spread out over several years and inflict only a temporary burden on the extraordinary budget. For the rest there was needed only simple administrative measures, with new organization, and limited credits added to the ordinary budget. This was a valiant exercise in cost-effective serendipity, whose "unhoped-for effects" lay far in the future. All the sages of scholarship and of scientific research supported Duruy's initiative, for, modest as the reforms of 1868 were, their potential was great for enlarging the scientific research establishment. Among those who offered to "lead young scientists on the road to discovery" were Bernard and Berthelot. The list of laboratories and seminars for the first year of the school shows that very few big names were missing from the roster of those giving their services freely. But the achievement would be that of another regime.

Justification for the new policy required an explicit recognition of the inability of universities to produce much research, except for the Sorbonne. Part of the reason for this anomaly, it was recognized, lay in the dual structure of higher education in France. Lawyers, doctors, engineers, teachers, and archivists were produced by the faculties of law and medicine, the Polytechnique and the Centrale, the Ecole normale supérieure, and the Ecole des Chartes; so the thirty-three faculties of letters (17) and of sciences (16) had nothing to do with preparation for middle-class professions. Pure science, left to occupationally meaningless faculties of science, was regarded of so little material significance that no qualifications were required for entrance to courses demanding little application or even attendance. Unlike in the German universities, in the French faculties of letters and of sciences there were no true students, only a general public demanding the titillation of *haute vulgarisation*, uncorrupted by the dullness and boredom of scholarship and scientific research. In a system officially consecrated to pure science, this elusive product of the mind was not being created.

Duruy's explanation of this institutional oddity, a clear embarrassment when one compared French universities with the lights of *Wissenschaft* beaming across the Rhine, was a sort of commentary based on common sense and traditional criticism of the French faculties.[73] Science is not only a body of doctrines but also an instrument to be manipulated; lectures do not make the whole scientist, because in order to know how to use science as an instrument one must be experienced in laboratory practice. In the case of preprofessional education, laboratory work was supposed to be an essential part of normal courses. But the faculties of science had only a small number of teaching laboratories, attached in proprietary fashion to professorial chairs and reserved for the preparation of his lectures and demonstrations. Before the budget of 1869 increased credits for expansion and more equipment, these "laboratoires d'enseignement" were closed to students. Duruy argued that this type of laboratory would also have an important function in the research system by providing support personnel for the directors of the research laboratories. In addition to the feeding of trained students from the teaching to the research units, there would be a connection in the person of the director, who would often probably be in charge of both faculty

46

laboratories in his discipline. The functions of two types of laboratory corresponded to two clearly distinct needs of public service: the diffusion of science to those who could understand it, and the progress of science, the task of those who, possessing the spirit of research and invention, could advance science if they had good facilities.

The theoretical virtue of university education would be left intact but supplemented by a "school" devoted to research and scholarship with the curious name of Ecole pratique des hautes études.[74] *Pratique* made sense when put in the context of the traditional explanation of the so-called theoretical –that is, oral – emphasis in the faculties. Having nothing to do with industrial utility, the word *pratique* referred to the activity of eyes and hands in providing the vital component in science to confirm and extend "the highest and subtlest of the conceptions of the scientific mind." *Pratique* referred to the means and instruments necessary for scientific research and scholarship: laboratories, observatories, libraries, archives, missions, precise direction, or even personal advice – all that was vital to establish "science-in-the-making" alongside the "science-already-done being taught in the faculties." Although the aim of the Ecole pratique seemed better suited for the physical and natural sciences than for mathematics, even students in mathematics could be fed into astronomy in order to form a much-needed group or "school," while others would pursue mathematical theory and its applications. Duruy hoped that the reputation acquired by the French in chemistry could be duplicated in the other sciences. A curious analogy, perhaps, for by 1869 it could be argued that the French had not kept up very well at all in organic chemistry. No doubt Duruy was thinking of the international renown of scientists like Sainte-Claire Deville and Berthelot. Duruy was also proud that the Ecole pratique promoted philology and history, then appendices of literature, from the obscurity of erudition to the ranking clarity of science. This novelty (for France) meant that the Ecole pratique included in its program "the full cycle of human knowledge," or would have if the plan to include economics had ever come to fruition. Whatever the area of knowledge, the school's aim was purely scientific. "Faire avancer la science, par un institut de jeunes gens d'élite, érudits reconnus, qui ont la vocation de la science pure, et aspirent au titre de savants."[75] Just as long as it was all *pratique*, of course.

Before the Ecole pratique was founded, Duruy thought that only students of well-heeled families could afford to come to a school that did not prepare its graduates for a public job. Not that this would be necessarily a disadvantage. The "Ecoles pratiques du Muséum," for example, could help in establishing a true Muséum faculty of agronomy, whose teaching of the laws of animal and plant production would attract the interest of property owners and heads of farm businesses. Nor would industry be closed to graduates whose diplomas showed that their education joined high-level practical skills to the highest theoretical knowledge. Others would be attracted by the possibility of jobs in secondary and in higher education. One of the key functions of the Ecole pratique was to supply higher education with teachers and researchers, thereby ending the system in which lycée teachers became faculty professors. This quasi-monopoloy of lycée teachers on University jobs was a carry-over from the days when they were the only people available to teach in the provincial faculties in many disciplines. Secondary education had different aims and different methods from higher education; so a stint of twenty years devoted to discourse or exposition to lycée students was just bad preparation for University positions that required research skills above all else. Any practice in teaching could perhaps be acquired through Duruy's other creation, the "cours libres," France's version of *privatim docentes*, of whom there were already forty milling around in the Ecole pratique of the Paris faculty of medicine and thirty-four authorized for the Sorbonne. Duruy even hoped that the establishment of research laboratories would attentuate some of the evils of the law on *cumul*, which no one contemplated abolishing; but directors of laboratories should be able to devote themselves to research and teaching in one job, thus encouraging the separation of chairs and of jobs. Milne Edwards and Berthelot were among high rollers in *cumul* who found this idea beyond their comprehension. It is clear that Duruy's reforms aimed at a long-term and long overdue revolution in higher education.

The Ecole pratique was immediately popular. Instead of the eighty or less registrations expected, there were four hundred, many of whom were *agrégés*, doctors, and Parisian lycée teachers. Only 264 could be admitted: twenty-seven in mathematics, seventy-five in physics and chemistry, ninety-four in the natural sciences and physiology, and sixty-eight in history and philology. Examinations made a typically French reduction in this group. By 1872 the number had doubled, with more than 500 people showing up for courses, but because one person sometimes took several courses, the number of registrations reached 700. Beginning with forty-two laboratory and lecture courses in 1869–70, the Ecole pratique grew to sixty such courses by 1872. Most were in Paris, but provincial operations grew from five to fourteen between 1869 and 1872. In 1872 there were in Paris twenty research laboratories and twenty-six "laboratoires d'enseignement," a flexible term including mathematics and the fourth section of the Ecole. In the provinces there were eight research laboratories, five "laboratoires d'enseignement scientifique," and one "conférence de philologie." The Ecole pratique gave a considerable push to publication by funding research and in stimulating new books and journals.[76]

France's most politically powerful scientists were consulted on the desirability of creating the Ecole pratique. No one is on record as having opposed the chance of getting more money for scientists. Dumas played a large role in the discussions of 1868. Wurtz and Le Verrier were also among the great powers consulted. Outside science, no one was more influential than the philosopher and guru of ethical wisdom, Félix Ravaisson. Each of the four sections was placed under the direction of three professors: Serret, Briot, and Puiseux for mathematics; Balard, Wurtz, and Jamin for physics and chemistry; and Milne Edwards, Decaisne, and Bernard for the natural sciences and physiology. The lists of the *commissions de patronage* read like a roster of the Parisian scientific establishment: Lacaze-Duthiers, Hermite, Bertrand, Troost, Duclaux, Moissan, Fremy, and Wurtz. The Collège de France, the Institut Pasteur, and the Muséum were represented, but the faculty of sciences was preponderant, especially in mathematics. In 1911–14 the mathematics section included Darboux, Appell, Picard, Poincaré (replaced by Hadamard), and Borel; the physics and chemistry section included stars of great magnitude as well in Lippmann, Bouty, Brillouin, Haller, and Roux; nor was the section of natural sciences deprived of talent in Perrier, van Tieghem (replaced by Lacroix), Bonnier, Guignard, and A.-M. Lévy (replaced by Delage). The great and near great dispensed modest sums to themselves and their colleagues without worrying about the ethics of it all. Who, after all, was better qualified to judge the deserving poor in scientific research, and who could have done it more fairly? No one, probably.

Obviously a great deal of scepticism existed on the need for government to fund scientific research. *L'Opinion nationale* (July 17, 1869) thought that the active French scientific movement showed that pure science was accepted as just as much a legitimate activity as the work of the practical man, the politician, or the merchant. "The scientist works for the common good, which is the improvement of humanity." But the paper regretted that there was a dangerous current of thought in a democracy that regarded as useless all research not producing a direct material result. Such narrow minds ignored the connection between discussions of systems of organic chemistry and new methods of printing fabrics or between the theories of Volta and Galvani and the electric telegraph. One of the main functions of Ecole pratique would be to provide for pure science a shelter from the rude currents of democracy and also to ensure that France shared in the immense practical benefits that eventually derived from science unsullied.

The Ecole pratique des hautes études was "essentially an administrative superstructure to dispense funds for advanced research and training . . ."[77] As late as 1888, the budget for physics and chemistry was not much more than that for the IVe Section; the size of the latter budget is probably a good indication of the influence and power of scholars and especially historians in the educational bureaucracy:

Ire section (Sciences mathématiques)	7,000 francs	
IIe section (Sciences physico-chimiques)	76,000	
IIIe section (Sciences naturelles)	112,700	
IVe section (Sciences historiques et philologiques)	75,300	
Ve section (Sciences religieuses)	30,000	
General expenses	28,600	
Total	329,600	

(There was also an annual subsidy of 36,000 francs from the city of Paris for *bourses d'études*.)

The funding of physics through the Ecole pratique promoted the rebirth of originality in an area in which the French had not kept up so well as in mathematics and chemistry. The research laboratory in physics at the Muséum was headed by Edmond Bequerel. Henri later joined him, and the laboratory changed direction as a laboratory of applied physics. The research physics laboratory at the Sorbonne was headed by Jamin, but the faculty was turning out an average of only one physicist per year in the 1870s. By 1882–3 Bouty was assistant director of the laboratory, and the new generation, among whom were Pellat, Lippmann, and Leduc, was working there. The laboratory was transferred to the new Sorbonne under Lippmann. In 1893–4 twenty-five researchers were in the laboratory, including Amagat, Leduc, and Marie Sklodowska. Foreigners and engineers were conspicuous. The physics laboratory at the Ecole normale also became important. Violle, professor at the Conservatoire and *maître de conférences* at the Normale, was director and Brillouin, professor at the Institut agronomique and *maître de conférences* at the Normale, was assistant director; the names of Abraham, Weiss, and J. Perrin appeared in the 1890s. Another brilliant period had begun in French physics.

Physics early adopted the device of separate research and teaching laboratories. In 1872–3 Desains, the director, had forty students in the Sorbonne teaching laboratory, where they watched basic scientific experiments, repeated them, became familiar with instruments and learned the art of manipulating them. They did no specialized research. Some research was done in the laboratory by the regular personnel, such as the *chef de travaux*, Branly, who had just finished several years of work on electricity. The "laboratoire d'enseignement" of Desains quickly became well known, even abroad, and was probably the only laboratory in physics to equal the former reputation of Regnault's at the Collège de France thirty years earlier, a period when scientists like Kelvin and Soret came to learn experimental physics from Regnault.[78] In 1882–3 there were 136 students under Desains and Mouton.

In 1893–4 Bouty had 112 students in physics, of whom 65 were regulars, including three young foreign women. In 1872–3 Schutzenberger's chemistry laboratory at the Sorbonne had 35 of its 49 students doing practical exercises, while the other 14 were doing serious research. By 1882–3, under Troost, the laboratory had in the practical exercise section 85 candidates for the licence and 11 *boursiers d'agrégation*, as well as 23 in the research section. By 1893–4, out of 101 students, 31 were in the research section, while most of the others were still licence candidates. Other laboratories with substantial sections devoted to teaching and practical exericses were in mineralogy, geology, and physiology. Even Balard's small chemistry laboratory (Collège de France), which was crowded with 18 students in 1872–3, had mostly students for the licence in physical sciences or pharmacy students, although research work was also done there.

More famous at the Collège de France was Berthelot's laboratory of organic chemistry, which had thirteen serious people in 1872–3, including Jungfleisch, Bouchardat, and Amagat. This laboratory continued to attract some of the best French scientific minds like Paul Sabatier and a sizable contingent of foreign scientists. Equally famous was Henri Sainte-Claire Deville's laboratory at the Ecole normale supérieure, where a dozen or so scientists of the calibre of Debray, Troost, Hautefeuille, Gernez, Mascart, Ditte, and Joly worked with the diector. Researchers like Gernez, Baubigny, and Lespieau continued to frequent the laboratory when Joly took over. Adding to the eminence of the Normale was Pasteur's laboratory of physiological chemistry. Schutzenberger's laboratory at the Sorbonne was attracting foreigners even by the early 1870s. Numerous publications resulted from the research done there. One of the two assistant directors was Armand Gautier, who became head of the laboratory of biological chemistry at the faculty of medicine. Foreigners were also attracted to Wurtz's laboratory at the faculty of medicine, whose activities showed up regularly in the *Comptes rendus* of the Academy of Sciences, the *Annales de chimie et de physique*, and the *Bulletin de la Société chimique de Paris*. Fremy's laboratory at the Muséum had 49 students in 1882–3. Some of Fremy's students went into industry, including sugar factories and Saint-Gobain. His numerous students formed an association that offered prizes for the best students. The support of chemistry by the Ecole pratique gave excellent results.[79]

It was not until late in the nineteenth century that there appeared on the scene of French physics a galaxy of scientists comparable to the stars of the earlier part of the century. Coulomb, Ampère, Fresnel, Dulong, Arago, and Sadi Carnot had honorable research offspring, but whether because of lack of funding and institutions or for genetic or accidental reasons, the importance of work during the July Monarchy and the Second Empire was not comparable to what was done before or after. A great deal of significant discovery and invention continued, of course, especially in heat, light, and the thermomechanical properties of bodies. The work of Pouillet, Despretz, and especially Fizeau, Foucault, and Regnault was carried on by Cailletet and then by Amagat. Experimental work in electricity was done by Pouillet, Antoine and Edmond Becquerel, Fizeau, Foucault, Desprez, d'Arsonval, and Mascart. Fizeau, Foucault, and Cornu did basic work on the speed of light. Most of the work was done within the unexhausted paradigms established by their French predecessors or by following discoveries made in Germany or England. This is not to deny originality to work like that of Foucault on the gyroscope, on the earth's motion, on light, and on electromagnetism. But it might be aruged that there was nothing revolutionary about this work; nor was there anything so fundamental as Maxwell's work in electromagnetic theory. Scientists were happy with the peaceful pursuits of normal science within these areas. It was possible to do excellent work within the existing paradigms. Such was Regnault's redetermination of the specific heats of many solids, liquids, and gases, as well as his finding that Mariotte's (Boyle's) law was only approximately true for real gases. Similarly, Cornu did a classic redetermination of the velocity of light by Fizeau's method to give a far greater accuracy to the results. It is tempting to conclude that the ideal type of science practiced in France during this period was "the experimental study of facts, without theory or preconceived ideas."[80]

Even late in the nineteenth and early in the twentieth century, there were famous practitioners of the type of reasoning centering on "certain relations between observable facts," as in Le Chatelier's clear statement of the laws of chemical equilibrium. Charles Fabry thought that Pierre Duhem was perhaps "the purest representative of this doctrine; in a long series of publications, stretching from 1890 to 1915, he developed in impeccable mathematical form the laws of all physico-chemical transformations."[81] This outlook can also be illustrated by the pioneering work of Raoult in cryoscopy and in tonometry, whose results were used by van't Hoff, Arrhenius, and others in working out a theory of solutions. But the idea of the completeness of science, implying that research could only be a matter of working out the next decimal place – physical theories would take care of themselves – was a general "low-grade infection" in scientific circles in the nineteenth century, not a specifically French disease.[82] In view of other types of French scientific activity, especially at the end of the nineteenth century and the early twentieth century, as well as before the period of "completeness," it would be silly to associate this activity with any putative French mental quirks.[83]

At least until well after the experiments of Hertz, French physicists did not pay enough attention to Maxwell's electromagnetic theory of light, which meant that no major French contribution was made to a key area of science. More characteristic was the criticism of Poincaré and Duhem about the sloppy logic and defective mathematics of Maxwell's system.[84] From the end of the century on, however, work of fundamental importance in magnetism was done by Pierre Curie, Langevin, Weiss, and Aimé Cotton.[85] Duhem's thesis on magnetic induction was also an important and influential work. But in the study of ionization of gas, following the discovery of X-rays by Röntgen in 1895, French research, emphasizing the atomic nature of electrical charges, was pioneering and profound, as well as extensive, especially in the work of Langevin, Benoist, Perrin, and Pierre and Marie Curie. Perrin's famous thesis dealt with the nature of cathode rays. Maurice de Broglie, brother of Louis, created a laboratory in Paris for the study of X-rays, thus making possible the work of Dauviller, Thibaud, and Trillat. Especially after Perrin's work on colloidal suspensions (1908), resistance to atomism among physicists became minor.[86] The discovery of radioactivity by Henri Bequerel led to the famous studies on radiation by the Curies and Debierne. In the 1880s Pierre Curie and Paul Desains worked on quantitative studies of the wave lengths of infrared radiation. Pierre also worked on piezo-electricity with his brother Jacques Curie. The early work of the Curies was done at the municipal Ecole supérieure de physique et de chimie industrielles but was relegated to the research and physical periphery in this industrially oriented school. The new burst of creativity in physics in France, itself part of the "second scientific revolution," was directly linked to funding through the Ecole pratique des hautes études. It was also connected to the revival and growth of the University, although many important laboratories and scientists were in other key institutions, like the Muséum and the Collège de France. In science, especially for an important faculty post, the possession of a state doctorate was vital, and this in turn enhanced the value of the degree and the University.[87]

In 1848 the basis of mathematical research was expanded from the confines of rational or celestial mechanics to permit a doctoral thesis in any branch of mathematics. The only restriction imposed was that the thesis contain a discovery. A second thesis was no longer required; there could be substituted a discussion of questions determined officially by the faculty. Given the remarkable development of mathematics in the first half of the nineteenth century, the Conseil supérieur de l'instruction publique decided that the new régime was needed to permit the University to participate fully in the new wave of mathematical discoveries. Although theses in the sciences between 1808 and 1820 were often a few unoriginal pages, there was a steady, accelerating improvement in quality thereafter. So high was the level of quality finally attained that it was unnecessary to change the doctorate. By contrast, growing dissatisfaction with the licence made it necessary to change that program completely by the end of the nineteenth century. Originality in scientific research became the touchstone of the doctoral program. "L'examen ... doit avant tout favoriser et mettre en évidence l'esprit d'invention, l'aptitude à ces recherches scientifiques qui constitutent la véritable fonction de l'enseignement supérieur."[88]

Nowhere was the increase in the quality of higher education more evident than at the level of the doctorate. (See Table 1.) Between 1847 and 1881 about fifty-seven theses were rejected as unsuitable or of too low quality for the doctorate in the faculty of sciences in Paris. The number of rejections in mathematics was twenty-one, chemistry twelve, natural sciences eleven, mechanics ten, and physics seven. In 1884–85 Lippmann rejected Duhem's thesis ("Le potentiel thermodynamique") in mathematical physics, on the basis of technical defects making it unsuitable for defense at the Paris faculty of sciences. A jury (Bonnet, Darboux, and Poincaré) that rejected the thesis ("Considérations nouvelles sur les diverses courbures d'une ligne tracée sur une surface et sur la courbure elle-même de cette surface") of abbé Issaly in 1886 explained the new situation. The scientific level of theses presented for the doctorate in mathematics rose considerably within a few years. Unlike many of the old theses, too often simple works of erudition, with a few demonstrations or a few new points of view, the new theses had to present true discoveries. Increasing emphasis was being put on real discoveries, as a result of the great progress made generally in higher education and specifically in the theory of functions, in higher geometry, and in mathematical physics. The new emphasis would play a great role in the future of mathematics in France and had to be encouraged at all costs. The thesis by Issaly was a good piece of work, done with order and method, containing a certain number of new and interesting results; it would have earned him a doctorate in 1870 but it did not fulfill the requirements of the 1880s.[89] A doctorate would be awarded only if the candidate presented a sufficiently important new discovery. The later institution of the University doctorate would make it possible for some people like Issaly to get doctorates, which, if they did not open up careers in higher education, at least served as a reward for a few who had spent long years in secondary

Table 1. *Number of Science Doctorates, 1810–85*

Faculty	Math	Physical	Natural	Total
Paris	157	218	174	549
Bordeaux	1	1	3	5
Grenoble	5	3	2	10
Lille	2	3	3	8
Lyon	3	1	8	12
Marseille	3	1	8	12
Montpellier	9	11	16	36
Strasbourg	15	18	8	41
Toulouse	14	4	7	25
Total	217	274	235	726

Note: Out of a total of 742 doctorates in letters in the same period, Paris produced 548. Other faculties were Caen – 23; Dijon – 21; Lyon – 20; Rennes – 26; Strasbourg – 36; Toulouse – 12. By 1910 the total for all science faculties was up to 1,300. In the period 1876–80, the faculties of science produced an annual average of 20 doctorates; in the period 1908–12, the annual average had risen to 37 doctorates. (A few small faculties included in the totals have not been itemized.)

Sources: *Enquêtes et documents*, 21 (1886); Terry Shinn, "The French Science Faculty System, 1808–1914," *Historical Studies in the Physical Sciences*, 10 (1979), 308–30; and George Weisz, *The Emergence of Modern Universities, 1863–1914* (Princeton, N.J., 1983), p. 236.

education and wanted the degree but were incapable of doing something original enough for the state doctorate.

In 1880 a report appeared in the *Revue scientifique* comparing biological science in Germany with that in France. Although in a country-to-country comparison of the higher educational resources of Germany and France, it was clear that France was far behind, it was also true that Paris was by far the most active center of biological study and research in Europe, indeed the greatest educational city in the world. What France lacked in 1880 was a group of good provincial universities comparable to those spread out over the imperial territory of Germany. But at the Muséum the professor of comparative anatomy had a laboratory, personnel, and resources equal to those anywhere, and far superior to most European institutions. Only Reichert's collection in Berlin could be compared favorably with the

Muséum's, and Reichert's was in a worse state than the Muséum's.[90] With such large and old collections the winner in a comparison of this sort would be the one in the state of lesser decrepitude. The state of the Muséum and its collections had become a hot scientific and political debate in France, perhaps taking second rank only to the debate over the future of medicine. Unfortunately for the Muséum, this debate took place at the same time as the tremendous growth of the University of Paris and the remarkable development of universities in several provincial cities. The Muséum still occupied an important place in French science, but it was no longer unique.

By the time of the memorable Senate debate over the budget for public education in 1907, it was clear that a definitive reversal had taken place in the scientific roles of the Muséum and the University of Paris. By the turn of the century the great Muséum d'histoire naturelle was literally falling down, at least in parts, because its pathetically low budget was totally inadequate for the maintenance of the physical plant. Lecomte, professor of botany at the Muséum, warned that the herbarium, the vaunted national treasure made up of the collections of Tournefort, Desfontaines, Lamarck, and Jussieu, was in danger of disintegrating. Senator Strauss also warned of the filth and lack of safety in the Muséum. New galleries of zoology and comparative anatomy went unfinished – zoology since 1889 – while personnel remained in the old ones, condemned by the Conseil des bâtiments civils. When a wall of the monkey house collapsed, the press amused the public with accounts of the incident, without bothering to emphasize the danger if some of the bigger primates decided to take off for the nearest café. The ceiling of the orangery completely collapsed into the room. Called in to repair hothouses, glass-workers were not permitted by their bosses to climb to the top. Over a million francs would have to be paid for the desperately needed repairs, but the Ministry of Finance would not supply the funds. And no financial autonomy would be permitted within the Muséum before a financial inspection was carried out. A reputation for nepotism and the Muséum's isolation from teaching did not encourage governmental alacrity in allotting it extra funds.

The educational darling of the Third Republic was the University of Paris, which became a showcase of competition against several – mostly *much* lesser – German splendors. In the same issue of the *Revue générale des sciences* lamenting the physical horrors of the Muséum, there was a glowing report on the latest marvels produced by Louis Liard at the Sorbonne, by now a term of medieval modesty totally inadequate to indicate modern glories, especially in the sciences. National and municipal governments joined forces to build the Institut chimique, eventually costing over 3 million francs. It took up part (9,000 square meters) of the vast establishment of the Congrégation des dames de Saint-Michel, and the rest of the property was bought for 1,850,000 francs by the University for its future expansion. A subsidy by the national government covered three-quarters of a million of the amount. The extensive scientific power and intellectual charisma of the Sorbonne made it, in French terms, a formidable fundraiser, although Liard and his cronies looked in vain for a French Stanford or Carnegie or Rockefeller. Part of the real estate deal associated with the Institut chimique was a sale of 1,000 square meters of land to the Prince of Monaco for the construction of his Institut océanographique. The price: 300,000 francs, which went into the University's fund to buy the old religious establishment. While the Muséum was decaying and new construction put in limbo, a new university city, complete with gardens, was being established in the Latin quarter. Georges Pouchet argued that the financial pinch damaging the Muséum could be blamed on the senseless luxury of new construction. Charles Richet agreed, adding the Ecole de médecine and the Sorbonne as similar cases showing the need to resist the Pharaonic projects of architects.

When Alphonse Milne Edwards, director of the Muséum, died in 1900 the scientific reputation of the Muséum was so low that scientists could call for a further reduction of its role in French scientific life by an abolition of its courses. Not that teaching had ever been of cardinal importance at the Muséum, except in explicitly research-related courses and popular public courses. From 1793 to 1840 the Muséum had been France's leading scientific institution, with a budget of 350,000 francs in 1825, while that of the faculty of sciences in Paris was 75,000. By 1905 the faculty budget reached the million-franc level of the Muséum and exceeded it in succeeding years. "By 1900 . . . the Muséum . . . faced a constant threat of annexation by the University of Paris. . . . " The factors involved in the Muséum's failure were also those explaining the success of the faculty. First, the attempt to turn the Muséum in the direction of experimental science was effectively foiled by the naturalists. Second, no effective teaching program succeeded at the Muséum; so it had no student body. Third, it really found no "vocation that would render . . . essential its naturalist and museological orientation, . . . recognized as socially useful and worthy of governmental support." Like its provincial sisters, the Paris faculty became a bastion of experimental science, catered to the career needs of hordes of students, and, not least important, became an important weapon in the ideological arsenal of the republicans, the politicians who funded science. While the Muséum was "relegated to the periphery of the system," the faculty became "central to the Third Republic's educational preoccupations."[91] The University's faculty of science was a great research factory of sixty laboratories – only twenty of which were at the Sorbonne – world-famous institutes like the Institut Henri-Poincaré and the Institut du radium, and a total of 530 researchers at all levels. Between 1926 and 1939 the student population of the faculty never fell below 4,000, including well over 1,000 women after 1929. After 1933 the annual production of doctorates rose to just over 100, of which eighty were the key *doctorat ès sciences* in 1938–9.[92] The Muséum's condition was much better by 1934, a year in which Paul Lemoine, the director, boasted of having about 12 million francs for capital construction, a gift of 2 million for a zoo, and another gift of 1 million from Paul Marmottan, the art historian.[93] But the shift of the center of science to the University of Paris was never reversed.

To some extent, at least, the faculties of science were funded by the state because, as Liard and Appell pointed out, the *grandes écoles* were no longer capable of fulfilling national needs, and the Polytechnique in particular could not produce the research scientists, mathematicians, and engineers vital for an industrializing society. Nor was the Polytechnique qualified or able to give civil servants the administrative and legal knowledge necessary for the modern bureaucratic state. From the 1890s on, the Polytechnique was subjected to heavy attacks on the backwardness of its curriculum and the incompetence of its graduates, who supposedly kept the more competent faculty graduates out of the choice positions in the *grands corps*. The Polytechnique was finally prodded into some reforms by 1911. The automatic entrance of its graduates into positions of influence and power meant that the Polytechnique was one of the means used by the *haute bourgeoisie* to maintain its slowly eroding power in the state. This displacement also occurred in the school. Terry Shinn estimates that between 1830 and 1880 nearly 30 percent of the students came from *propriétaire* and *rentier* families; between 1880 and 1914 this group had dropped to just over 12 percent of the students. The student increase was from entrepreneurial and middle-rank civil servants, both of which made up 37 percent of the students in 1910. So the school reflected the process at work in society at large; the lower segments of the bourgeoisie were also increasingly using the *grandes écoles* as a means of upward mobility and as an entry to choice state jobs. In 1910 nearly 13 percent of the graduates of the Polytechnique were entering the civil service. The competition between the *grandes écoles* and the faculties cannot be simplified as rivalry between social classes, for no institution was the preserve of any one class by the late nineteenth century. The *haute bourgeoisie* was simply not numerous enough to monopolize the growing number of positions available in government and in commerce and industry; the upward thrust of the middle and lower bourgeoisie coincided with a significant expansion of job opportunities. In any case, Shinn found that class origins had no influence on intellectual achievements in the Polytechnique, which permitted those who achieved high ranking in graduation to have first choice in the job market.[94] It was the growth of an industrial society that led to the growth of the faculties and their institutional competition with the *grandes écoles*.

By the first decade of the twentieth century, as a result of the introduction of courses of an industrial nature, the faculties of science were undergoing a transformation that would become characteristic of the entire University after the Second World War. As Prost has pointed out, the old ideological functions of the University receded as the development of the economy gave the University new functions. Faced with a threat to the social hierarchy through the "school population explosion" and the demand for equal educational opportunities, the ruling classes considered and partly pursued another plan, that of giving the economic function priority over all others in the University; this had the advantage of ensuring social stability through economic progress. But the demands for the improvement of techniques and the transmission of technical knowledge by the University could only be implemented by sacrificing some of the traditional aims of the University as the producer and transmitter of abstract and theoretical knowledge. The lack of research and development in French business meant an increased pressure on the faculties of science for more "applied" research at the expense of "pure" research.[95] This has been recognized as a serious problem in the provincial faculties. The faculties of science, because of their importance for industry and commerce, had thus to face a problem at the end of the nineteenth century that other faculties could avoid for nearly another half century.[96]

2

FATA MORGANA

POSITIVISM IN NINETEENTH-CENTURY
FRENCH BIOLOGY

"La vie, voulez-vous que je vous la définisse scientifiquement?
C'est de l'inconnu qui f . . . le camp."
                    — Dr. Fornerol, in Anatole France, *L'anneau d'améthyste*

O NE of the most striking areas of nineteenth-century scientific growth
was in biology. Apart from anti-Darwinism, two of the most
intriguing issues in French biology are the influence of Auguste
Comte and of Claude Bernard, the odd couple who laid down nearly
opposite royal roads to the great research empire that the French believed to
be their birthright. The scientific community, backed by republican
enthusiasm, worshiped Bernard as its great culture hero. Dumas put it so well
that Pasteur quoted him: "Ce n'est pas seulement un grand physiologiste,
c'est la physiologie elle-même." On Bernard's death in 1878, the black-
banded pages of grief in the *Revenue scientifique* morosely proclaimed his
reputation even in Germany. "Claude Bernard était de tous les français celui
que l'orgueilleuse science d'outre-Rhin osait le moins méconnaître." The
Chamber of Deputies paid 10,000 francs for his burial from Saint-Sulpice in
the first public funeral for a scientist. Yet Comte's influence lurks in the
inspirational ideology leading directly to biology. Some even believe that it
has left its potent traces in the Bernardian gospel itself.

Did Auguste Comte and the divided band of positivists significantly
influence the development of biology in nineteenth-century France? Many
historians of biology, perhaps even most scholars, would not hesitate to
answer yes. Because most of those who argue for the positivist influence do
so for solid ideological reasons, it would be surprising if my negative brief
changed any minds. Sir Impey Biggs explained a similar problem brilliantly:
"A person who can believe all the articles of the Christian faith is not going
to boggle over a trifle of adverse evidence."[1]

The argument of this chapter can be outlined simply: (1) the role of
positivism in the Société de biologie, long regarded as the most enduring
monument to positivist ideology; (2) the lack of consequence in the
temporary happy congruence between positivism and biology in opposing
evolution, especially Darwin's doctrine;[2] (3) the ideas of Comte and other
positivists on the nature of biology and the direction it should take; (4) the
putative influence of positivism on Claude Bernard; (5) the philosophy of
science of positivism's "black pope," Grégoire Wyrouboff. A dissonant

60

voice in the scientific community, Wyrouboff could not provide an ideological rallying point for any scientific group, let alone in biology, which was outside his own specialization. My balance sheet puts the positivist influence on the debit side: the case is strongest on an institutional level in the founding of the Société de biologie, but only superficially; there is little evidence of much influence on biological disciplines or on the leading scientists. The positivist philosophy of science could have little basic appeal to scientists, especially in the late ninteenth century, when the old positivism was regarded, perhaps unjustly, as a quaint relic of ancient times.

The role of positivism in the founding of the Société de biologie

The Société de biologie was conceived right in the midst of the Revolution of 1848, in May, the relatively peaceful month of the feast of concord. The Society was certainly positivist in one regard, its indifference to revolution in 1848. By contrast, Virchow waxed enthusiastic about his hope for a German republic in 1848.[3] Brainchild of Charles Houel, the Society originated when he collaborated with Eugène Follin and Charles Robin to organize regular meetings in Paris of doctors, chemists, naturalists, and physiologists who were interested in the "phenomena of life." Pierre Rayer accepted the presidency. Among the other founders may be counted Claude Bernard, Charles Huette, Alexandre Laboulbène, and Lebert, and perhaps some others who were omitted from the honor roll of "founders" by a wanton Clio. The first volume of the proceedings appeared in 1849; so on the reasonable grounds that the life of a scientific group consists chiefly in its publications, French logic decreed that the jubilee of the Society be celebrated in 1899. Certainly the signifiance of this Society was to a large extent determined by the possibility of quick publication of scientific results, although rapid publication was not its only claim to greatness and may have been a direct result of the high scientific caliber of the members of the Society, men who were not used to waiting for publication while their research became old news.

The first four presidents of the Society were Rayer, Bernard, Bert, and Brown-Séquard; the first vice presidents were Bernard and Robin; Huette was treasurer and Brown-Séquard, Follin, Lebert, and Segond the secretaries. M. D. Grmek has made clear the importance of the Society in the career of Claude Bernard.

> In this period (1849) Claude Bernard went faithfully every Saturday, at three o'clock in the afternoon, to the sessions of the Society of Biology ... This society brought together after January 1849 the best

61

physiologists and naturalists in Paris...The Society of Biology regularly published the proceedings of its meetings. Printed as fascicles, the proceedings had the advantage of appearing quickly. For Bernard the society was an ideal audience. During the entire year 1849 he and his rival Brown-Séquard spoke at nearly all the sessions. Bernard drew on his old laboratory notebooks or spoke of his contemporary research.[4]

Like anthropology, biology in France was a lusty daughter of medicine. Follin and Houel were two young surgeons in 1848; Pierre Rayer was a member of the academies of science and of medicine, a doctor at the Hôpital de la Charité, and later dean of the faculty of medicine; Charles Robin was a professor at the Paris faculty of medicine; and over two-thirds of the regular members of the Society in 1850–1 were connected with medicine. Both Dubois Reymond and Virchow were among the fourteen German and Austrian foreign correspondents. At first members met at the top of the "Ecole pratique de la Faculté de médecine," and the deans of the faculty, smiling benevolently on the Society, helped out when possible. When the commission of Balbiani, Berthelot, Brown-Séquard, Dareste, Guillemin, Robin, and Broca was charged by the Society to investigate the issue of the reviviscence of dried animals and settle an ancient squabble, the last evidence of which was a polemical exchange beween A. de Quatrefages and Louis Souleyet, the experiments carried out in the laboratory of Jules Gavarret in the faculty of medicine formed the basis of Paul Broca's report of 1860. The medical connection was really consecrated in the declared aim of the Society: to study life in its normal and pathological states. "La Société de Biologie est instituée pour l'étude de la science des êtres organisés, à l'état patho-logique."[5]

In 1864, thanks to the diplomacy of Rayer, an imperial decree elevated the Society to the select company of those organizations whose existence was declared to be in the public interest. Although Claude Bernard was the "star and favorite" of the Society, we should not ignore the part played by Rayer in assuring the success of the young organization. Because traditionally the history of science has, with ruthless logic, installed in its pantheon only the intellectual creators, an organization man like Rayer quickly falls into oblivion. He deserves a short paragraph in the history of science: They also serve who only encourage others to create. Rayer died in 1867, but when Paul Bert became president in 1878 he still remembered the work of Rayer in building up the Society. In a speech of 1898 to the Academy of Sciences on Brown-Séquard, Berthelot harkened back to Rayer's presidency. What impressed Berthelot was Rayer's friendly encouragement of young scientists; the support of Rayer was valuable because of influence deriving from his work in medical science and the medical care he had given to some members of the establishment. Bernard, Robin, and Berthelot were among the recipients of Rayer's numerous favors. Anticlericals had another reason to like Rayer: In his youth he had been a victim of Restoration discrimination when he was refused permission to take the *agrégation* examinations because he had married a Protestant. Rayer was not an innovator in science or even in medicine, but this had not prevented him – indeed, it may have encouraged him – to play the role of good fairy in the lives of many young scientists.

If Comtean positivism was an ideological inspiration for the Société de biologie, perhaps in much the same way that Baconianism was for the Royal Society, it was in a rigorously scientific form elaborated by one of the founders himself. Yet as soon as Charles Robin had expounded his organization of the sciences and the road for the society to follow, an ungrateful science, and biology in particular, found new gospels to follow, including some never even imagined by mid-century positivists. Thus one is reduced to accepting statements about the Gallic equivalent of *Geist* to describe the relation between positivism and French biology. Such was the funeral rhetoric of Berthelot when he inaugurated the statue of Claude Bernard: "La Société de Biologie, fondé sous l'espirit profond de son règlement, rédigé autrefois par Charles Robin." In giving his sketch of the history of the Society up to 1899, Gley identified the positivist spirit as "the true cause of the fortunate development of the society; the essence of that spirit animated the society throughout the second half of the nineteenth century, in spite of many modifications positivism underwent during that period." What was this powerful gospel? None other than the well-known classification of six fundamental sciences – mathematics, astronomy, physics, chemistry, biology, and social science (sociology) – fulminated by Comte and expanded and modified by Robin for biology.[6]

Robin was superbly qualified to fulfill the request of his colleagues that he draw up the guidelines of the Society. Segond, Littré, and he left the Positivist Society in 1852 to perpetuate the scientific purity of positivism, which had become contaminated by Comte's subjectivism. He later collaborated with Grégoire Wyrouboff as coeditor of *La Philosophie positive* and was one of the editors of Littré's famous medical dictionary. As a result of an accident in childhood, Robin had a glass eye, with which, his enemies said, he looked into the microscope. He was solidly entrenched in the outer corridors of influence in the Second Empire. Robin did the autopsy, ordered by the emperor, on the duc de Morny when it was rumored that he had been poisoned. Although the report has disappeared, Robin could only show, as Prosper Mérimée put it, that the duke had died in good health. Along with

Broca, Vulpian, Charcot, and others, Robin was accused of teaching materialism in public courses. A petition was deposed against them in the Senate, but that body gave them its approval by a vote of 83 to 40. Perhaps a Catholic denunciation of Robin's atheism helped. Some of the faithful of the *dîners Magny* – Taine, Renan, Flaubert, Robin – were accused of eating sausage and *boudin* at Saint-Beuve's on Good Friday, 1868. Robin's views probably had some impact on republicanism through Gambetta and Ferry, and although he did not leave much of a positivist imprint on Clemenceau, one of his famous students, Robin wrote an introduction to his medical thesis. Not everyone believed that the philosopher had woven the cloth of science with irremovable threads of gold.[7]

Within his positivist framework, Robin thought that each science can be considered from abstract and concrete viewpoints. An abstract designation refers to general science, the laws of the different classes of phenomena each science is concerned with; a concrete classification means from a descriptive and specific viewpoint, involving the application of general laws to the history of each specific object. The concrete sciences are the so-called applied sciences, of which a large number may derive from one principal science. The abstract part of each science is the basic part, for the others, being derivative, depend on pure science for any advances they might make.[8] Progress might be science's most important product but it is a product of pure science.

According to Robin, biology studies phenomena that are special and also more complicated – perhaps he meant complex – than those studied by other sciences. "La biologie est la science qui a pour objet de ses études les corps organisés, et pour but la connaissance des lois de leur organisation et de leur activité." The highest aim of biological knowledge was an increasingly precise conception of the organization and acts of an organic order.[9] He was certain that one can observe in living bodies the mechanical, physical, and chemical phenomena occurring in inorganic ones, but these living bodies become so complex that it is impossible to do a direct physical analysis of nervous, sensible, intellectual, or moral phenomena – they are *vital phenomena*.

If instead of considering the totality of organized bodies, the scientist of the mid-nineteenth century got down to concrete or individual examination of them or of their parts, whether from the viewpoint of anatomy, biotaxy, physiology, or environmental science, this concrete or individual approach implied a division of biology into two parts. Abstract biology was an instrument that permitted scientists to consider organized life from two points of view: the static – having a tendency to act – and the dynamic – acting. The Comtean categories, first stated in de Blainville's *De l'organisation des animaux* (1822), bifurcated again. Static biology covered

64

two of the four basic branches of biology: anatomy, which used comparison as its chief method of intellectual investigation, studied the organization of life; biotaxis studied the laws of the arrangement of life in natural groups, according to the similarity of their organization as seen in corresponding modifications of external organs. Dynamic biology ("bionomie") covered the other two basic branches of biology: the science that studied the influence of environment or external agents on living beings ("mésologie"); and physiology, which studied directly the functions of each organ. When Gley mentioned Robin's classification he put physiology first under the dynamic group, a change that may have had the effect of playing down the strong emphasis that Robin put on the reciprocal influence of environment on life and the resulting science of hygiene. Robin admitted that this "science" was the first part of physiology and could be rejoined to it. Indeed, all of the branches were intimately connected, with each branch dependent on the one preceding it. The entire "science of organization" was based on chemistry, a point that Robin made twice in one page, in spite of an earlier salute to the vital nature of biological phenomena: "La biologie dérive de la chimie." In spite of this widely advertised dogma, there was little research in biological chemistry in France. Some blamed this deficiency on biologists' ignorance of chemistry and physics. When Gautier's work took a physiological direction, Wurtz warned him to get back to pure chemistry – "Elle seule étudie des faits exacts, définis, relativement simples, seule elle permet les généralisations et déduit des lois" – for the chemistry of living beings was still too imperfect a science in the 1860s.[10]

In his further classification of types of anatomy, Robin did not list comparative anatomy, for he thought that it was simply an irrational transformation of one of the methods of biological investigation into a distinct science. Just as physics and chemistry developed to the highest possible degree the arts of observation and experimentation, so biology had developed comparison as its chief intellectual method of investigation; the comparative method was not a separate science but the methodology of all the branches of biology. Biology also had the capacity of developing to the highest degree the art of classification, just as chemistry had the property of carrying the art of nomenclature to its pinnacle. The comparative approach was of key importance in the study of man, whose physiology could not be well known except with the help of a serious study of animals, "which show us separately what is united in him."

Although Robin did not devote so much attention to biology from a concrete viewpoint as he did to it from an abstract viewpoint ("considérant les êtres en général"), he did give it strong emphasis, especially in its connection with medicine. There were two branches of biology envisaged from the concrete point of view ("c'est-à-dire individuel, descriptif ou d'application"). First, there was natural history, which considered each species of being successively from the four viewpoints of anatomy, biotaxis, the science of milieux, and physiology – all of which also applied to being in general. Secondly, there was pathology, "nonnatural history," a science complementary to natural history, extending the four study-areas of natural history to accidental states. With the help of precise knowledge of the normal state one got to know the changes that organs can undergo, and this knowledge led to the reestablishment of the normal state. Pathology was the indispensable complement to experiment: It was a sort of spontaneous experimentation that resulted from a methodical comparison between the different anormal states of the organism and its normal state. (Following Broussais, Robin did not think that the pathological differed radically from the normal, which meant that a precise and rational idea of the healthy state must inevitably form the indispensable point of departure for any non-fictitious pathological theory.) One defect of pathology was that it concentrated on man, ignoring the information furnished by studies of animal diseases. To each science corresponded a technique; in biology, vegetable culture and animal domestication corresponded to natural history, and medicine to pathology. Hygiene corresponded to the science of environment and obstetrics to one of the parts of physiology.[11]

Robin's belief that medical technique had been the past source of knowledge in physiology and pathology, evident in the necessity for sciences like anatomy, did not stop him from warning that the development of these sciences had reached the point where they had to be considered independently of all applications. Fortunately, a stage had been reached in which separation of science from its applications led to more applications. This development took place spontaneously in most sciences, but biology, because of its complexity and the irrationality of its methods of study, required that the path of progress be mapped out in advance. Only by a planned separation could medicine benefit as much from biology as industry had from physics and chemistry. The aim of the Société de biologie – and biology studied plants as well as animals through the same sciences—was general but could be broken down into specifics. Not being a society of pathological anatomy or of pathology, the members studied anatomy and the classification of beings in order to explain the mechanism of functions, and they studied physiology in order to find out how organs can change and within what limits their functions can deviate from the normal. At the core of this knowledge was the natural arrangement of things; knowledge of the anormal should lead to knowing the normal state. If anormal cases did not establish relations between the normal structure and function, they were

66

useless. This safeguard would ensure that medical applications would flow spontaneously from "pure science." In any case, other societies existed for such activities as the direct study of pathology. Paradoxically, then, the separation of medicine and biology was inspired by a desire to increase the number of medical applications that would issue from advances in biology. Also Robin did not neglect the love affair between doctors and plants.[12] No doubt the medical scientists could live quite comfortably with these ideas in the mid-nineteenth century, for, after all, they were the biological community. It is unlikely that they would have proceeded so sanguinely had they been able to foresee the inevitable independence of the new discipline a short generation away.

Positivistic scepticism on Darwinism

The founding and history of the Société de biologie provide a certain amount of prima-facie evidence of the influence of positivism on French biology, especially in the lip-service given to Robin's pronunciamento. These guidelines proclaimed from on high probably exerted little lasting influence on the development of biology. A generation later most of the ideas were as peripheral to the discipline as Robin himself became. Yet for a while positivism had the illusion of an intimate association with contemporary science; it was a luxury few ideologies could boast of, although some, like Marxism, tried to profit from a contrived congruence. This association seemed doubly cemented by the alliance of positivism and French biology in rejecting Darwinism. The rejection was not unusual in the nineteenth-century international scientific community. In the pre-Darwinian choice offered scientists between Lamarckian evolutionism and Cuvier's fixism, Comte, with the reservation that the issues be left open, picked Cuvier solely on the strength of purely formal arguments in contemporary science and for "the logical perfection of science" evident in the theory of fixity of species.[13] There is really not much difference between the sceptical attitudes toward Darwinism littering the pages of *La Philosophie positive* and the general doubt and even hostility of French specialists in botany, zoology, and the social sciences. Positivist scepticism was derived not so much from subservience to the pope of positivism, who had biological evolution on his index of erroneous doctrines, as from positivism's own blinkered view of the nature of science. When the biological community shifted its allegiances from a Cuvierist position to a Darwinian one, or more likely to a Darwinian-Lamarckian one, positivism's previous hostility and scepticism toward evolutionary theories made it quite unsatisfactory as a philosophical doctrine for biologists. In the third volume of *La Philosophie positive* (1868), the physician Charles Letourneau, a notorious materialist, conceded that it was legitimate, in the true scientific sense of the word, to accept the doctrine of the variability of organic species. In the same year Clémence Royer, the controversial translator of the *Origin*, who turned Darwin into a bête noire of the clerical camp, domesticated Darwin in assuaging the amour-propre of the French in the area of scientific originality by telling them that Darwin had really only added two principles, two laws, to the theory of Lamarckism, an old taboo that became a new totem. Lamarck, long forgotten, was then being resurrected as proof of French scientific prowess, in the face of the greatest threat from Perfidious Albion since Newton, although his definitive rehabilitation would come only in the cultural nationalism engendered by the defeat of 1871. The two laws were those of the struggle for existence ("concurrence vitale") and of natural selection, which derived from the first law. "And the innovation was certainly important." No doubt, but the concession might have been more gracious if not added as an afterthought. These two series of articles were the only really unreservedly favorable reactions to scientific evolutionary theory to appear in *La Philosophie positive*.[14] In light of the criticism of both Letourneau and Royer that would later appear in the journal, it is clear that the contributions of both were in basic disagreement with the positivist position on evolution.

That both Letourneau and Royer were leading French materialists undoubtedly did not help the Darwinist cause in the positivist camp. In the course of a comparison between Darwin and Agassiz, E. Jourdy noted that Royer's book on the simian descent of man (*Origine de l'homme et des sociétés*, 1870) accepted Darwinian theory in its entirety, including spontaneous generation, with the natural result of a materialist system explaining everything.[15] But until 1879 the young French materialist school of the nineteenth century did not have any more than a few articles and studies to bolster its intellectual stature; then André Lefèvre published a two-volume materialist bible for the guidance of the public. This "complete code" or "general treatise" of the materialist school was, Wyrouboff conceded, well thought out, well organized, and well written, but totally false. And it gave Wyrouboff great pleasure to show with ease and his usual relentless rigor the rotten scientific and philosophical basis of materialism.[16] The philosophy of positivism was that science has no philosophy.

The intimate relation of positivism to science was succinctly stated by Littré: Positivism is founded on the idea that the basis of science is experiment. On the rock of this cliché, so dear to the hearts of worshipers at the empiricist shrine, the high priest of the positivist heretics could build "a philosophy that was both relative and the expression of all positivist knowledge." Positivist philosophy is not deist, pantheist, atheist, or materialist. No individual science came to any such conclusion; neither could the positivist philosophy, for "the general conclusion is only the expression of all the individual conclusions established by positive knowledge."[17]

The counteroffensive against acceptance of evolution as a solid scientific doctrine was clearly under way by 1869, the third year of *La Philosophie positive*. Wyrouboff thought that discussion of the mutability or fixity of the species showed a low level of understanding of the nature of science itself. Every classification must be artificial in the sense that it is constructed not with real facts but with abstract mental concepts. Biologists easily forgot this capital point.[18] Antievolutionists did not show any stronger reasoning powers than their opponents. Wyrouboff found that Agassiz's philosophy was that of the childhood of humanity, presumably the theological first phase: All of organized nature shows, in each of its details, the existence of a creative, intelligent, omnipotent, and omniscient God. This weak reasoning often reached the heights of a sublime naïveté in Agassiz.[19]

Positivists also shared in the general French *Angst* arising from the cloud of vagueness shrouding the *Origin*: The book itself is not clear because of a prolixity of style, aggravated by the expression of ideas by an embarrassed scientist who has published in a hurry the result of notes accumulated in disorder for many years. Even praise was damning: Darwinism was so ingenious and grandiose in its concepts that it would prevent anyone from ever being puzzled by a biological fact. Concentrating on the logical form of Darwinism, the positivists thought that Lyell's theory of big effects through small causes had seduced Darwin into forgetting that science has certain rigorous requirements, that hypotheses must be verified, and that doctrines must be sanctioned at each step of the way. "Mais la fantaisie va son train; le lecteur sent bientôt qu'il perd pied dans ce tourbillon vertigineux et qu'il quitte la terre ferme pour s'enfoncer dans le monde des rêves où nous refuserons de suivre le naturaliste anglais." Agassiz was guilty of a different type of fantasy, but his conception of the intimate nature of man was the same as Darwin's: Agassiz raised nature up to man, and Darwin lowered man down to nature. Starting out from opposed viewpoints they ended up with the same result.[20] André Sanson took much the same attitude toward the two chapters on pigeons in Darwin's work *The Variations of Animals and Plants under Domestication* (1868). A tropistic French complaint about style, a logical lament over contradictions, but worse, a complaint that there was nothing new on evolutionary doctrine. Looking at Darwin's work in his own area of specialization, zootechny, Sanson declared Darwin's work to be inferior to experiments done before and after on variability in cross-breeds to show the fixity of specific characteristics. Darwin's experiments had little scientific value, as one might expect of experiments similar to the empirical experiments of breeders aiming at creating new breeds through *croisement* and *métissage*.[21] Judged by the sacred criteria of style, logic, and utility, Darwinism was deemed unworthy to be admitted to the body of positivist doctrine.

As late as 1879 Littré still stressed the hypothetical nature of spontaneous generation and of evolution. He had always believed that in spite of the support evolution found in embryology and in paleontology, there was no experimental fact on which to base the transformation of one species into another. On the basis of this common scientific view, he therefore rejected evolution as a candidate for incorporation into positive science. It did, of course, have the status of a biological hypothesis. Littré's view of evolution was fairly sophisticated, certainly more intelligent and complex than the simple views of those who just rejected evolution on naïve empirical grounds, barely hinting at an awareness of the new philosophy of science developed by Jules Lachelier and Emile Boutroux. "Le transformisme ... [est] un artifice logique." To call evolution a logical construction was not in any way to deprecate the idea. "Les artifices logiques ne sont aucunement à dédaigner." In physics the hypotheses of the ether and of the wave theory of light showed their scientific power by the number of inferences verified; they were useful as means of specific research and provided a general view while remaining amenable to the controls of observation and experiment. The atomic theory in chemistry, although it continued to receive constant confirmation and had not been contradicted, still belonged to the region of unverifiability. Evolution was far from being as advanced a hypothesis as any of these, but Littré still did want to dismiss it out of hand. If only an experimental proof in the form of the production of living species could have been established! The technical scientific difficulties rather than any philosophical hang-ups explained Littré's reservations.[22]

Robin, labeled by Conry as a positivist histologist, voted against admitting Darwin to the elect of the Academy of Sciences on the grounds that Darwin was not first-rate and was inferior to Bischoff in scientific production. Having reduced Darwinian theory to Lamarckism, Robin was unable to appreciate Darwin's originality; such was the baleful consequence of accepting Comte's pontification that Lamarckian transformism was a zoological projection of embryological evolution. This pre-Wolffian sense of development was taken up by Littré: "Le transformisme procède par voie d'évolution, et se met de la sorte en opposition avec l'embryogénie, qui, elle, ne procède que par l'épigenèse ... Pour aborder l'idée de la production des espèces vivantes nous avons deux principaux instruments: l'embryogénie et la série organique, y compris les fossiles. Il est fâcheux pour le transformisme de n'être pas évidemment épigenétique." *La Philosophie positive* later hailed the "revolutionary" doctrine of Wolff. Dr. Ernest Martin, in an analysis of "La Production artificielle des monstruosités; recherches de M. Dareste," wrote that the doctrine of Wolff (epigenesis) completely changed the idea of life and consequently the whole of biology.[23] Conry sees this misinterpretation of Darwinism as the twisted offspring of the Comtian sketch of development. In the fourteenth edition of the *Dictionnaire de médecine* (1878) "darwinisme" is demoted to a classification under "transformisme." Anyone who adopted the systematic thought of Comte and Littré could dismiss Darwinism as doubly superfluous. "Le vrai problème biologique consiste à articuler structure et fonction."[24]

Comtean and positivistic parameters for biology

Comte's concern for defining life was solidly based on biological research in the early nineteenth century. This interest in seizing on a single definition of the elusive properties distinguishing organic matter and behavior from the inorganic was the culmination of a concern of "biologists" for two centuries. Comte participated in the last epistemological quest for this Holy Grail before the search was relegated to a distinctly peripheral activity in biology. A definition of life or of death was not of much use in biological research, although it had some significance in medicine, as it still does, especially in terms of the old paradigm of Bichat. When the French queen of the biological sciences, physiology, consecrated her reign to experimentation, attempts to understand life in essentially philosophical terms became far less important, for one can't do an experiment on life, but only on life-related, limited problems having experimental solutions. This was equally true of zoology, which underwent a rebellious separation from physiology in the second half of the nineteenth century. This reorientation of research priorities, a departure in a new direction, led to one of the most fruitful periods in the history of the life sciences in France. Rarely has there been a time of more fertile, first-rate scientific minds: Bernard, Bert, Lacaze-Duthiers, Giard, Delage, Dastre, d'Arsonval, Henneguy, and Richet. While these men were creating biological science the positivists remained wedded to a sterile logomachy.

During much of the nineteenth century, biologists were not sure exactly what biology was and therefore tried to conquer their doubts by imprisoning the elusive subject within the comforting confines of definition. The French positivists were excellent in this game. First, they were French intellectuals, who were generally addicted until recently to achieving a sort of Cartesian

71

clarity in discourse. Second, they were positivists, who especially liked to have the illusion that they always knew exactly what they were talking about in science. Comte, in a moment of excessive generosity, attributed the invention of the word "biology" to de Blainville, but it had been used by Burdach in 1800 and by Treviranus and Lamarck in 1802.[25] Perhaps Comte should be given credit for giving biology some impetus toward its independence from physiology, although in the end this was more a matter of different research paths than of definitions.

Comte's aim in life was to create sociology, the new science of society, thereby putting this intellectual activity on a level of certainty with the queen of the sciences, mathematics, and her immediate royal followers: astronomy, physics, and chemistry, with biology holding up their trains. Biology had evolved rapidly in the eighteenth century and during Comte's lifetime in the early nineteenth century. This science was not only a closer chronological example of the evolution of a science than the others, but also in the group of sciences already in the positive state it was the one directly before sociology. It is not surprising that Comte devoted most of the third volume of the *Cours de philosophie positive* to biological philosophy. Comte's biological philosophy was essentially an intelligent manipulation of the ideas of Lamarck, Bichat, Cabanis, and especially de Blainville. Based on his knowledge of the work of contemporary biologists, Comte's biological philosophy later struck Paul Tannery as worthy of high praise: Comte's synthetic exposition of the mathematical, physical, and natural sciences is a historical document of great importance on the state of the sciences and scientific ideas at the beginning of the nineteenth century. Tannery knew, of course, that Comte's real influence would be on sociology, not biology. "Peut-être est-il permis de se demander si, dans la sociologie de l'avenir, la trace de l'oeuvre de A. Comte restera marquée plus profondément qu'en biologie." With his Cartesian mania for breaking up a subject into its basic parts, Comte divided biology into a trinity of anatomy, physiology, and biotaxy – indeed, biology was a name conveniently designating the "ensemble total." "Ensemble" of what? "L'ensemble de l'étude réelle des corps vivants . . ." Comte insisted that the word biology should be carefully reserved for the truly basic part of such an immense subject: the part "in which research is both speculative and abstract."[26] It is not surprising that, in contrast to the fate of Comte's comparatively moribund pronouncements on mathematics and on physics and chemistry, his biological philosophy became a subject of some controversy in the definition of research programs, rather than remaining solely a matter of tedious commentary by pious positivists or of jejune banquet rhetoric by scientists and politicians. He also had the luck to be writing when the subject was in a period of formation, when a powerful

72

intelligence superbly informed about contemporary research directions could fulminate potentially influential dogmas, especially when he echoed what the scientists themselves were saying. The philosopher of science is most influential when he is a clone of the scientist.

What were Comte's main ideas on biology? Comte accepted de Blainville's definition of life as a sort of peristaltic movement of composition and decomposition: "le double mouvement intestin, à la fois général et continu, de composition et de décomposition, qui constitue, en effet, sa vraie nature universelle." Comte rejected Bichat's brilliant tautology ("La vie est l'ensemble des fonctions qui résistent à la mort"). Although he applauded Bichat's severe mauling of the mechanistic explanation of life, Comte did not accept the consequences of Bichat's dogma that there is no connection between the two sets of laws governing the organic and inorganic worlds. Inspired by Cabanis, he believed that harmony between the living being and its environment is the basic requirement for life. Comte put it in his best entangled prose: "Elle [life] présente ainsi l'exacte énonciation du *seul phénomène rigoureusement commun à l'ensemble des êtres vivants* considérés dans toutes leurs parties constituantes et dans tous leurs divers modes de vitalité, en excluant d'ailleurs, par sa composition même, tous les corps réellement inertes. Telle est la première base élémentaire de la vraie philosophie biologique."[27] In Comte's opinion, the ideas of life and of organization were osmotic. So he could easily accept the determinism implicit in de Blainville's definition, for which a determinate organism and the special environment of the process characteristic of life were axiomatic.

Harmony between organism and environment is the primordial condition of life. The great problem for the positivist biologist must be to establish, according to the smallest possible number of natural laws, the link between the double idea of organ and environment and that of function. This guiding idea of Comte's philosophy of biology meant that biology was not exempt from his trivial and famous binary law dividing phenomena into static and dynamic categories, meaning the disciplines of anatomy and physiology in this case. It was in strongly declaring himself a supporter of the doctrine of the determinism of vital phenomena that the First Positivist showed that he was in the camp of the moderns, or at least on the side of a future winner – Bernard – as opposed to that of a past loser – Bichat – who insisted on the regularity of the physical phenomena of the inorganic world in contrast to the instability of the physiological phenomena of the world of living beings.[28] Many commentators judge the value of a philosopher according to the degree to which he can predict the successful research strategy of some science or other – perhaps not a bad idea, at least as long as the philosophers insist on talking about the future of science.

The positivists were as obsessed as Descartes and Bacon had been with method; their pronouncements on it were no more fortunate than the classic statements of the seventeenth century. Ordinary minds are no more immune from error than great ones. Like Bichat, de Blainville, and H. Milne Edwards, Comte thought that the microscpe had usurped too large a role in biological research. Obviously it is very dangerous to follow scientists in charting the future of science. In following the biologists, Comte often imitated their mistakes. Comte was not optimistic about an experimental future for biology. The chief difficulty in experimenting on organisms was in the complexity and changeability of biological phenomena. For Bernard this was a challenge to overcome, for Comte it was a limitation on the applicability of a technique having its true domain in physics rather than in the study of living bodies. What was the key to physics could be only of minor use in biology. Of more significance was the use of chemical reactions in the analysis of the structure of tissues, but again the demon of complexity made the precision generously attributed to chemistry an impossibility in biology.[29] Could biology be a science in the sense that physics and chemistry were?

Yes: the comparative method to the rescue. This "basic method, so suited for biological investigation," had long been operational in chemistry, but Comte seemed more inspired by the work done in comparative anatomy, perhaps because he saw more organization and structure in this subject than in any other area of biology. The ideas of life and organization made up a basic unity that could not broken up: All organisms possess certain general characteristics as well as a host of secondary variations. All living beings have a certain common vitality, and their similarities are in practice necessarily more important and more fundamental than their distinguishing characteristics. The comparative method would raise biology to a scientific stage as the science dealing with all organized beings. Out of the confused would come forth the hierarchically ordered.[30] Cuvier, Dutrochet, and Comte all trumpeted the need to apply the comparative method to physiology; the fact is "that during the first four decades of the [nineteenth] century a large number of dissertations had been written in Paris on such questions as observation and experimentation in biology, and the relation of physiology to physics and chemistry and to philosophy."[31] It seems that all of this is regarded as significant by most historians of science because its culmination was the brilliant experimental physiology of Claude Bernard. Alas! This seems to be another case of nonexistent precedents, where the "origins," so beloved of historians, turn into fantasy: Bernard "connected his ideas concerning the nature of experimentation not with any of those authors, but with Tiedemann and Réaumur, two of his most eminent predecessors in the study of digestion."[32] Comte might pontificate at length, but the *libido sciendi* of the scientists came from science itself rather than from an injection of positivistic philosophy of science, and Comte might not even have been the agent of transmittal.

Without doubt, the leading positivist in the biological sciences was Charles Robin. Not only did his lifetime extend back into the period of the creation of positivism, but he knew the creator himself. He was one of the founders of the Société de biologie, the only genuine scientific organization basically influenced by the creator of the science of society. In 1849 Robin published a treatise on the microscope for its users in the biological sciences.[33] For Robin "life did not depend on a rigid structure but on a 'state of organization' – in fact on 'a particular molecular state.'" Believing that microscopic research was a prelude to chemical analysis, which would establish the molecular organization of life, Robin could not accept the idea "that the cell could be the single fundamental component of organized beings." But in the mid-nineteenth century, the chemical analysis of organisms could lead only to the dead paradigm enshrined in the three-volume *Traité de chimie anatomique et physiologique, normale et pathologique*, done in collaboration with the chemist Verdeil, for, "given the contemporary state of chemical knowledge, the superiority of a morphological approach was undeniable."[34]

In the first edition of *Du microscope* Robin made the second part a magnificent monument to the role of positivist ideology in biology: "De la classification des sciences fondamentales en général, de la biologie et de l'anatomie en particulier." Especially convenient for the reader in search of the positivist influence is the indiscreet abundance of rigorous definitions of elusive entities like law, life, and biology.[35] The definitions are pretty much the same as those Comte had lifted from contemporary scientists and covered with a patina of positivism. Law: "relations ou rapports constants de similitude et de succession qui rattachent les uns aux autres tous les phénomènes que présentent les êtres qui composent l'univers." Then he recites the standard litany on the aim of science and its method. Hypotheses in science are part of the positive method; we sigh in relief that they were not banished into the limbo of theological or metaphysical fancies. A definition of life is pure de Blainville and Comte: "on donne le nom de VIE au double phénomène de mouvement moléculaire, à la fois général et continu, de composition et de décomposition, que présentent les corps organisés placés dans un milieu convenable."[36] Biology studies the organs of organized beings from a functionalist viewpoint. "La biologie est la science qui étudie les êtres organisés dans le but d'arriver par la connaissance des organes ou des modifications organiques à connaître les fonctions ou actes, et réciproquement." This was essentially the party line Robin delivered in his address to the Société de biologie. It was innocent enough and unlikely to damage anyone's research program, even if he took it seriously. The edition of *Du microscope* published in 1871 did not contain the section on classification. *Autres temps, autres méthodes?* The practicality of the subjects treated is given prominence in the title.[37] Comte would have approved of this, no doubt, but probably not of the dropping of the scientific catechism exuding positivist virtues.

One disadvantage inflicted upon positivism was that its biological definitions were immortalized and preserved from corruption in Littré's great mausoleum of the French language, instead of being given the typical ephemeral half-life characteristic of scientific publication – unless the dogma had the luck to be given intellectual virility and attired in a seductive style by Claude Bernard. And every discipline seems to have its genetic quota of at least one great stylist per era. In his *Dictionnaire de la langue française* (1863– 78), Littré was unambiguous in defining biology: "Science qui a pour sujet les êtres organisés, et dont le but est d'arriver, par la connaissance des lois de l'organisation, à connaître les lois des actes que ces êtres manifestent."[38] Littré also defined physiology: "Science qui fait partie de la biologie, et qui traite des fonctions des organes chez les êtres vivants, chez les végétaux et chez les animaux."[39] A later orthodox positivist was not at all happy with heretic Littré's supererogatory definition, for, like most attempts at precision, it produced uncertainty and confusion rather than conviction and clarity. Porfirio Parra argued that if one penetrated beyond the abstraction of Littré's formula (arriving at knowledge of the laws of action by the laws of organization), it is clear that the characteristic aim of biology is to know the functions through the organs, but because that is the only way to know functions, Littré really said that biology deals with the functions of organs – the same definition he gave to physiology.[40] Obviously a circle that was not necessarily vicious, but certainly not an enlightening argument.

Other deviants from the positivist camp were also excoriated by Parra. Bernard was guilty of confusion in his *Leçons sur les propriétés des tissus vivants*: "... la physiologie est la science qui étudie les phénomènes manifestés par les êtres vivants; elle est, par conséquent, la science de la vie, la biologie, comme on l'appelle également."[41] Physiology was mingled indiscriminately with zoology, botany, and even anatomy and embryology, among other branches of the life sciences. The meaning of a word varies inversely with its scope of meaning: to make a word comprehensive can lead to meaninglessness; Parra made a valiant and misguided effort to strangle Bernard's definition in a burst of panlogistical analysis.

Even Bernard's determinism was not without its obscurities for the positivists, at least those as fanatically logical as Littré. Bernard's expression "les phénomènes métaphysiques de la pensée" ressurected ghosts long exorcized from the thought process. Any role for metaphysics was an idea horrendous enough to strike atavistic terror in the hearts and minds of properly hatched positivists. Littré's explanation of Bernard's use of this proscribed word was not unreasonable: It was the unconscious product of reminiscence, which he had used without thinking that there were hardly any phenomena less compatible with physicochemical determinism. Also disturbing was Bernard's distinction between physiology and psychology. "Considérer le cerveau comme l'organe de l'âme est nuisible aux progrès de la physiologie et de la psychologie."[42] Littré's equation of both was the characteristic position of a whole brilliantly innovative school of French physiological psychology: "If it is true . . . that intelligence is the function of the brain, psychology is nothing else but this same function." Littré thought that Bernard's basic problem was the conflict between two doctrines, both of which he seemingly accepted. One doctrine got its theory of the knowledge of living bodies with the help of general physics, chemistry, and mechanics; the other doctrine assigned special properties to living bodies – a species of vitalism. With these two doctrines warring within the confines of a single breast, Bernard's method and his somewhat obscure determinism could not sate the appetite of positivists for consistency and agreement with the biological gospel of Comte and Robin. Unfortunately for the positivists, Bernard had what they lacked: creative genius in science.

Bernard's confusion passed into French scientific consciousness. The positivists found it prominently displayed in the *Dictionnaire de physiologie*, edited by Charles Richet, in which Henri de Varigny wrote the article on biology. Recognizing the difficulties in using the word, de Varigny noted that two kinds of studies were brought together under the same rubric: biology *stricto sensu* and the biological sciences. Modern textbooks like Wilson's *General Biology* (1889) presented the morphological sciences (anatomy, histology, taxonomy, distribution, embryology) as subdivisions of biology, which was consequently all of the biological sciences together. Biology included the group of sciences dealing with the phenomena of living matter. After a historical and critical analysis, de Varigny arrived at a definition: "La biologie est la science des relations des organismes avec le milieu ambiant et avec les organismes présents ou passés."[43] Parra found de Varigny's formula defective even from the denotative viewpoint, for it did not give biology the attributes necessary to distinguish it from the biological sciences as a group.

The word *relations*, with its two meanings, was also ambiguous. The word could mean the acts of an animal toward another or those acts of which it is the subject; in the concrete language of physiology, these acts made up the functions of relation. Understanding its mechanism was the area of classical physiology, including even an animal's sexual life. This was the area of the individual.

To move beyond the individual to the collective results of actions and reactions that living beings have upon one another as groups was to move out of classical physiology to a vast new area, one that had been worked in by Lamarck, Geoffroy Saint-Hilaire, and Darwin, the area formerly called zoological philosophy and then biogenesis, although it was not the distinct and independent science one could call biology. Relations also included similarities and differences discovered in the objects of the environment. These objects make up the unity of knowledge. In the midst of the variability of phenomena, they make prominent the contrast between the single and immortal idea and the fleeting and individual fact. This type of study of relations led to taxonomy (Comte's *biotaxie*) but is the domain of the sciences of classification – zoology and botany. De Varigny dealt with various types of relations – hereditary, that is, with the past, and nonhereditary, that is, between organism and milieu. The fact that relations belong to pure biology and to concrete physiology caused the strict positivist Parra to lament that de Varigny's definition allowed biology to include questions falling within the provinces of zoology, botany, and general and special physiology.[44] To ensure that biology would have the necessary scientific autonomy, it was not enough to say only what biology does; behind the research must lurk a philosophy.

Parra's way of ending the putative confusion in the mater of definitions, which presumably implied some confusion in the area of biological research itself, was charmingly naïve in its simplicity. A science cannot be characterized exactly by the phenomena it studies, for different sciences study the same phenomena with different methods. We may be grateful that positivism did not inflict upon us yet another verison of the scientific method. General and special pathology both study disease but are distinguished by the different points of view they start out from. The capital question was, then, What should be the viewpoint from which biologists start out? It was no less important than the viewpoint that separates the abstract from the concrete. The "grand philosophe" Comte had decreed that some sciences are abstract, general: They aim at discovering the laws governing different classes of phenomena while taking into account all the individual cases that can be imagined. The other sciences, called concrete, individual, or descriptive, are sometimes also called natural sciences: Their activity consists in the application of the laws of the abstract sciences to the real history of the different beings that exist. On the basis of Comte's opinion and Parra's knowledge of what had happened in biology since Comte, Parra proposed his own definitions for the modern positivist. Physiology is the concrete science of the life of a determined species. Biology is the abstract science of life.[45] *Lucus a non lucendo?* Perhaps. Whatever the philosophical value of Parra's critical analysis of the working definitions of contemporary life science, the orthodox positivists were convinced that some of the leading scientists of the day were floundering in conceptual confusion because they had not followed Comte's philosophy of biology.

Limited role of positivism in the Bernardian *Weltanschauung*

Claude Bernard was certainly a thorn in the mind of positivism. What was his philosophical camp? Perhaps Schiller was right: Bernard is all things to all men. A spiritualist to Sertillanges, Chevalier, and Lamy; an idealist to Caro, Ravaisson, Renouvier, Pierre Janet, Boutroux, Brunetière; and an agnostic to Renan. No big-name system builder has been able to get much out of Bernard, at least in the way Schopenhauer was able to gut Bichat. This does not indicate Bernard's philosophical vices, and it may be a sign of his scientific virtue. Paul Bert had said essentially the same thing. "Ses écrits ont put servir, à tour de rôle, à tous les souteneurs de thèse." This knowledge did not prevent the physiolgoist Mathias Duval from comforting the positivists with the assurance that Bernard was more in their camp than anyone else's. Writing in *La Philosophie positive*, Duval concocted a clever argument: Bernard's quest for determinism resumes his scientific philosophy. A general study of his work and its tendencies led Duval to the welcome conclusion that positivism is the only philosophical framework that corresponds exactly to the doctrine of determinism rather than to spiritualism or materialism. But Duval admitted that no explicit declaration or profession of faith could be found in any of Bernard's numerous publications.[46]

There is no agreement between scholars about the influence of Comte on Bernard. In the camp of believers may be placed Gley, Canguilhem, Charlton, and also Grmek, although the issue is quite minor for him. In spite of Bernard's condemnation of positivism as a philosophy, Charlton maintains that Bernard's "assertion of neutrality is deceptive: his arguments reveal over and over again his adherence to the positivist theory of knowledge." Holmes is among the nonbelievers: The similarities between

Comte and Bernard are superficial and the differences fundamental, chiefly because of the disagreement of Bernard's idea of the nature of physiology with that of Comte's. Virtanen concludes that "Bernard is a positivist only in a vague and general sense" because of the differences between them in philosophical outlook and in biology. Most of the problem comes from the elusiveness of a universally accepted definition of positivism and from Bernard's ambiguities and potential contradictions. Annie Petit gives an excellent analysis of the different views of Comte and Bernard on the value of the history of science, or rather its lack of value for Bernard. "Le passé est, pour Bernard, fait d'erreurs, de vérités partielles et provisoires; celui de Comte est fait de tâches nécessaires." Bernard's philosophy of history is not all black, for he has to make allowances for the creative achievements of scientific geniuses like himself. As Petit cleverly put it, "Bref, Claude Bernard hésite entre le hasard et la nécessité."[47]

Was Bernard a philosophical positivist, who believed that a sound epistemology can be built only by accepting science as the model for attainable knowledge? Did Bernard really care? In a manuscript now ironically published under the title *Philosophie*, Bernard cast philosophy outside the scientific pale, although he conceded that it was a useful amusement for the mind to discuss philosophy after work – or, we may assume, in the Beaujolais, at any time. There is only experimental science; beyond experiment nothing is known; philosophy can discover nothing new because it neither experiments nor observes. To follow Bacon would be to "become as sounding brass, or a tinkling cymbal." "Quant à Bacon c'est une trompette et un crieur public . . . "[48]

Science was not the epistemological revelation that positivists imagined. Comte's mistake was to believe in something called positive; by accepting positive philosophical generalities, he thought in vain he would get rid of metaphysics. All scientific theories are metaphysical abstractions. Bernard echoed Chevreul in recognizing that even facts themselves are only abstractions. Nor could science eliminate the search for causes: The scientist is always looking for first and final causes; indeed, science is the consideration of an infinity of proximate causes leading to the unattainable first and final causes. The end can never be reached because the end of the search would mean that man knew everything. Ignorance is a stimulus to knowledge. When man knows everything he will be annihilated. Having echoed Pascal, Bernard could do no better than to quote him: "L'homme est fait pour la recherche de la vérité et non pour sa possession."[49]

Bernard declared Comte wrong in assuming that the positive state would destroy the theological state; they would only separate as science developed. An even greater objection to Comte's scheme was the assumption that it could suppress the moral and sentimental part of man. "Ce n'est pas la tête, c'est le coeur, c'est le vague, l'inconnu qui mène le monde."[50] Whatever the validity of Bernard's curious exegesis of Comte, his works contains more antipositivistic than propositivistic statements.

How convenient it would have been if positivists had been able to unfurl the banner of Bernardian physiology in the positivist camp. Little matter that his philosophy was a stylish, scientifically inspired hash of no epistemological power – just as long as it was positivist.[51] The important thing was to connect the teachings of Comte directly with research producing good science. Not an easy task, and probably a hopeless dream apart from some second-rate medical research, which was not what the philosopher and would-be legislator of research programs had in mind.

The scientific domination of the Société de biologie by Bernard obscured for a long time the physiologist's lack of commitment to the ideology that Robin had injected into the official aims of the Society. One could hope for Bernard's guilt by association. The secret of his innocence might never had arisen but for peevish complaints by materialists about Bernard's philosophical neutrality and for rejoicing by the enemies of positivism that "the leading physiologist of the nineteenth century" had a philosophy of science diametrically opposed to Comte's.[52]

In 1879 Charles Letourneau, who had become one of the haughty guardians of nineteenth-century French materialism, proclaimed the impossibility of Bernard's refusal to make a decision for either materialism or spiritualism. Such blasphemy was soon punished by an authorized pen of official science. Charles Richet, perhaps France's best-known physiologist after Bernard, began his defense with the classic tactic of smear. Was it not remarkable that the religious camp said the same thing as Letourneau? Presumably if the materialists and Catholics agreed on a proposition, this was a double a priori proof of its falsity. The flaw of Letourneau's position was that metaphysics was bad only if it led to spiritualism; if it led to materialism, it was excellent. Should not the author of an elementary book on biology have understood that at the core of Bernard's method was the epistemological axis of experimental determinism? Richet emulated Bernard in proclaiming *urbi et orbi* the independence of physiology from philosophies dead and living. "La physiologie s'est enfin dégagée du vitalisme, du spiritualisme religieux ou philosophique, mais ce n'est pas pour se heurter au matérialisme, car ce serait de tomber de Charybde en Scylla."[53] Philosophy might make a good handmaiden but a poor mistress. Positivism, even though endowed with the Circean charm of the science of yesteryear, could hardly hope for a much kinder judgment. The scientific child had become father of the philosophical man.

In France old quarrels never die or even fade away; they flicker tediously, erupting from time to time with renewed vigor. The metaphysical heat recently generated by partisans and enemies of the ideology of "chance and necessity" is a fascinating illustration of this intriguing characteristic of French intellectual life, especially the component that survives by being hooked up to the mind support system of scientism. So well into the twentieth century, intellectuals – and not only historians, who are, after all, condemned by the nature of their discipline to perpetuate old quarrels – have deemed it of the highest importance to declare Bernard free from any taint of positivism. In the 1920s and the 1930s Pierre Lamy effectively argued that Bernard, far from having lent any support to Comte's positivism, had in fact dealt it a death blow by striking at the heart of its most fundamental and naïve illusions. Comte's idea of fact was derived from Magendie; Bernard transformed it by highlighting, if not discovering, the primordial role of reason in the experimental process itself. For Lamy, then, Bernard took on a new significance in the history of thought: he restored the value of reason through an analysis of the very idea of fact while invoking the genesis of experiment.[54] The fertile paths of discovery that Comte had put off-limits to science were reopened by Bernard. What better rhetorical flourish than to note with smug glee that the philosophy priding itself on being the most scientific of the century was rejected by Pasteur and turned on its head by Bernard? Comte had wrapped positivism in a mantle of tomorrow's science only to find that it soon became a shroud of yesterday's philosophy.

It was certainly a grave injury to the positivist cause that Bernard's ideas on biological research were vastly different from those of Comte; it was also very irritating to positivists that Bernard insulted them by keeping ominously silent on the garrulous if penetrating philosopher of science. This insult was officially consecrated when the participants in the celebration of the centenary of Bernard's birth omitted any reference to the influence of the philosopher on the scientist. Perhaps scientists could be forgiven for not knowing any better, but the faithful to Comte's cause were especially wounded when Bergson also ignored Comte. To counteract this omission, the positivists took desperate action. First, the citing of honorable and official precedents: the close connection between Comte and Bernard had been noted by Ravaisson in his famous report of 1867 on *La Philosophie en France au XIXe siècle*. Second, the strategy of showing Bernard's perfidy toward Comte: Bernard used the law of the three states of the human mind without giving his obvious source. The amusing accusation that Bernard plagiarized Comte's law may have seemed plausible in 1914, but in the 1860s Comte was well known, and it was quite reasonable to make a liberal application of this convenient model to the history of medicine without citing the obvious source of a sort of historical folk wisdom. Bernard was an intellectual prodigal son of Comte – "un fils intellectuel de notre maître, un illustrateur, sous beaucoup de rapports."[55] A third reason positivists used to explain Bernard's silence concerning his debt to Comte was his craving for official glory. Bernard did not wish to jeopardize his candidature for the Académie française, as Littré had done in 1863, by recognizing his debt to Comte.[56] No point in provoking the Catholic rhetorician Dupanloup by acknowledging a debt to one of the scientistic Molochs of the nineteenth century. Although it seems simpler just to recognize that Bernard did not cite Comte because he was not influenced by positivism, the accusation does have a certain spurious appeal. The later entrance of Charles Robin into various public bodies elicited squeals of protest from Catholics. But his notorious anticlericalism and his publicly declared positivism sharply separated him from the Bernardian *Weltanschauung*.

Georges Canguilhem has argued that the historic turning point at which modern medicine began is found in the idea that *experimental* medicine was a declaration of war against Hippocratic or clinical medicine. Hippocrates, the founder of observational medicine, was admitted by Bernard to the Valhalla of genius, but medicine had simply evolved beyond a sort of Comtean stage or state that stretched from Hippocrates to the early nineteenth century. There was a historical division, then, between the positivistic position on the aims of biology, which was based on the medicine-biology of a generation – Cabanis, Broussais, Magendie – committed to clinical medicine, and the new experimental position emphasized by Bernard. Experimental medicine – a "conquering science" – should separate itself from observational medicine, which would continue to fulfill a justified clinical role; but scientific discovery would be the domain of the experimental. The biologist would join the physicist and the chemist in an elite of scientific sorcerers. "A l'aide de ces *sciences* expérimentales *actives*, l'homme devient un inventeur de phénomènes. . . ."[57]

Bernard believed that experimental physiology was the science that carried in its womb the entire future of biology.[58] (Zoologists, when they invented a discipline and themselves, would reject this stifling imperialism on the part of an aggressive new arrival on the scientific scene.) Physiology would therefore have to be the solid basis on which scientific medicine was founded. "La médecine scientifique ne peut se constituer définitivement que par la physiologie, et le problème physiologique contient aujourd'hui le problème médicale tout entier." Medical science could acquire the certainty characteristic of other sciences if clinical observation went back experimentally, through physiology, to the immediate causes of the phenomena of health and disease. Bernard did not give any credit to Comte for the new science; he brazenly took full credit himself: "C'est moi qui fonde la Médecine expérimentale, dans son vrai sens scientifique; voilà ma prétention."[59]

The new science was founded by its greatest practitioner, not legislated by a philosopher of science. Nor could one expect that a man who believed in a version of the modest proposition *La Science, c'est moi* would tolerate rivals in revealing the new *Logos*. On a more fundamental level, Bernard's idea of experimental medicine was "the negation of systems," for it proposed "recourse to experimentation in order to verify medical theory." "Le grand principe de la médecine expérimentale, qui est en même temps celui des toutes les sciences expérimentales, c'est de ne marcher que d'expériences en expériences, et de ne pas faire de théories qui ne soient établies par l'expérimentation."[60] Positivism could only have been as acceptable as the old-fashioned science on which it was based, even if Bernard had deigned it worthy of serious consideration.

The aim of science was to arrive at experimentally based theory, the source of the certainty and power of physics and chemistry. Bernard never tired of preaching the experimentalism he practiced, the royal road to the theory from which practice derived. Clinical medicine grouped in the dark, far from that degree of perfection. By following the Bernardian program experimental medicine could achieve the certainty and power of the sciences based on theory.

> D'abord, qu'entendons-nous par médecine scientifique?
>
> Nous voulons désigner par là une médecine dans laquelle la pratique se déduira avec certitude de la théorie. N'est-ce pas là d'ailleurs le caractère de toutes les sciences faites: savoir, c'est pouvoir. La connaissance dans les sciences expérimentales a pour sanction la puissance. Le physicien et le chimiste dirigent à leur gré tous les phénomènes naturelles dont ils connaissent les conditions d'existence; dans ces sciences la pratique se déduit toujours rigoureusement de la théorie. La médecine est encore loin de ce degré de perfection ... Au fond médecine clinique tatônne et marche dans l'obscurité.[61]

Bernard was not alone in believing that he would lead French medicine into the promised land of science, just down the experimental road. The Société de biologie also believed this – the prevalence of the Bernardian research imperative in the "life sciences" in France would have serious implications for the French biological community. Given the "ideology" of the Société presented by Robin and the nods of respect often given to Comte's memory by members, the triumph of Bernard's ideas is quite ironic. But scientific societies do not live by ideology alone, but more by the examples of research done by the master, whose inspiration is often more a matter of his genetic nature than of philosophic nurture.

Even for Robin, no stranger to experimental procedures, experimentation was only one of three procedures of the art of observing, precariously perched between observation as direct, concrete, and synthetic contemplation of the arrangement of the phenomenon in a natural state, and comparison, whose essential characteristic consisted in conceiving as similar all the cases considered in biological research. "La méthode comparative est . . . fondée sur le fait de la comparaison très-prolongée d'une suite non interrompue ou du moins fort étendue de cas analogues, où le sujet se modifie par une succession continue de gradations régulières, croissantes ou décroissantes."[62] Experimentation was more important in dynamic biology, especially in physiology, than in static biology. The basic characteristic of Robin's experimentalism was that it placed observation outside natural circumstances; this was achieved by putting bodies in certain artificial conditions expressly instituted to facilitate the examination of the movement of the phenomena being analyzed in a determined aspect. "Le caractère fondamental de l'expérimentation réside surtout soit dans l'institution, soit dans le choix des circonstances du phénomène pour une exploration plus évidente et plus décisive."[63]

Physics, rather than chemistry, was the science in which experimentation triumphed totally. The great generality and relative simplicity of physical experiments gave physics a great experimental advantage over chemistry, where an ostensibly wider field for experimentation was in reality considerably reduced by the difficulty of isolating different influences and determining conditions of chemical phenomena. In biology the situation was worse, for it was extremely difficult to obtain two experimental cases exactly alike in all respects. Robin aped the pessimism of Comte in regarding the difficulties as insuperable barriers rather than follow the optimism of Bernard, who saw them as challenges to be met in establishing physiology as the experimental queen of the biological sciences. This view, in combination with untenable physiological ideas, led Robin to adopt a fruitless research strategy for part of his own career. Rejecting Virchow's formulation of the cell theory, he believed that because the "real seat of life was constituted by the humoral parts of the organism," there must lie "beyond the fixed anatomical elements . . . a molecular organization that explained the morphology. In his opinion, therefore, microscopic investigation was only a stage of biological research that must be followed by chemical analysis."[64] But Robin's collaboration with the chemist Verdeil to study the chemical compounds composing an organism led to little more than a fat treatise because of the comparatively primitive state of the chemistry of life.[65] At that time "the superiority of a morphological approach was undeniable."[66] Most of the biological community prudently kept their distance from Robin's approach, with its implication of a definite subordination of biology in the hierarchy of the sciences.

On a very general level, Bernard defined physiology as knowledge of the laws of life. (Cf. Balbiani: Physiology studies the vital manifestations of beings and the functions of their different organs.) Experimental physiology is solidly based on an indispensable tripod: the physicochemical sciences, anatomy, and experimentation on the living organism. Lavoisier and Laplace established the basis of the physics and the chemistry of physiology: "que les actions physico-chimiques qui manifestent et règlent les phénomènes propres aux êtres vivants rentrent dans les lois ordinaires de la physique et de la chimie générales." After Bichat it was possible to define the aim of general physiology: "déterminer par l'analyse expérimentale les propriétés physiologiques élémentaires des tissus, afin d'en déduire ensuite d'une manière nécessaire l'explication des mécanismes vitaux." But Bichat did not move beyond the great task of establishing general anatomy. The path to general physiology was further opened by Magendie, who took a key step toward creating physiology as a science when he introduced the method of the experimental sciences, meaning experiments on living organisms. Yet this advance was limited to *empirical experimentalism*.[67] This fortunate shortcoming of Bernard's master let the pupil advance the theoretical progress of physiology by adding interpretation and reasoning. General physiology could only be definitely established as a science when its direction was determined in a rational manner by a clear conception of the problem it proposed to solve. A few perceptive and Buffonesque remarks made the antiempiricist point. "L'empirisme peut servir à accumuler les faits, mais il ne saurait jamais édifier la science. L'expérimenteur qui ne sait point ce qu'il cherche ne comprend pas ce qu'il trouve."[68] Physicists would soon be making similar statements, a reluctant admission that even the nature they revealed was at best an imitation of art.

Physiology was the supreme biological science because it was nearer the goal than any of the others; it led the way in researching the great problem tantalizing man, the explanation of life.[69] Most definitions of life – and its very existence was a challenge to scientists to attempt a definition – given by the biological community were a bit more edifying than that of Anatole France's Dr. Fornerol, although they were necessarily more circular in logic. Often the science invoked to substantiate positions leaning toward vitalism or toward a physicochemical explanation depended upon the philosophical or religious beliefs of the scientists. Bernard rejected vitalism but did not believe that physics and chemistry could explain life.

Bernard agreed with the idea that life is a conflict between the physicochemical properties of the external environment and the vital properties of the organism, each reacting upon the other. (Cf. Bichat's definition of life.) General physiology had to take both factors into account before moving on to experiments on living organisms. In putting general physiology on a firm scientific basis, Bernard imperiously declared his aim of ending the era of sterile controversies over vitalistic and mechanistic theories. But if life's cause is in the power of organization that creates the living machine and replaces its continual losses to external physicochemical forces, vitalists were wrong in thinking that life and those forces are not compatible; there is "perfect and necessary harmony, for the causes that destroy organized matter are those that make it live, that is show its properties." The living matter of organic elements is not intrinsically spontaneous and reacts like inorganic matter to such external excitants as air and light. Do the vital manifestations of the living machine have special laws, or do they follow the same laws as inorganic matter? Bernard believed that all phenomena have one science. Vitalists were wrong: By their nature and the laws governing them, living beings do not differ from inorganic bodies. "Sous le rapport physico-mécanique, la vie n'est qu'une modalité des phénomènes généraux de la nature . . . " Bernard did not intend to fritter away the heritage of Lavoisier in the hope of some insight from vitalism.[70]

And yet life is unique. What precisely is unique about it? Bernard's answer: the instruments and procedures employed by living beings, life's organic tools, which belong to it alone. This special morphology of the apparatus of an organic being meant that the chemist could create the products of living nature in his laboratory but could not imitate its procedures. Organic matter possesses vital procedures, inorganic matter does not. "Je conclurai donc que, bien que les phénomènes organiques manifestés par les élémènts histologiques soient tous soumis aux *lois* de la physico-chimie générale, ils s'accomplissent cependant toujours à l'aide des *procédés vitaux* qui sont spéciaux à la matière organisée et diffèrent constamment sous ce rapport des procédés minéraux qui produisent les mêmes phénomènes dans les corps bruts."[71] The mistake of the physicochemists was that they did not make this distinction between laws and processes, upheld by Bernard as a basic proposition in physiology.

For Bernard the special foundations of physiological science were to be sought "only in the organic structure of living beings," rather than in the vitalist hypothesis or in the exclusive views of physicomechanists. He would use the term *vital force* to describe the phenomena of organic renovation unique to living beings; it had no resemblance to a marvelous and extraordinary essence outside the understanding of science. Even the word *force* was "only an abstraction or a form of language"; other words ("*phénomènes organotrophiques* ou *nutrifs*") might be substituted for this vague phrase. In physiology, as in the sciences dealing with inorganic matter, the scientist must never "look for the explanation of things in the hypothetical attributes of imaginary properties of any hidden force."[72] As Gohau points out, Bernard belongs to the methodological camp of antivitalistic thought. In the dualistic exegesis of antivitalism, the methodological form of the doctrine is an antimetaphysical defense used by physiologists and chemists or biochemists like Bernard and Berthelot. A second form of antivitalism is substantial in nature, interpreting life as a set of physicochemical phenomena. The second level of antivitalism, basically mechanical, incorporates a subtle reductionist viewpoint, which, irreducible to the Loebist perspective, recognizes the irreversibility of facts about heredity and evolution – the historical character of biological mechanisms.[73]

Positivism's unacceptability as a philosophy of science for the scientific community

Probably the most important reason explaining the absence of most scientists from the positivist camp is that many nineteenth-century scientists did not want to have anything to do with philosophy and especially with a *Weltanschauung* that was proud of its successful mixture of science and philosophy. The whole trend of science in the nineteenth century was to take refuge in an elaborate but useful fiction separating these two human activities. In French biology, the separation was viewed as especially important, so important, in fact, that it led some scientists, who ought to have known better or to have read more of Darwin than they did, to assume that what was true in Darwin was trivial and what was false was philosophy. German biologists did not have this hang-up about philosophy, which was both good and bad, depending on the nature of the philosophy and the use to which it was put. Emulating the experimental nature of physics and chemistry was the way in which biology had to become a science, or so it was widely assumed by biologists. This meant a ruthless purging of all ideas that could be used as a hypothesis to guide experiment. Both physics and biology would lose their phobia about philosophy toward the end of the nineteenth century; to some extent, positivism would benefit from the return of the sciences to a healthy relation with philosophy, but it would not be the Comtist variety. The positivist program of giving philosophy its rightful position as a synthesis of the sciences ran counter to the trends in nineteenth-century science, especially in biology.

It must be emphasized that the only philosophy allowed into the sanctuary of Isis by the positivists was their own concoction. Metaphysics was a horrendous error. No one could put it better than Littré: "En résumé, la métaphysique est la théorie des principes de l'esprit, d'où l'on tire les principes des choses, ce qui est impossible. La philosophie positive est la théorie des principes des choses, d'où l'on tire les principes de l'esprit, ce qui est possible."[74] But the most sophisticated philosophy of science was developed by Grégoire Wyrouboff, inorganic chemist, collaborator of Littré, redoubtable polemicist, and from 1904 to 1913 holder of the chair of the history of science at the Collège de France.[75] In 1867, Wyrouboff lamented, science and philosophy were in two enemy camps: The hatred of the scientist for the philosopher was an atavistic survival originating in a sane reaction against an imperialistic metaphysics that tried to explain the world without worrying about experiment or observation.[76] But since the old scholasticism had been long destroyed, there was no need to inhibit the human mind from practicing its need to generalize, without which science would be no more than a tedious catalog of wonders. Only one limitation could be placed on philosophy: All its speculations must be based on the study of reality. Truth was no longer defined as what was logical but what was real. Whereas Thomists labored to place science under the umbrella of philosophy, the positivists proudly proclaimed the reverse program. Thus the scientist could be a positivist philosopher without fear, for he could generalize without leaving the safety of the bosom of Isis.

Wyrouboff's laudable aim was to correct the baleful tendency of modern scientific education to develop an excessive taste in young scientists for a specialization that produced rather meager results. "Rétrécir arbitrairement l'horizon intellectuel ne peut jamais être utile ni pour le penseur, ni pour le savant; car si le premier a besoin de savoir, le second a besoin de penser. Sans un système, sans un aperçu général de la science qu'on cultive à chaque phase de son développement, on ne peut avec fruit creuser aucune de ses parties; on ne peut rien produire de sérieux ni de durable."[77] Comte would have been pleased with Wyrouboff's upholding of his opinions – if only it had come from the orthodox camp! Meanwhile science was wantonly abandoning itself to the evil of specialization, source of its greatest successes, and nowhere more than in zoology and physiology, the double cutting edge of biological advance.

In the first number of *La Philosophie positive* (1867), Wyrouboff set forth in magisterial terms the main tenets of the positivist philosophy of science.[78] Under the seductive rubric of "certainty and probability," he explained the ramifications of the limiting idea or fundamental dogma of positivist philosophy: Only reality can be true. All reality is not necessarily true because a real fact is one that is recognized by the senses, notoriously fallible. Truth can only be a relative agreed-on idea, the result of experiment and observation practiced according to the special rules of science. Reality is not, then, a condition sufficient to endow a scientific result with certainty. Seeing a fact does not mean that it exists: miracles, animal magnetism, and witchcraft had all been seen. But all man's knowledge comes through the senses. How then can one escape from this dilemma? How is science possible? The second condition of certainty must therefore be in the means of regulation or control we have over experiment and observation, permitting us to relegate the senses to less than a primary role and to make verification possible for everyone, a community experience. These conditions are called law. Separating the constant from the accidental in studying a phenomenon leads to certainty, to the formulation of a law; the law is the fact reproducing itself exactly the same in the same circumstances. Error is no longer possible, for observation and verification become a community function. Science is always infallible because it is a group of laws, although the scientist is often mistaken. The affirmations of science are certain, whereas those of the scientist are only probable. Law represents the relation connecting reality to truth, the indispensable condition without which we cannot be sure that a fact is true.

Wyrouboff did not remain satisfied with this reasonable certainty but went on to establish "an even greater certainty, an even truer truth for the conclusions of science. How could one be sure that scientific law would always be the same?" One takes into consideration not only the conditions accompanying the phenomenon but even the other known laws governing the body in which the phenomenon is observed. Because these laws have a determined relation to one another, the rejection of one would mean the rejection of all, a possibility Wyrouboff clearly considered absurd. Necessity, not chance, emerged triumphantly from this argument based on a convincing use of the logical trick of reductio ad absurdum. And law was elevated from a lowly empirical status to a rational one. The law of gravitation is true not only because scientists observe it governing falling bodies but because it explains all astronomical phenomena. Wyrouboff failed to note that the Newtonian statement of this law was not an explanation of reality at all but the presentation of gravity as "a mathematical, not a mechanical force." Newton did not know the *cause* of gravity; cause was scientifically irrelevant.[79] In the nineteenth century it became increasingly clear that however much science might love certainty, she would marry probability. And although the scientific community might reluctantly acquiesce in this morganatic union, the mid-nineteenth-century positivists could not accept it. But this was more important for physics than for biology. The positivistic emphasis on the guarantee of certainty implicit in rationalism, with its correlative downgrading of empiricism as a weak and limited doctrine, would have aroused the suspicions of the average biologist more than any emphasis on certainty. In fact, it is certain that most biologists would have chosen certainty over probability had they given much thought to the issue at all. Robin argued that the great virtue of positivism as a philosophy was that it showed that the criterion of truth in science is prevision, confirmed or weakened by action. Comte had stated the basic dogma: "All science has prevision for its aim," meaning, as Lévy-Bruhl put it, that all science tends to substitute deduction for experience, rational for empirical knowledge. This is not surprising when one recalls Comte's definition of natural laws in terms of the constant relations between phenomena.[80]

Rational law was for the positivist the greatest certainty in science. Doubt had a minor role in arriving at truth, but carried to an extreme it became a fruitless *jeu d'esprit*. One could not doubt the basic axiom, the limiting conception of all philosophy, which, for the positivist, was reality, understood through rational law provided by science, the only means of acquiring certainty about this reality. His enthusiasm for achieving this certainty led Wyrouboff to consecrate reality as the absolute – only law has the privilege of expressing an absolute truth. But science as a system of rational laws could never change any of its achievements – a scientific truth is eternal. Mathematical and astronomical laws discovered in antiquity are still true, absolute certainties. No possibility for a switch of paradigms here!

How does science progress, then? Science also contains relative truths, a perfectable part that can be modified with the development of investigative procedures. True, the supreme aim of the progressive movement of the relative in all philosophies has been toward the absolute, and science is no different. "Les vérités relatives des sciences exactes tendent incessamment, elles aussi, vers cette connaissance parfaite de la réalité, qui est notre absolu et notre idéal."[81] Positivist alchemy came close to its philosopher's stone of certainty by means of an odd linguistic algebra in which relative truth led to an absolute one.

Pronouncements about absolutes, even in science, and flattering remarks about the eternity of scientific truth were not likely to lure many nineteenth-century scientists into the positivist camp. It all sounded like lycée philosophy or ideology, and positivism contained substantial amounts of both ingredients. A scientist who spoke and wrote about atheism, pantheism, and materialism was likely to be regarded suspiciously by many scientists. Biologists had more reason than any other scientists for remaining aloof from religious and philosophical issues, for their young science was suspected in some quarters of being laden with philosophy and ideology anyway. No point in giving aid and comfort to the enemy! The positivists believed that Comte would have been justified in boasting, "Ils m'ont appelé l'Obscur mais j'habitais l'éclat," but most of the scientific community thought exactly the reverse.[82]

Perhaps the fundamental positivist principle of a hierarchical ranking of the sciences precluded any significant positivist influence in biology. In contrast to Bernard's innovative role for experimental physiology, Robin's scientific study of living bodies was "by its nature, essentially reserved for the general development of the universal art of coordinating and classifying as well as of the art of comparing." A complex science like biology was based on more general and simpler science (physics and chemistry), although the complex could by some kind of cultural feedback influence the simpler. For the biologist, the striking thing was the Comtean "principle of the subordination of the more complicated sciences to those that are less so," the principle establishing not only the "true scientific hierarchy" but also the "vigorous unity of the system," although the unity of the system was widely ignored.[83] Hardly a viewpoint that would appeal to the creators of the essentially new discipline of biology in the mid-nineteenth century. Men of big egos, discipline builders do not accept philosophies implying that their activity is not quite a first-rate intellectual activity with the potential of achieving the same certainty and objectivity traditionally associated with mathematics, astronomy, and physics. The scientist must be "le Poète qui rentre seule avec les Filles moroses de la gloire." Nor does he usually forget to pay "hommage, hommage à la véracité divine!" – of which he is the discoverer.[84]

3

BIOLOGY IN THE UNIVERSITY

THE SUCCESS OF MARINE BIOLOGY
LEADS BUT TO PARADIGM LOST

Incorruptible Mer, et qui nous juge! . . . Ah! nous avions trop présumé de
l'homme sous le masque!
Qui donc es-tu, Maître nouveau? Vers quoi tendu, où je n'ai part? et sur
quel bord de l'âme te dressant, comme prince barbare sur son amas de
sellerie; ou comme cet autre, chez les femmes, flairant l'acidité des
armes?

– Saint-John Perse, *Amers*

A T about mid-century, research and teaching in natural history and
physiology in both the Muséum and the Collège de France were
much stronger than in the faculties of science. In 1855–6 it was
possible to classify the faculties into three categories according to the level of
their teaching of natural history. In the first group, with a full faculty,
including three professors in natural history, one for each "order" of the
natural world, were Paris, Lyon, Montpellier, and Toulouse. The second
group, with two professors of natural history, one of whom taught two
different courses, included Caen, Dijon, Strasbourg, Besançon, Grenoble,
Rennes, and Bordeaux. The third group, made up of new faculties having
only one professor for all of natural history, included Lille, Nancy, Marseille,
Clermont, and Poitiers.[1] But less than a quarter century later the faculties had
improved considerably in the whole range of biological science; in the last
decades of the nineteenth century a great deal of new biological research was
done in the faculties. In areas like marine biology, the faculties were the most
productive institutions in research and in training researchers and teachers.
The growth of biological science is one of the most striking features in the
scientific empire of late nineteenth-century France.

One nice thing about the growth of knowledge is that it leads to more
jobs, especially for the group creating the new knowledge. After the early
1870s zoologists were clever enough to spread the word that their science
was growing faster than the other sciences and therefore needed more
University positions. The provincial faculties of science made clear to the
Ministry of Education that one professor could no longer teach all natural
science. The subject had fissioned into a number of well-developed
subspecialties: histology, comparative anatomy, paleontology, and descriptive zoology, an area leading to general classification and physiology. If the

93

label of higher education was to have any meaning, the new scientific specializations could not be left out of the programs in the faculties of science.[2] The dizzy bifurcation of modern science was well under way. Its initial result of a great increase in scientific knowledge, with the promise of an improvement in the human condition, made it eminently acceptable to the nineteenth century. Because science also seemed to produce greater material wealth, even those who paid the bills did not complain too loudly.

As late as 1873 it would have been difficult to predict a great future in biological research for the faculties of science, even in Paris. Part of the problem, if such it can be called, was the tidy division of functions among various institutions, with the Collège de France leading the research effort in experimental biology. Claude Bernard emphasized the primary, indeed only, role of the Collège when he gave his inaugural lecture in 1873. "Our only tradition at the Collège de France must be the tradition of progress."[3] Institutional reforms giving the professors of the Collège an autonomy denied during the Second Empire, when a governmental administrator ruled, were meant to consecrate the rule of the research imperative, or at least this was the favorable interpretation given by the scientific community. Ultimate power was concentrated in the assembly of professors, who delegated power to carry out administrative tasks to an administrative council and drew up a list of three professors from which the minister of education would pick the administrator. When a chair became vacant the assembly decided whether to keep it in the same subject or give it to someone in another area answering better to the "new directions of the human mind." Dropping subjects that had become of little research importance in order to add ones that were more promising allowed the Collège to escape – at least this was the theory – from the trap of having to pass on the chair to the former professor's heir apparent. Such switches were possible because the Collège did not have to conform to official programs in order to prime students for examinations, or so it was believed, and often practice conformed to the belief that the Collège de France did not have the responsibility of teaching known areas of science. That lowly activity was the role of the faculties. The Collège should teach areas of science in the process of creation. Thus its teaching had always to change as it left known areas for those being explored. To fill a chair with someone who would continue his predecessor's course would run the risk of dealing with accepted science, thus encroaching on the territory of the faculties. It seemed that the Bernardian research imperative had been institutionalized, with a monopoly for the Collège de France.[4] But scientists and scholars hate the monopolies of others, and science and learning were too vast to be kept in the Collège's gilded cage. In the doctorate the faculties had the perfect research weapon for breaking down the division between teaching and research that seemed so logical but was so impossible in practice once a decision had been made to produce increasing numbers of high-quality research personnel.

Even in physiology the faculties were not wont to leave research initiative to the Muséum and the Collège. Bernard's first chair was one of general physiology at the faculty in Paris, created in 1854 especially for him by the government, but when Magendie died in 1855 he became professor of medicine at the Collège. In 1868 the Sorbonne chair went to the Muséum with Bernard's succession to Flourens. Flourens's chair emigrated to the Sorbonne for Bernard's disciple, Paul Bert. When Jules Béclard was named to the chair of physiology in the Paris faculty of medicine, he inaugurated a sort of competitive game with Bernard and the Collège. On a teaching level he did well, for he had a great talent for popularization, a valuable skill in teaching science to medical students. By 1870 his *Traité de physiologie* was in a sixth edition. Posterity is notorious for forgetting teachers and remembering discoverers and inventors; so after his death in 1887 little was heard of Béclard's valiant effort, which fell into certain oblivion when the faculty of medicine became a bastion of clinical rather than experimental medicine. But there was no shortage of physiology in Paris, with a keen competition between the holders of chairs in the different institutions.

The arrival of physiology in the provincial faculties was a generation later than in Paris, coming only with the reform of the universities in the 1880s. In 1883 Paul Bert asked the Chamber and the Senate to ensure that physiology be taught along with other sciences in big provincial university centers by qualified professors. After Bert's famous report Saturnin Arloing was appointed to the first chair of physiology in the faculty of sciences in Lyon. When Chauveau ("son maître") got the chair of comparative pathology in the Muséum in Paris Arloing replaced him in the faculty of medicine. The course in the faculty of science was then changed to general and comparative physiology, which was closer to the knowledge deemed useful to naturalists and the problems they tried to solve. When he gave his opening lecture in this course in 1893, Dubois professed satisfaction with the division of the course into two equal parts and with the function of his subject in the faculty.[5] The "service course" provided many a provincial faculty of science with its raison d'être. Better to exist in service than never to exist at all.

Within the University, physiologists were quite defensive about their jobs – if they had one. Thirty years after the first chair of general physiology was created in the Paris faculty of sciences for Claude Bernard, Arloing felt obliged to defend the independence and importance of his subject when he gave his first lecture on becoming professor of physiology in the Lyon faculty of sciences in 1884. The establishment of chairs of physiology in the faculties of science showed that there was a recognition – somewhat grudgingly given, it is true – that physiology had a place in the system of scientific education. Arloing could blithely establish its role in medicine in a few words because there was little need to engage in much propaganda for the teaching of physiology in the faculties of medicine, although even there it required defense from the danger that medical educators would turn it into a completely practical, applied activity. It was assumed that progress in pathology is correlated with that in physiology because understanding the difficulties of a function requires a deep knowledge of the function in its normal state. So Arloing proudly proclaimed physiology to be one of the solid bases of medicine. But physiology could not live by medicine alone; it also needed its place in the palace of pure science.

According to the high-minded view of its practitioners, physiology was not limited to the study of the functions of man and the furry relatives whose organization is closest to him. Arloing raised its status to the "science of life, dynamic biology," consecrated to the study of organic life in its search for the laws governing it as well as to predicting the nature and progress of phenomena not yet brought within the range of scientific scrutiny. Three solid reasons justified the teaching of physiology in the faculties of science. It is the science of the physicochemical phenomena of life; it cannot be separated from the other branches of human knowledge; and it is a valuable part of a general education.

An autonomous role for the teaching of physiology in the faculties of science had not been accepted by several of the great historical figures in French biology. This burden of the past may explain part of Arloing's defensive attitude. A view long dominant in the faculties was that physiology should be part of zoology. This dubious view had probably been hatched in the Muséum. Who could be more convenient to blame than Cuvier, guilty of advocating the transformation of physiology into a contemplative science absolutely subordinated to the progress of zoology and comparative anatomy. And if Milne Edwards was more generous in merely favoring the union of anatomy and comparative physiology, he was more dangerous because of his position as dean of the faculty of sciences at the Sorbonne. Arloing could not accept the argument that anatomy provides full knowledge of certain functions in many areas like respiration and the liver. Only Bernard's position of independence, based on "the method of experimental determinism," was acceptable to Arloing, whose program included general and comparative physiology. Bernard was Arloing's great source of disciplinary inspiration. "Je m'inspirerai constamment du déterminisme expérimental de Claude Bernard, 'ce principe nécessaire de la physiologie' qui nous révèlera 'les rapports entre les phénomènes et leurs conditions, la seule et la vraie causalité immédiate, réelle et accessible.'"[6] Arloing was rather eclectic, for he also accepted many of the pronounce-ments of positivists on physiology and the connections between sciences, including Littré's definition of physiology. If the role of science was to enable man to rule over animals and plants, to conquer the chief riches of the planet, and to prepare for the proper government of men and societies, physiology had an important social role in making known the instruments of feeling, intelligence, and will, thus furnishing a solid support for the study of esthetics, logic, and ethics. Pure positivism or some sort of model for a grant in the social sciences.

The tension that had existed between Bernard's physiology and clinical medicine had more than intellectual consequences – although they are always the primordial and most serious, of course – when it was inherited as a conflict between Bernard's followers in the faculties of science and their opponents the faculties of medicine. The conflict was more a matter of dispute over the content of official programs stipulating the material to be covered in courses in physiology for medical students. Charles Richet was strongly critical of a publication by Jolyet, professor in the faculty of medicine of Bordeaux, dealing with the proper way to teach physiology to medical students. Jolyet's clinical view was that any theoretical approach was useless; the currently taught courses, drowning students in theory, should be replaced by practical exercises. For Richet this would be to prescribe ignorance. All students need practical and technical knowledge, but all must also know its scientific basis, from which come the results making serious medical knowledge possible.[7] The spirit of Claude Bernard may have been pleased at the new staging of the drama he had written and in which he had once played the leading role. Reforms would soon lessen the tension but not eliminate the dispute; indeed, it is endemic in any system of medical education. With the growing power of faculties of science, the potential for clashes became greater. A notorious quarrel developed in Lille, and although its malignity was to a large extent the personal joy of Giard, the issues were very general.

In his presidential address in 1883 to the British Association for the Advancement of Science E. Ray Lankester lamented the dreadful state of biological research in England, where public support was far less than in Germany and France. After enviously referring to Germany's hundred or so biological institutes with 300 places for research workers, Lankester warmly admired the set of scientific institutions clustered in Paris, especially the Collège de France, which he set forth as a model for a similar institution that the British government should build in London. The English also wished for

97

French salaries. But most of all Lankester wished for an institution comparable to that worthy of the talents of the "four most brilliant biologists of our period: Brown-Séquard, Marey, Balbiani, and Ranvier." It was bad enough that England, with a population of 25 million, had only 38 biological posts, compared to Germany's 300 chairs for a population of 45 million, but it also did not have a single permanent laboratory of marine biology, which was quite astounding in a country where the fisheries played such an important role in the economy. To reach a level of biological activity comparable to Germany's, Lankester estimated that England needed forty new chairs with laboratories in the various areas of biological science, located in eight big cities, and the new model Collège de France for London. The chairs and laboratories alone would cost the equivalent of 50 million francs.[8] But how else could the English match Dohrn's laboratory in Naples or the French laboratories in Roscoff and Concarneau?

As the republican faculties of science developed into quite respectable centers of teaching and research, the biological sciences came to occupy an important place in them. Biology's dependence on experimental medicine – so clearly illustrated in the foundation of the Société de biologie in 1848 – was considerably reduced by the growth of biology in the faculties and training of biologists quâ biologists. When Henri de Lacaze-Duthiers challenged the claim of Claude Bernard that physiology was the true experimental science he was trying to raise the position of zoology in the intellectual hierarchy of the sciences.[9] Observational science belonged to the age of the amateur and remained the preserve of the bird watcher; experimental science was the craft of the high priests of the new science, and it showed surprising strength even in the Muséum, that stronghold of the naturalists.[10]

Bernard's usurpation for physiology of the cutting edge of explanatory power in the biological sciences alarmed the few thinkers of the French zoological community who were solidly entrenched in scientific institutions and in possession of the few perquisites of scientific power existing in the Second Empire. The notorious declaration in the *Rapport* seems to have put the fear of the physiologists in the mind of some zoologists and perhaps speeded their awakening from a long Cuvierist slumber. Bernard's contention that "physiological science must master the phenomena of life, as physics and chemistry master the phenomena of material bodies," was cleverly linked to the flattering claim that France was always a promoter of new ideas, which eventually spread to other countries. Unfortunately, unlike gastronomy, these ideas sometimes developed more there than in France. This claim was part of Bernard's strategy for obtaining governmental support for new laboratories and an extension of the role of physiology in teaching. He was

98

also proud of the clarity of his definition of physiology. "L'idée que la physiologie expérimentale est une science autonome, distincte des sciences naturelles, une science expérimentale distincte: *physique et chimie vivante . . .* "[11] In two short, brilliant paragraphs Bernard raised the threatening standard of the small but élite corps of physiologists – especially himself – who would be the leaders in the biological sciences.

> Nous établirons tout d'abord que la physiologie n'est point une science naturelle, mais bien une science expérimentale. Les sciences naturelles et les sciences expérimentales étudient les mêmes objets (corps bruts ou corps vivants); mais ces sciences se distinguent néanmoins radicalement, parce que leur point de vue et leur problème sont essentiellement différents. Toutes les sciences naturelles sont des sciences d'observation, c'est-à-dire des sciences *contemplatives* de la nature, qui ne peuvent aboutir qu'à la *prévision.* Toutes les sciences expérimentales sont des sciences explicatives, qui vont plus loin que les sciences d'observation qui leur servent de base, et arrivent à être des sciences d'action, c'est-à-dire des sciences *conquérantes* de la nature. Cette distinction fondamentale ressort de la définition même de *l'observation* et de l'expérimentation. *L'observateur* considère les phénomènes dans les conditions ou la nature les lui offre; l'expérimentateur les fait apparaître dans des conditions dont il est le maître.
>
> La physique et la chimie ont conquis la nature minérale, et chaque jour nous voyons cette brillante conquête s'étendre davantage. La physiologie doit conquérir la nature vivante, c'est là son rôle, ce sera là sa puissance.[12]

Georges Canguilhem argues that nineteenth-century physiology was distinguished from that of the eighteenth century by the construction of instruments and the systematic use of techniques of detection and measurement, rather than by an increasing emphasis on experimentation. In adopting physics and chemistry as model sciences, physiologists were led to adapt to biology the analytical and measuring techniques that had been so fruitful in the model sciences. So nineteenth-century physiology was constituted "by the solid alliance of these two styles of research."[13]

Bernard thought that the ability of physiology to conquer nature, to snatch secrets from it, would benefit humanity. Just as physics and chemistry gave man domination over inorganic nature, physiology gave him control over living nature. Zoology and botany could be useful servants of physiology in their study of vital mechanisms in new species. But the vague or purely empirical data supplied by the natural biological sciences could not give them the real power over living nature possessed by experimental biology. Some zoologists saw only anatomy in physiology, others saw only chemistry; a point of view including both approaches was needed. There were still some physiologists who were too close to physics (Dubois Reymond), some too close to chemistry (Dumas, Liebig), and some too close to anatomy (Milne Edwards). "Il faut être physiologiste et fondre les deux points de vue en seul. Le côté physico-chimique est *conditionnel*. Le côté anatomique est *formel*. Il n'y a pas de transformation *directe* des forces physico-chimiques en action."[14] It was Bernard's legitimate proud claim to have instituted "an autonomous life science based on physicochemical determinism, which nevertheless takes account of the specificity of living beings."[15]

Bernard's general physiology had moved a long way from Auguste Comte's prescriptions and expectations. Positivists might have hoped for some help from zoologists defending their bastion against the physiologist's claims, but the nature of the zoological counterattack centered on an equal claim to the glories of experiment, which thus made assimilation to the Comtean scheme unfeasible.

Wounded amour-propre ensured a speedy reply from zoologists: J.-J.-V Coste led the pack of scientists stung by the classification of their work as contemplation rather than conquest.[16] Since science was an exclusively male activity, insult was certainly added to injury in Bernard's categories, which might have conjured up out of the depths of the male ego the horror of a relegation of zoology to an area of female activity. If nature's secrets were to be penetrated, Coste, Daubrée, and Lacaze-Duthiers did not wish to be deprived of their epistemological phallus (experimentation).[17] More important issues were at stake. Lacaze-Duthiers's assertion of the experimental rights of zoology, the best known of the defenses against Bernard's attack, is not much of a surprise when one considers that he was a product and leading representative of the disciplinary matrix being challenged. Nor did it hurt his situation to defend the work of his powerful *maître* Henri Milne Edwards, old-style zoologist, anatomist, and experimental physiologist, whose versatility probably irritated Bernard to some degree but not too deeply, for he could take comfort in Edwards's lack of originality.[18] Bernard also believed that zoologists wanted to usurp some of the professorships belonging to physiology. Milne Edwards had emerged as the triumphant successor to Cuvier's "Macedonian Empire"; this meant the domination of the school of "physiological zoology" over the other three squabbling research factions: (1) pure morphology and "transcendent anatomy," with Geoffroy as its culture hero; (2) zoology concentrating on the relation between the center of an organism and its peripheral parts and also concerned with substantiating de Blanville's "série des vivants"; (3) Cuvierist zoology, founded on the subordination of characters, which assumed a model of the organism as a set

100

of anatomico-physicological structures – it venerated the techniques of comparative anatomy.[19] Even if he knew them, Lacaze-Duthiers did not refer to any of Comte's lucubrations on biology. His concern was to convince his audience that zoology was not stuck in the classificatory bog of Linnaeus.

The zoologist, especially in France, had to enter resolutely "upon the new and fecund path of experimentation."

> Why should we oppose this title *experimental zoology?* By enlarging the narrow circle to which some have tried to restrict him and by breaking apart the shackles and barriers set in his way, the zoologist readies himself for the most brilliant successes and the science which follows will then enjoy the right to speak of herself as a science explicative of natural phenomena and one to a certain degree the conquerer of organized matter. We admit, however, that our hopes are not as great as those of M. Bernard and we dare not hope as he does, to see new organisms created by man. Let us finish:
> The day of descriptive zoology is over. It is now but one of the parts, indispensable, certainly, but not co-extensive with *general zoology.* Thus, in order to hold incontestable value, the results collected by general zoology must be
> – supported by the precise laws of *morphology,*
> – deduced from the most minute researches of *histology,*
> – demonstrated by the lengthy and continuous studies of *development,* and
> – submitted to the *test of experiment,* which must always prepare, assist, and conduct the studies of *morphology* and *development.* Things must be such that, in a word, general zoology merits also the name experimental zoology.[20]

The distinction made by Bernard in his report in 1867 between the two categories of science plagued and irritated French zoologists for a quarter of a century or more. Nothing could wound a Frenchman more than to be told he is devoting his life to a second-rate intellectual activity. As late as 1905 one of France's leading zoologists, Alfred Giard, took the trouble, in the columns of the *Revue scientifique,* to explain away the shameful dichotomy that Bernard had introduced into the life sciences as a result of his fight to gain equality for physiology in the pantheon of the life sciences. Giard summed up the basic problem: Certain sciences, including astronomy and the natural sciences, were categorized as sciences of contemplation and of observation, able only to achieve the prediction of facts; the other sciences – physics, chemistry, and physiology – were alone the sciences described as explicative, active, and able to conquer nature, in short, the experimental sciences. "Car l'observateur considère les phénomènes dans la condition où la nature les lui offre; l'expérimentateur les fait apparaître dans les conditions dont il est maître. Le naturaliste est un descripteur, le physiologiste est un créateur."[21] In 1908 Maurice Caullery also attempted to kill off the distinction between physiology and morphology by showing that the latter had come to acquire an explanatory character – it was "the historical register of the transformations that living matter has undergone and is undergoing under the influence of physicochemical forces" – and could not be opposed to the experimental sciences. "On peut dire que la physiologie est la biologie limitée dans le présent, tandis que la morphologie y associe le passé. La forme organique est inséparable d'un immense passé."[22] What scientist worth his salt would have tolerated a definition of his discipline that deprived him of the masculine divinity that derived from his creative power? The scalpel of the physiologist looked like the dread tool of the great castrator.

In Giard's opinion, what had made Bernard's view untenable was the triumph of the doctrines of Lamarck and of Darwin – no one would have been more surprised than Claude Bernard at this turn of events. A shadow of the Comtean spectrality lurked in Giard's description of the change in points of view in natural history: The categories of the static and the dynamic were not obsolete.[23] In his speech to the congress of German doctors and naturalists in 1877, Haeckel expressed the same viewpoint: The theory of descent had raised biology, zoology, and especially botany to a new conceptual level, equal to that of physics and chemistry. Giard added that the theory had also introduced a common aim for all sciences, eliminating the issue of supremacy for one or the other science. Anatomy, physiology, ethnology, geonomy, systematics, and paleontology were all parts of an indivisible whole tending toward the realization of the same program: "retracer d'une façon aussi exacte et aussi complète que possible l'histoire des manifestations de la vie sur notre planète en laissant aux métaphysiciens et aux poétes le soin de chercher les origines premières ou d'en célébrer les finalités." If Bernard had read this, he might have been puzzled as to how a restatement of the gospel of Milne Edwards slathered with the frothy icing of Lamarckism–Darwinism achieved a synthesis of what his dialectic had rent asunder.

A definition of biology given by the Medawars is comfortingly ecumenical: "Biology is a general term comprehending all the sciences (the 'life sciences') that have to do with the structures, performances and interactions of living things."[24] What is striking about the nineteenth century is the proliferation of areas of specialization in studying and experimenting on living things. One of the most successful of these specializations was marine biology. Like any new area of research, marine biology immediately concocted for itself a pedigree based on great practitioners of the past. In the beginning there was Aristotle, the founder of zoology, who did comparative observation of the structure of internal organs and their functions. Pliny the Elder, a writer rather than a naturalist, was nothing compared to Aristotle. And there was not much in Lucretius on marine life. With the rebirth of zoology in the Renaissance, the level of research took a quantum leap upward. In France Pierre Belon made his mark in the study of marine animals before being assassinated by robbers in the Bois de Boulogne in 1564. Rondelet wrote a history of fish. The seventeenth and eighteenth centuries were by no means devoid of interest for the study of marine life, but it was with the advent of Georges Cuvier that marine biology began to take on its cognitive shape. Cuvier had the good sense to spend the Terror preparing his famous memoirs on molluscs, ascidia, and corals while he was a tutor in the Château de Fiqainville, near Fécamp. Jean Ardouin and Henri Milne Edwards, accompanied by their wives, continued the great tradition while frolicking at the seashore. Milne Edwards's investigations produced the classic three-volume *Histoire naturelle des crustacés* (1834–40). With de Quatrefages and Blanchard he also did some zoological exploration in Sicily. Before it could become a full-fledged subdiscipline marine biology needed a special budget; only then could it become, in French eyes, "une science officielle." Henri de Lacaze-Duthiers, more than any other scientist, deserves the credit for starting the process of institutionalization of marine biology.[25] But the story is by no means exclusively French.

The laboratories of marine biology: a great success for French biology

The second half of the nineteenth century, and particularly the last quarter, was the golden age of the founding of laboratories of marine research. To the Belgians go the laurels for priority: In 1843 P. M. van Beneden, a great naturalist, professor at Louvain, founded the first permanent marine laboratory at Ostend, where he did the work published as *Recherches sur la faune littorale de Belgique*, which appeared in the memoirs of the Académie de Belgique (1850, 1860, 1862, 1866). Edouard van Beneden, son of the founder, professor at the Université de Liège, later brought pupils and disciples to Ostend and created the Belgian tradition of university maritime zoological field trips.[26] Others might pioneer, but the Germans built the biggest and most famous monument. The Neapolitan zoological station with an aquarium, established in 1872, was universally regarded as the very model of a modern marine biology laboratory.

103

The success of the incomparable Neapolitan zoological station was the result of the efforts and money of Anton Dohrn. Equally important for its international reputation were the substantial publications of Dohrn and his collaborators, whose attention was devoted chiefly to marine invertebrates, part of an intensive study of the flora and fauna of the gulf of Naples. The special facilities of the station permitted a research emphasis on embryology, including the most complex technical work. Between 1874 and 1884 about 300 scientists worked at the station and 250 publications were researched there.

There were thirty tables at the station, most of which were occupied or reserved. Each of the tables was a complete, perfectly equipped "laboratory." The cost of maintaining such an establishment was partially covered by the practice of having different governments and scientific societies rent tables annually for 2,000 francs each. Four tables were taken by the government of Italy, three by Prussia, two by Russia, one each by Holland, Belgium, and Switzerland, and one table each by different organizations like the British Association for the Advancement of Science and institutions like the University of Cambridge; one table was reserved by the Berlin Academy of Sciences, which also provided the station with a small steamboat. Renting a table gave foreign governments the right to send their scientists to Naples on research missions. For this impressive research institution Dohrn spent the equivalent of 270,000 francs from his personal funds, including gifts by relatives, and a 100,000-franc subsidy by the German government. English naturalists donated 25,000 francs. The land was contributed by the city of Naples. Annual expenses rose from 20,000 francs in 1874 to 100,000 francs in 1882.

The total scientific personnel of the station was thirty-four in 1882. Dohrn, the director, had six assistants: one for administration and replacement of the director when he was absent; one for the administration of the aquarium, the conservation of collections of gulf fauna, and the methodical listing of the flora and fauna of the gulf; one for the systematic formation and classification of collections; one to oversee the preparation and conservation of specimens sold to museums, laboratories, and universities; one for the running of the botanical laboratory and herbarium; and the sixth as a research assistant and librarian. The library of 4,000 volumes, worth between thirty and forty thousand francs, was strong in embryology. A laboratory of experimental physiology was built later. There were three series of publications put out by the station: *Mittheilungen aus des zoologischer Station zu Neapel*, made up of notices and memoirs, published every trimester; *Fauna und Flora des Golfes von Neapel*, a series of monographs that could be in German, English, Italian, or French; and the *Jahresbericht* and annual *compte rendu* subsidized by academies, societies, and learned institutions. True, by the end of the 1880s only one monograph had been produced, but it was of great importance and so big that it cost 50,000 francs. Alfred Giard recognized that the Neapolitan station, incomparably superior to all other marine biology establishments, had an enormous influence on the progress of zoology during the last quarter of the nineteenth century.[27]

Certainly, marine biology was one of the areas in which French scientists could justly boast of their striking achievements in the nineteenth century. The oldest of French laboratories in marine biology is that founded in the 1850s by the embryologist J.-J.-M.-C.-V. Coste (Collège de France) at Concarneau, in Finistère, on the south coast of Brittany, a small town loved by sardines, artists, and tourists.[28] Coste was clearly aware of the virtue of self-advertisement, especially in the matter of priority. In 1868 he declared to the Academy of Sciences that he had opened up a new field of exploration to the energies of the nineteenth century by creating a vast model establishment, visited by scientists from all countries, the center of important scientific work from its beginning. Among the early research was that of Robin on the electrical apparatus of rays, Moreau on the formation of gas in the swimming bladder of fish, Gerbe on the metamorphosis of crustaceans, Legouis on the pancreas of fish, van Beneden on the *Phillobotrium*, and Coste on the fertilization and the incubation period of different species. Equipped with seventy aquariums and six large reservoirs for the study of marine life in its "natural" habitat, this "ocean observatory" was carefully planned according to data available from long experience; it was not surprising that it produced such excellent results in higher educational programs. Coste was proud that the unique combination of reservoirs or sea-ponds and aquariums made his laboratory "an improved instrument of investigation unprecedented in the history of science."[29]

Coste's laboratory, founded with the support of the ministries of the navy and of public works, provided facilities for work on the condition of the bay and for the measurement of water temperatures, research done by the scientist-commander of the schooner *La Perle*. At first intended solely for research, the laboratory was soon visited by students working for the university degree of licence in natural science. An early connection was also established with the Muséum. Some personnel from the medical faculties of Lille and of Paris found it useful to do research at Concarneau, which was both a laboratory of zoology and of marine physiology. Chabry, who did work on the mechanism of the swimming of fish, the mechanism of jumping, and the physical laws of diffusion, became associate director of the laboratory. Some of the better-known scientists associated with Concarneau were famous for their work in physiology: Robin, Marey, Ranvier, and d'Arsonval. In fact, it reflected the special interest of the Collège de France in physiology and in pisciculture. For a few years in the 1880s there was a laboratory devoted solely to marine physiology at Le Havre under the direction of Paul Bert. Regnard, Blanchard, and Richet worked there, but it closed when Bert's jump into politics ended in his death while on an administrative mission in Indo-China in 1886.

Although the laboratory at Concarneau united two friendly rival disciplines in its title, the Laboratoire de physiologie maritime of Le Havre made no secret about its concentration on physiology. Milne Edwards had nearly got the laboratory started before the Franco-Prussian war, but he was unable to raise enough money to buy an aquarium that had cost 200,000 francs for the exposition of 1868. In 1882 Doctor Gibert of Le Havre bought the decaying aquarium, restored it, and offered it to the Ministry of Education along with a laboratory building, all for an annual rental of 1,000 francs, which was also the amount given by the municipal council for heating and lighting. Paul Bert became the director of the laboratory, Paul Regnard the assistant director, and Raphael Blanchard and R. Dubois assistants.

How did this laboratory differ from the other laboratories in marine science? As Bert noted, a zoologist could not do much on the Norman beaches because of their lack of a rich marine life. This did not matter to the physiologist because generally one member of a species was sufficient to experiment with – "the laws regulating the life of one mollusc are evidently those that regulate all the others." Since physiology drew on physics, chemistry, and even surgery, the physiologist needed apparatuses and complex instruments and the means to repair them found only in cities. One of the five large rooms of the laboratory contained the delicate equipment in conditions vital to protect it from the ravages of coastal climate. Regnard and Blanchard did work on the respiratory capacity of the blood of diving animals, and Richet did some of his important work on the digestive liquids of fish. Much of the work was published in the same journals and outlets as the work of the other laboratories – *Comptes rendus* of the Academy of Sciences, *Bulletin de la société zoologique de France, Comptes rendus* of the Société de biologie – but Richet's work on digestion appeared in the *Archives de physiologie* (1882). Perhaps the work done in this laboratory shows that the sibling rivalry with physiology exhibited by Lacaze-Duthiers in the 1870s became far less evident once zoology had shown its experimental strength and was clearly established as a scientific discipline.

The continuing success of Concarneau was intimately connected with the direct utility of much of its research for the important local fish industry. For several years in the 1880s the laboratory carried out studies on sardine fishing in the Bay of Concarneau. For some reason sardines did not show up in any quantity between 1880 and 1882 and again between 1884 and 1886; the year 1883 was one of only middling success for fishermen. Politicians and bureaucrats became alarmed at the fickleness of the sardine; a commission of enquiry met at Brest to discuss a new type of net. Georges Pouchet was able to tell the commission that, speaking from a strictly scientific point of view, he believed that fishermen should be left completely free to determine the best methods and equipment to use. Science supported need and greed, and in 1887, after three years of scarcity, nature collaborated by supplying an abundance of fish in the Bay of Concarneau. In 1888 Marcel Baudouin explained the decline of the catch in terms of the pathology of the sardine. This surprising news also appeared as a pamphlet in 1894. Baudouin took it all back in 1905. Although it was useful to know the parasitic diseases of the sardine, the crisis of 1905 was not really the result of the disease but the natural successor to the crisis beginning about 1882–3. Following the recommendation of the Brest commission, a governmental decree created a consultative committee on fisheries with the deputy Gerville-Réache as president. Curiously, chambers of commerce and manufacturers of sardine oil were not enlightened enough to be interested in the report prepared for the Ministry of the Navy. Little matter. Scientific publications pullulated, and Pouchet's article on the use of sardine nets, published in the *Phare de la Loire* (November 3–4, 1887), was translated in the *Deutsche Fischerei-Zeitung* – the ultimate accolade: for French science to speak in German.[30]

Of some significance for the fisheries was the project carried out by Chabry to measure the annual winter temperatures in the Bay of Concarneau. A great number of new facts on the life and behavior of fish – species migration and their regular return and annual periodicity in the manifestation of ocean life – could be used in the exploitation of the riches of the sea. So meteorology was harnessed to solve the problems of general biology. In order to devise instrumentation that would withstand the pressures of the deep, close collaboration was required with the constructor of scientific instruments, Richard Frères in this case. Funding came from the Association française pour l'avancement des sciences and the results were published in the *Comptes rendus de l'Académie des sciences*. Truly a national effort. The work was also explicitly recognized as part of a general issue of universal concern, overfishing of the world's oceans, for which there were few scientific data. The establishment of tables showing the quantities of certain species of sardines caught as early as 1815 and especially for the period after 1855 was an admirable beginning to the research for this vast inquiry.[31] Not everyone was able to see the importance of measuring the testicles of sardines – a developed testicle might have a diameter of 0^m 060 – and noting the condition of the eggs of spawning females – most eggs would be about 0^m 10 in diameter but could be more than 0^m 50 – but not the least of the virtues of the French government was that it did.

In the history of marine laboratories, the laboratory of experimental zoology at Roscoff occupies a dominant position, partly because of its achievements and partly because of the justly deserved praise Henri de Lacaze-Duthiers heaped on himself for creating it. Even Giard gave the old scientist his due, once defending Lacaze-Duthiers against the faint praise of Delage that the master of them both was the man who had carried to its limits what could be taught by forceps and scalpel, and magnifying glass and microscope applied directly to tissues: He also solved important morpho-logical problems and so moved beyond purely descriptive anatomy.[32] The director of higher education, A. Dumesnil, asked Lacaze-Duthiers to organize the laboratory because there was nothing comparable in France. A few aquariums existed in Le Havre, Arcachon, and Paris, and, of course, there was Coste's operation in Concarneau. It is probable that the Ministry of Public Education wanted its own laboratory, which, under the aegis of Hautes études, could be its own official showcase. Ministerial jealousy was not without its benefits for research. At first devoted exclusively to research, the laboratory soon succumbed to an invasion of candidates for University degrees and assumed an important teaching role in the natural sciences in France.

The faculties of science had three categories of students: the medical students, who were the most numerous and took the most elementary courses; the students studying for the licence; and students who would be the scientists of the future. It was this last group that interested serious scientists like Lacaze-Duthiers, who rated them as the most important and the most in rapport with the nature and the very spirit of the higher education given in the faculties. Unfortunately, the organization of the faculties could not accommodate the three levels of students at the same time. Research laboratories offered a refuge for serious students from the rest.

Financial support for the station was always insufficient. Has there ever been a scientific institution that did not cry poverty? The initial annual subsidy was 3,000 francs, with an extra 3,000 given for the construction of reservoirs in 1875. In 1876 a beautiful estate was bought for 23,000 francs and improved for another 24,000. The cost of a park was 3,000 francs, and a sea-pond (*vivier*) carried the high price of 24,000 francs. Lacaze-Duthiers failed in his attempts to get subsidies from the departmental government of Finistère, which, unlike many departments, was uninterested in supporting science. The absence of any tradition of higher education and ignorance of its possible benefits to the local economy probably explain the indifference of the regional authorities to the laboratory. The Association française pour l'avancement des sciences and a generous faculty professor contributed a total of 1,100 francs. This hardly made up for local indifference, which extended to the commune level, but Roscoff gave the establishment the right to an old and useless path after a year's hard bargaining. Lacaze-Duthiers had the good idea of beginning a sort of yearbook, the *Archives de zoologie expérimentale*, for the publication of the research done at Roscoff. He had to pay for it himself: The first volume, in 1872, cost 6,800 francs. Lacaze-Duthiers also believed in the need to show foreigners, who were not so numerous in French schools as before, that one could still do serious work in France. After the Franco-Prussian war he also thought it his moral duty to make personal sacrifices. As the experiences of Dohrn and Giard showed, science was an expensive mistress who could not be supported in style solely on governmental subsidies. Fortunately, Lacaze-Duthiers, an old bachelor content with the sexual activity of marine life, could afford her. The more she was paid, the more she produced, or so it was argued by her admirers, who were also her procurers.

Roscoff became the most important of French marine laboratories. Under the directorship of Yves Delage the principal building was constructed by Charles Pères; the laboratory, named for Lacaze-Duthiers, was part of the University of Paris. (A new *national* direction in science funding is evident in the building of Le laboratoire Yves Delage, finished in 1953, which belongs to the Centre national de la recherche scientifique.) But Roscoff became such an important research center because of the support of the Ministry of Public Education, the University of Paris, and private funds. The construction of the *Pluteus*, a boat of eighteen tons for fishing and dragging, with sails and a ten-horsepower engine, would not have been possible without a grant of 2,000 francs by the Conseil de l'Université de Paris and contributions from Prince Roland Bonaparte, the prince of Monaco, the Rothschilds, and others. The University of Paris Council also had to vote 2,500 francs for repairing aquarium machinery in 1904–5. The engine for the boat was bought with a gift of 2,000 francs in 1905 by the Société des amis de l'Université de Paris. Ten contributors gave 32,000 francs. Such generosity emphasized the need for many thousands more. The justification for the gifts was the scientific significance of the type of science done at Roscoff. This was perceived with an exaggeration entirely appropriate to the occasion: The laboratory had no equivalent anywhere, for at Roscoff professors and students could carry out their studies on living creatures in their natural environment, rather than on preserved specimens deformed by alcohol.[33]

In spite of all the difficulties in getting Roscoff under way and the failure of governments to implement completely the grandiose plans of Lacaze-Duthiers, the laboratory was a great success. In the summer of 1883 forty people applied for admission and twenty-six showed up, among whom were four Englishmen and three Swiss, including Hermann Fol, professor of comparative anatomy at the University of Geneva. Among the French was Charles Richet, then professor in the Paris faculty of medicine and editor of the *Revue scientifique*. Fol continued his work on the embryology of polyps during the year of 1883; thirteen others, including four Englishmen, also continued their research.

Beginning in 1879 Lacaze-Duthiers undertook to establish a new laboratory for Mediterranean zoological studies. There were three basic reasons why he undertook the struggle finally resulting by 1881 in the establishment of the Laboratoire Arago at Banyuls-sur-mer (Pyrénées-Orientales). First, the "sea of the Pyrénées-Orientales" was a vast, unexplored area: It was the traditional challenge of the unknown to science. Second, zoological research was quite different in the Mediterranean from what it was in the ocean: The education of a naturalist would be incomplete without direct knowledge of both. Third, work was difficult at Roscoff between October and May: Winter on the Mediterranean would mean no interruption in research. In 1879, when Perpignan erected a statue to Arago, the minister of education and the director of higher education showed up to honor a local glory who was now enshrined in the Third Republican pantheon of deities. Lacaze-Duthiers seized the opportunity to raise the issue of a winter biological station. Ferry and Dumont were in favor of a laboratory on the coasts of Roussillon comparable to Roscoff. With the blessing but not the financial support of the government Lacaze-Duthiers started negotiations. Difficulties fabricated by the army engineering corps, which was working on wet docks at Port-Vendres, threw the advantage to its competitor, Banyuls, whose municipal council offered land and money to get the laboratory. But 1880 passed without any definite decision.

In 1881 a subsidy of 20,000 francs to help construction was approved by the general council of the Pyrénées-Orientales; work began in the fall. Lacaze-Duthiers's influence and reputation got him supporting funds from the Academy of Sciences. Total expenses in 1883 had risen to 26,500 francs. The Ministry of Public Education gave 18,000 francs for furniture, books, and instruments. But the total cost by 1884 was 137,000 francs. Only a great fundraiser like Lacaze-Duthiers could undertake such projects. Local support counted for a great deal. What a happy contrast with Finistère, whose general council had turned down his requests for funds on four occasions and where the commune of Roscoff had given him 300 out of 100,700 francs it had received from Paris for its schools. The department of

110

the Pyrénées-Orientales gave 27,000 francs and the commune of Banyuls 30,000 francs. Lacaze-Duthiers's friends, "devoted friends of science," were also generous with their help. Support came from the municipal councils of Perpignan (1,000 francs) and of Toulouse. In the case of Toulouse, Professor Barthélemy got the council to vote 4,000 francs. Even the railways of the Midi did their bit in giving reduced fares to researchers; Banyuls being twenty hours from Paris, this was not a small concession.[34] The Arago Laboratory showed what could be done, even in biology, by a top national scientist with strong regional support.

Confidence in Banyuls was justified from the beginning. Even in 1883 researchers settled in, while as many as eighty construction workers were still doing their job. Seven *licenciés ès sciences naturelles* came from Paris to begin their doctoral theses. Two military physicians came to do doctorates in science. Henri Wegmann of Zurich did work at Banyuls and Roscoff as well as Paris. Three Americans came for the winter and part of the spring, and George Dimmock wrote a flattering article on the laboratory for an American journal. One of the features of the Banyuls operation was an annual Easter expedition for the University – the laboratory was early annexed to the University of Paris. This type of scientific jaunt could provide an opportunity for collaboration between universities, as in the case of trips to Barcelona. In 1898 the visit of the French to Barcelona was continued to the Balearic islands along with the Spanish students of Odón de Buen, professor of natural history at the University of Barcelona. The stations of marine biology were a remarkable minor but basic means of international scientific interaction.

No less important than the international function of the marine stations was their educational-diversionary role in French society. The Laboratoire Arago was important to the faculty of sciences at Toulouse, both as a possible place of research and as a source of supply of living animals from the Mediterranean, but it also served as the object of excursions by the Toulousain naturalists. With Professor Barthélemy in charge, an excursion could be sufficiently scientific to attract lycée and collège teachers from Auch, Foix, Figeac, and Castelnaudary, students from the faculty of sciences, as well as members of the Société d'histoire naturelle de Toulouse. Other groups using the facility later, in November 1883, were the Ecole normale des jeunes filles and the Société d'histoire naturelle de Béziers. Lacaze-Duthiers had evolved into a strong supporter of the popularization of science, one of the basic republican dogmas.

> I earnestly want the Laboratoire Arago to extend the services it can give
> in popularizing zoological knowledge. These services will undoubtedly

111

be of a lower order than those we generally claim to provide . . . but they are not less important. In developing a taste for the natural sciences through the viewing of the curious creatures abundant in our waters, we can be certain of assuring greater final progress in zoology than solely by the studies of the overrated programs with which education is today burdened.[35]

In 1887, when he visited the Laboratoire Arago along with the Congrès scientifique de Toulouse, P. van Beneden wrote a flattering note in its visitor's book. "No doubt the laboratory in Naples is more luxuriously housed, but it does not have all the apparatuses and riches of the establishment of M. de Lacaze-Duthiers. The construction of the laboratories of Roscoff and of Banyuls marks a new period in the history of science . . . " He was right. By 1910–11 Banyuls, under Professor Pruvot, housed sixty workers, of whom twenty-one were foreign. Unfortunately, by then the buildings were thirty years old and totally insufficient for their function. But this deficiency was the result of success, not of failure.[36]

One of the most creatively productive of France's laboratories of marine biology began in difficult circumstances at Wimereux (Pas-de-Calais). When in 1873 he was appointed to the chair of natural history at Lille, Alfred Giard found himself in the material poverty typical of a French provincial faculty of the period. But Lille already had a tradition in zoology and the natural sciences established by Lacaze-Duthiers, Dareste, and Gosselet, and it had a small group of researchers in the faculty of sciences. Giard believed that the natural sciences should be taught in nature: Several months of faculty courses could be replaced by excursions or a stay in a laboratory. Giard had done his doctoral thesis at Roscoff on *Syncarida* when Lacaze-Duthiers was in the process of founding his zoological station; so he had been initiated quite early into the cult of zoology at the seashore. The dynamic Giard immediately undertook to get a research station established in the Nord. In spite of promises by Dumesnil, the minister of public education could not find the money to help Giard. It was impossible to expect more help from the municipal government, which had just given him a big building for laboratories of histology, anatomy, and physiology – including everything the Germans called an institute of zoology. Giard decided on Wimereux as the location for the station, rented a house, and paid for the construction of aquariums with his own money. (Giard married, but later.) Wurtz was at this time president of the Association française pour l'avancement des sciences and got the council of the organization to give Giard a subsidy of 1,500 francs. The subsidy was renewed several times. In 1875 the Laboratoire de zoologie maritime became part of the Ecole pratique des hautes études; this meant an annual subsidy of 3,000–4,000 francs. Giard continued his own annual contribution of 1,000 francs. This laboratory probably produced more research for less money than most nineteenth-century scientific enterprises.

Giard began with the hope of forming at Lille a group of zealous workers "like one meets in most foreign universities." His passion was to see zoological work done which would be the vehicle for the propagation in France of the "admirable doctrines that Darwin, Vogt, Claparède, Kowalewsky, and Haeckel had spread among the peoples who had made the most rapid progress in science," for "these doctrines had produced in the biological sciences a revolution comparable to that formerly achieved by the hypothesis of Newton in the astronomical sciences; these doctrines have impressed upon the natural sciences the character of *grandeur* and simplicity that the mechanical theory of heat and the wave theory formerly did for the study of the great physical laws of nature."[37] Most of Giard's students strongly supported his efforts at Wimereux. Backing up H. Leloir, Charles and J. Barrois, Dutertre, and de Guerne was Giard's assistant, Paul Hallez. By 1884 twenty-one students had successfully prepared their examinations for the *licence ès sciences naturelles*; six doctoral theses had been done in the laboratory and two more were in progress – those of Hallez, Ch. Barrois, J. Barrois, who did two, Moniez, Bonnier, and Billet. Over ninety publications, many of them by Giard, resulted from the work done in the laboratory between 1874 and 1882–3. Two systematic publications depended on Wimereux: *Travaux de l'Institut zoologique de Lille et de la station maritime de Lille* and the *Bulletin scientifique du Nord* (later *de la France et de la Belgique*). The *Bulletin* was especially interested in evolution. "Sans négliger aucune des parties des sciences biologiques, la direction s'attache surtout à publier des travaux ayant trait à l'évolution (ontogénie et phylogénie) des êtres vivants. Les recherches relatives à l'ethologie et à la distribution géographique dans leurs rapports avec la théorie de la descendance occupent aussi une large place dans le *Bulletin*." Authors frequently paid their own publication expenses, and Giard himself funded the *Bulletin scientifique*. Wimereux was well known outside France, especially in Belgium. Among the foreigners who worked there were E. Ray Lankester, A. Kowalewsky, Horst, C. Vogt, F. Monticelli, Ch. Julin, P. Pelseneer, and G. Hermann. Brains, enthusiasm, and hard work offset material poverty, provided researchers were not unwilling to use their own money to offset the most serious disadvantages of the absence of official subsidies. Giard did build up a good library at Wimereux. The laboratory also seems to have a had a strong attraction for students of the Normale.

In 1889 Giard was appointed to the newly created chair of the Evolution des êtres organisés given to the Sorbonne by the city of Paris, and Wimereux was integrated into the University of Paris. One of the results of the curious system of funding men rather than institutions was that the institution followed the man; not that it made much difference to a discipline in the end. When Louis Liard was director of higher education he made it possible for Giard to ask parliament for 30,000 francs for his program in marine biology, but this was not the time to expect such large sums for studying sponges, parasites, and the genital organs of crustaceans. Then, in 1898, Maurice Lonquéty built Giard a new laboratory at Pointe-aux-Oies, near the race track of Boulogne. With the support of the University of Paris, where Darboux was now dean of the faculty of sciences, the new laboratory of 2,500 square meters was respectably furnished with scientific equipment. In the old days at Wimereux researchers had to provide their own simple instruments, but at a time when ideas in biology were at least as important as instrumentation, this was not a barrier to doing good work.[38]

A catalog of the laboratories of marine biology would include the establishments at Marseille (1869), Le Portal (1888), and several others. Le Portal was Hallez's private creation. The proliferation of these establishments cannot be explained by the hoary model of centralization usually invoked by commentators on science in France. It would perhaps have made more sense to have had fewer laboratories but better funded, or at least this is what was said by the professors who wanted money; it might have been better to have cut back on some operation to fund Giard better, but this is by no means certain. First, it is difficult to see how this could have been done, for departmental and municipal support was usually important in the founding and operation of the stations. It was not a matter of coordinated decisions emanating from Paris. Even where Paris was involved through ministries or Hautes études, influential and productive professors could and did use centralization to proliferate laboratories. The production of a great deal of high-quality research can be interpreted as a justification for the existence of the laboratories. The relative cheapness of the marine biology operations as compared, for example, with some of the applied sciences laboratories in Nancy made possible a curious embarrassment of riches in this area of French science. Local pride and, more important, belief in the practical significance of marine biology, combined sometimes with the presence of professors and their friends in local government, were powerful stimuli to the growth of marine biology. There were some notorious exceptions to local support. Concarneau was surpassed in its indifference by the bourgeois philistines of Wimereux – "des goîtreux!"[39]

Most of the laboratories we have not dealt with here were basically similar to those treated in some detail. Departmental and municipal governments generally supported the ventures. In 1880 the general council of Calvados voted to establish a laboratory at Luc-sur-Mer on the property of de Caumont, valued at 30,000 francs, given by his widow to perpetuate the archeologist's name; the state gave 9,114 francs in 1883, when Delage took over after the death of Deslongchamps. At Villefranche the commune gave a yearly allocation of 200 francs, which brought the annual credit by 1884 to 2,100 francs, 500 francs of which came from the Ministry of Agriculture. A pattern of financing by local, departmental, and state governments was sometimes augmented by private donations or contributions by associations. The biological station at Arcachon was created by the Société scientifique d'Arcachon in 1867. Marseille gave the Laboratoire de zoologie maritime, run by Lespès from 1869 to 1872, an annual credit of 3,000 francs. Under A. F. Marion, Charles Lespès's successor, the new Station zoologique d'Endoume began with a municipal contribution of 20,000 francs in 1884, the year of completion. Armand Sabatier's laboratory at Cette found support in both Cette and Montpellier and by the general council of the Hérault because of the clear utility of the laboratory to the fisheries, especially the oyster industry. On Sabatier's death in 1910, his widow gave 4,000 francs to the laboratory, which had a reputation for having the latest technical equipment. Biologists found that the funding of science on an adequate level was much easier when it was directly related to the regional economy.

When he wrote a report on the zoology laboratory at Marseille in 1884 Marion contrasted the research functions of foreign institutes with the teaching functions of French institutes. Establishments in Naples, Trieste, and Sebastopol, for example, were "exclusively reserved for higher scientific research," whereas French laboratories were "teaching establishments subject to the complex task of university education." The demands made on the laboratories for training licence students certainly took up a great deal of time: Delage reported that out of seventeen persons working at Luc-sur-Mer in 1883, thirteen were candidates for the licence. Since there was not loud protest on the part of the professor, in contrast with the later protests against overcrowded laboratories in all the sciences after the PCN students inundated the faculties, we may assume that the situation was tolerable and that biologists could get their work done. (The *certificat d'études physiques, naturelles et chimiques* [PCN] was introduced in 1893, ostensibly to give medical students some basic science but also to provide needed bodies for faculties of sciences.) Indeed, Marion attached to his report a list showing that between 1869 and 1872 he, Lespès, and G. Moquin-Tandon had published nine memoirs. Between 1873 and 1884 the number of publications based on work done in the laboratory rose to seventy-eight, of which over twenty were by Marion himself. One advantage of the combination of teaching and research functions was that it provided a feeder system to the doctorate. There was also good justification for upgrading the natural science programs in the collèges and lycées by initiating the secondary teachers into the mysteries of high-level scientific research.

It must be emphasized that the research ideology was quite strong among French biologists, so strong, in fact, that no marine biology laboratory failed to display proudly at the end of each official report the list of its researchers' publications. Naturally this list did not preclude inclusion in the report of the nearly tropistic complain of those bitten by the research bug that funds were insufficient and facilities deficient, and a warning that government should pay attention to the glorious achievements of foreigners who were far ahead of France. Even where the list of publications was not long, such as that of Cette for 1881–4, it could be significant: Carl Vogt did research there for his *Traité d'anatomie comparée pratique*, and the eight items listed by Sabatier made up the first part of a long series of studies on the nature and origin of animal sexuality. Even Hallez's little operation at Le Portal (Pas-de-Calais), founded in 1888 when he was professor of zoology at Lille, produced the research for the *Revue biologique du Nord de la France*, which he, Barrois, and Moniez edited between 1888 and 1896. An official publication boasted in 1955 that the nine laboratories of the Institut de biologie marine de l'Université de Bordeaux, which issued from the ancient womb of the Arcachon station in 1949, were only for research. The roots of this research ideology stretch deep into the nineteenth century.[40]

The establishment of the laboratories of marine biology may have been the most significant development in nineteenth-century French biology because of their importance as centers of research and of training for a whole generation of scientists who definitively made biology into a subject of substance in the faculties of science. They were so important that the French established about twice as many as they probably should have. Lacaze-Duthiers argued in 1894 that ten better-funded ones would have been better than twenty competing for limited funds. But it is by no means certain that this fit of excess in establishing laboratories was anything but a stimulus to science; in the end it probably produced more and better science than might have been the case had some "overlord of science" imposed an exemplary rational plan for marine biology.

The laboratories also deserve a note in the history of psychology. Giard was a strong supporter of the Institut général psychologique, welcomed psychological researchers in his laboratory, and participated, with Delage, Mlle Goldsmith, Bonnier, Perrier, Manouvrier, and others, in the formation in 1905 of the Group d'étude de psychologie zoologique. Delage set aside a room at Roscoff for the work of the Institut. The study of the psychology of fish and animals, and of the behavior of microorganisms, was well under way in the late nineteenth century.[41]

Those who worship the reality principle as the greatest of pedagogical virtues have seen in the laboratories of marine biology a symbol of their crusade against the excessive verbalism infecting French education and so damaging in the teaching of the natural sciences. Maurice Caullery and many others were delighted to quit their books in Paris for the joy of finding out about things in the world of nature at Wimereux. Caullery viewed the experience available at locations of the laboratories as a valuable antidote to the notoriously abstract nature of all French education. "Sweet is the lore which Nature brings," of course, but more important for the French scientist was the fact that a sort of self-education produced a quick, healthy, and penetrating judgment, the principal gift of all education.[42] This was not a fully Romantic interpretation of Wordsworth's bucolic lines in "The Tables Turned": "Come forth into the light of things, Let Nature be your teacher."

Whither biology in France? Giard versus Delage

At the end of the nineteenth century, French biology was not seriously Darwinist, at least not more than biology in other countries. Sentimentally Lamarckian, devoted to a cult of observational and experimental techniques, it was split into embarrassed factions by a sustained attack by one of France's two leading biologists on the other: Alfred Giard against Yves Delages. Perhaps much of Giard's ire came from his falling out with his dictatorial old master Lacaze-Duthiers, who passed on his Parisian scientific mantle and great influences to Delage rather than to Giard. Esau raged against Jacob, yet Jacob made no groveling submission or substantial gift. When Giard finally did get a chair in the faculty of sciences in Paris, it was a present from the anticlerical municipal council rather than the result of a recognition of his seminal contributions to biology by the faculty, which had not opposed the malign hostility of Lacaze-Duthiers to any attempt to move Giard to Paris.[43] But Giard's antics at Lille probably caused enough alarm in the camp of staid higher educational bureaucrats to spoil his chances for a Sorbonne post. He could expect no unusual promotions: His record was mediocre at the Normale, and he never took the *agrégation*, which had not existed for the natural sciences when Giard was there.[44] At least this was the bureaucratic reasoning within the higher educational section of the Ministry of Education, which was thus in agreement with the Parisian biological

117

establishment against a maverick who had a higher opinion of foreign than of French biology.

It was not entirely reprehensible from the radical republican viewpoint that the whimsical Giard wrote a burlesque on the new Catholic faculty of medicine in Lille. Dubious taste but good politics, and good for stimulating the state into establishing a full medical faculty in the University. But then Giard, no respecter of sacred secular objects, attacked the state medical facility in a spoof entitled "Note sur l'existence temporaire de myriapodes dans les fosses nasales de l'homme, suivie de quelques réflexions sur le parasitisme inchoatif." The medical professors, not noted for their sense of humor, expressed their collective outrage at this parody of their research. Nonplussed by a Voltairean in his scientific ranks, an astonished rector could only babble his astonishment to the medical dean. Neither faculty nor students wanted Giard in the medical faculty, which he started to neglect as a result of his preoccupation with the faculty of sciences. His transfer was logical and therapeutic. Giard's transfer to the faculty of sciences, where he got a justly deserved chair as a sort of unacknowledged part of the shift, did not end his war against the beleaguered doctors, whose scientific weakness was advertised anew in a note by H. Lescoeur published in Giard's *Bulletin*. Letters in the *Echo du Nord* achieved the ultimate horror in any scientific quarrel – going public. An investigation by the rector produced a ministerial reprimand of the culprits. Wannebrocq, the medical dean, was goaded into the maneuver of getting one of Giard's former collaborators on the *Bulletin*, Jules de Guerne, who had become a vulnerable medical student, to criticize Giard in print. Giard was able to increase the power of his attack on his medical colleagues from a radical pulpit on the municipal council. It did not improve matters that the dean of the faculty of sciences, Violette, had gone down to defeat in his bid for a council seat because of Giard's competition. The squabble was snuffed out only when Giard went off to Paris as a deputy for three years.[45]

Giard was admitted sixteenth in rank to the Ecole normale supérieure in 1867, became a *préparateur* at Hautes études in 1872, took the place of Dareste in zoology at Lille from 1873 to 1875, when he started teaching natural history at the Ecole de médecine et de pharmacie de Lille, which would shortly be evaluated to the status of a faculty. In 1881 Giard finally became professor of zoology in the faculty of sciences at Lille, where he stayed until he became a *maître de conférences* at the Normale in 1887, a year before he was given the post (chair in 1892) established in the faculty of sciences by the city of Paris. For several years it looked as if Giard, like Bert, might be lost to science, but after some years in municipal politics, not a full-time occupation, and a term (1882–5) in the Chamber of Deputies, Giard returned to his university job. Giard put Lille on the map of zoological research by his own work and by the training of well-known French naturalists: Charles and J. Barrois, Paul Hallez, Moniez, L. Dollo, J. Bonnier, and E. Canu. Certainly he was a Lamarckian, but that component of his research program did not preclude his introducing into France strong currents of Darwinism, reinforced by Haeckel's doctrines, by his teachings, writings, and polemics. This act of reconciliation was not unusual. Félix Le Dantec thought that neo-Darwinism applied to plastids while neo-Lamarckian theories applied to pluricellular beings. "Les deux systèmes se complètent et se concilient."[46] Even a rector frightened by Giard's violent attacks on his enemies and by his radical politics admitted that he was a remarkable researcher and initiator who filled his students with a passion for science. "C'est un 'fanatique' de la science, brillant, convaincu . . . " On his return to Lille for a brief period, after his defeat in the election of 1885, he could turn his attention to attacking the practice of sloppy science by the biological establishment in Paris. A convenient figure for attack was provided in 1885 when Yves Delage became professor in the faculty of sciences of Paris.

Marie-Yves Delage was born in 1854 in Avignon and died in 1920. His official functions in the French university system ranged over a typical spectrum of positions: "préparateur du laboratoire de zoologie expérimentale à l'Ecole pratique des hautes études" in 1878, a similar function in the laboratory of maritime zoology at Roscoff in 1881, "maître de conférences de zoologie à la faculté des sciences," Paris in 1882, and "chargé du cours de zoologie et physiologie animale" at Caen in 1883. Then in 1884 he achieved the elevated security of "titulaire," only to leave the next year to become "chargé du cours de zoologie, anatomie et physiologie" at the Sorbonne. The next year, 1885, at the tender age of thirty, he became professor in the faculty of sciences at the Sorbonne, a marvelous model of the brilliant, circumspect, safe university professor. The gods had smiled on Delage's career and, more important, Lacaze-Duthiers had approved. In 1888 Delage became adjunct head of the research laboratory of experimental zoology (Hautes études) and in 1901 the head of the Roscoff laboratory of the Paris faculty of sciences. By 1919 his annual salary had reached the first-class category, 25,000 francs. His visual problem, dating from 1904, officially recognized in 1909, forced him to take leave after 1913.

While Delage was at Caen Louis Liard was rector. The logician and philosopher who would become the administrative head of higher education in France praised Delage for his understanding of its true function. In 1884 Liard's report was a veritable panegyric: "I could not praise M. Delage too highly. This scientist and professor is always solely preoccupied with his work and his teaching. He organized extremely well the laboratory at Luc and the practical part of zoology teaching." Liard knew that Delage was not destined to remain at Caen, but he noted that Delage had left his mark on the faculty. While on his tour of inspection, E. Zevort concurred with Liard's assessment: Delage was full of zeal and very capable, especially as shown in his organization of the laboratory at Luc-sur-Mer, used by a handful of people each year to prepare for examinations and to carry out some original research. At Caen in 1883 Delage had six students and eight to ten auditors, and in 1884 four regular students and six to twelve auditors. The toleration of such small classes, not atypical of smaller provincial faculties, was striking evidence of the support of the government for science. Professors, deans, and rectors were uneasy, however, and made every effort to attract students to the faculties of science. The advantages of the blessings of officialdom became clear in 1889, when Delage took a tour, quaintly designated for fiscal purposes as a "voyage d'études," of many European cities, including Naples, Rome, Copenhagen, Berlin, and Hamburg.[47] (Most successful French academics were treated to at least one of these educational trips; some of them even produced interesting reports on the facilities visited. The custom was made famous by Wurtz, though few wrote such penetrating reports.) But Delage's career was soon to be called into question, quite without any material effect, when Giard incorporated him into his pantheon of incompetents.

Early in his career Giard advertised his break with the establishment in French biology by publishing an attack on the evil influence on French science of the prizes of the Academy of Sciences. In 1878 Giard attacked E. Oustalet's dissertation, *Recherches sur les insectes fossiles des terrains tertiaires de la France*, which had been given the Thore prize on the recommendation of a report signed by Emile Blanchard, a professor at the Muséum who became another victim of Giard's ironic barbs. Giard was also irritated by the seeming intrusion of a political factor in determining citations. Two new works by Kunckel d'Herculais and by Jousset de Bellesme did not cite – Giard said that they took care not to cite – J.-B. Robineau-Devoidy's *Recherches sur l'organisation vertébrale des crustacés, des arachnides et des insectes,* a work Giard believed to be relevant though published fifty years earlier. The author had made the mistake of dedicating his work to the old doctor and revolutionary, Raspail, whose name was not part of the iconography of the academic right and whose very mention might jeopardize getting a prize of the institute. Giard lamented the domination of scientists like V. Ardouin and Blanchard in the Muséum and the Academy of Sciences. What a situation for a country that should be proud of the achievements of Léon Dufour, Robineau, and Fabre![48]

In 1889 Giard returned with a vengeance to the topic of the vicious effects of the protection of scientific incompetents by the institute. His argument was ingenious and logical even if he exaggerated the importance of the actual biological squabble. "Errors are like heavy bodies: the higher the point from which they fall, the deeper they establish themselves in the minds they penetrate, and then they propagate themselves rapidly." So a mistake that one would not even dream of exposing in the work of an obscure worker must be carefully corrected when it is published by a man in a high scientific position and has received the approbation of judges of the highest competence, namely the Institut de France. The object of Giard's attack was the recent action of the Academy of Sciences in giving a prize to the author (Delage) of a memoir, *Sur la circulation des Crustacés édriophthalmes*, published in Lacaze-Duthiers's *Archives de zoologie expérimentale* (Vol. 9, 1881). Giard perpetuated Delage's fictional obscurity by not naming him in the article – a traditional ploy for diminishing the importance of an opponent, like Berthelot's relegation of Duhem to the anonymous category of persons indulging in some calculations when the physicist blasted the chemist's thermochemistry. By drawing attention to his own ignoring for several years a memoir important enough to be published by his old master's yearbook, Giard also relegated a good deal of Sorbonne biology to scientific limbo. Once the institute had given its recognition to Delage's work, Giard had to counter the influence such an accolade bestowed upon work and author. This was not an easy task in a prestige-oriented system like the French University. Giard believed in overkill, for he found nothing to praise in Delage's work. Since the work was Delage's doctoral thesis, condemnation included also Lacaze-Duthiers, whose official account of the defense waxed ecstatic on Delage's new discoveries, technical innovations, and talents for drawing and exposition.[49] The general conclusions, proudly advanced by Delage as the justification for the study of his obscure crustaceans, were given an airy dismissal by Giard; to clinch the case, Giard cited the work of a young zoologist, E.-L. Bouvier, who also had just shown the feeble nature of Delage's conclusions.[50] Nor did Delage get his details right: on the basis of the thesis Giard concluded that Delage had never seen the type of isopoda (crustaceans with seven pairs of thoracic legs) in their habitat enjoying natural relations with their hosts. An error of position was admitted by Delage but blamed on a printer's error. Delage refuted all the other technical criticisms.

Delage refused to be drawn into a squabble with Giard because he was convinced that Giard was engaged in a systematic campaign against him, carefully combing his works for mistakes and inventing them when he did not find any. Rather than a scientific discussion of serious works, Giard aimed at insulting the writer by couching his criticism in as insolent a form as possible.[51] Perhaps Delage was not wrong, and Giard was indulging what he imagined to be his righteous indignation against an opponent, necessarily and wrongly imagined incompetent, who was occupying a place Giard really believed should be his. And it might have been his had he not fallen out with Lacaze-Duthiers, one of the dispensers of scientific patronage in Paris.

Giard was also cruelly critical of Delage's memoir on the "Evolution de la sacculine," which had been published in the *Archives de zoologie expérimentale* in 1874, and which was, according to Giard, full of nonsense, some of it dangerous. It was not so serious that Delage did not seem to know that the sacculina was never an internal parasite in the proper sense of the word. What did matter was the denial of any correspondence between the sacculina and the *Cypris*, for that way of looking at things really denied the whole of comparative anatomy and of general embryogeny. Delage had cooked up a fantastic theory of the migration of Rhizocephala. Simple knowledge of this interesting type was sufficient to demolish the novel constructed by Delage and the gigantic mystification "imposed" on many zoologists.[52]

The culmination of Giard's attacks came after the publication by Delage of his fat tome in general biology on *La Structure du protoplasma* (1895), which Giard took as a rejection of the type of work he and his students had been doing for a generation.[53] Under a heading that announced the "conversion of M. Yves Delage," Giard wrote one of his ferociously critical and ironic reviews, devoting over twenty-five pages to Delage's introduction alone. Giard's heading for the review, imitating Delage's introduction, showed the issue clearly: the direction of biological research in France. Delage lamented the ignorance of his fellow biologists concerning the new paths of general biology, presumably those blazoned by Haeckel, Weismann, Lankester, Van Beneden, the Hertwig brothers, and W. Roux, among others. This odd accusation did not even apply to Delage's own faculty at the Sorbonne. Over a period of four years Giard himself dealt in his courses with Haeckel's *Gastroea-Theorie*, Weismann's theory of heredity, and Roux's mechanistic embryology. Giard thought that Delage had placed French biologists in a ridiculous position, from which they could be removed only by a destruction of Delage's arguments. Like Hugo's preface to *Cromwell*, Delage's work was dismissed on the grounds that what was true in it was not new and what was new was not true.

About eight pages of Giard's critique was made up of a series of parallel quotations from Delage's *Structure de protoplasma* and from his own works showing that Delage's call for a turn in French biological research in the direction of general biology ("conditions and causes of the great manifestations of life in the cell, in the individual, and in the species") was an echo of similar statements made by Giard over the previous twenty years. Delage was really a member of the "school" Giard had been attacking since his early days at Lille; his conversation to Giard's "school" was a little late, even in the area of biomechanics, which Delage made such a fuss about in his book. It was only by ignoring the work of Marion, Balbiani, Naudin, Sabatier, Millardet, and Maupas, that Delage could claim that the French biological community was outside the contemporary scientific movement.

Perhaps the claim of Delage that most excited Giard was that histological technique was mostly a foreign import in France, especially from Germany. "Eh, quoi!" What about Bichat? After merely citing Bichat's name, Giard advanced as French heroes in histology the names of Robin and Ranvier. With Raspail, Robin was one of the founders of parasitology and biological chemistry. Robin's work drew attention to the applications of the microscope, but he left an original mark on several areas, and although his name could not sum up the "histological period," he should not be left out of a book on *general biology*. For Giard it was Ranvier who ruled the domain of histology from his small laboratory at the Collège de France, frequented by generations of European physicians and scientists in search of the teachings later enshrined in the successive editions of the famous *Traité d'histologie*, a work described by Lieberkühn as the "gospel for technique." Giard did not limit his appreciation of the work of Ranvier to his technique, but his emphasis was on the mistake Delage had made in limiting Ranvier's role in science to that of a simple manipulator.

Nor did Giard agree with the friends of Delage who thought that the book was unoriginal, a presentation of other people's ideas, and unreliable where it was original but still a useful book for physicians and candidates for the licence. The difficulty was Delage's inaccuracy in stating the views of well-known scientists. This failing might be forgiven in an exposition of the fluctuating opinions of Weismann, but not in the cases of Wiesner, His, or Th. Eimer. Delage had even managed to state Th. Eimer's notion of "Génépistase" as the opposite of what Eimer really meant. Worse, the original ideas of Delboeuf, available in excellent French, as Giard exclaimed with heavy irony, and discussed by R. Baron and F. Houssay, were not understood by Delage any better than those of the Germans. The worst travesty of authors was Delage's statement that Wallace supported Darwin's theory of sexual selection, which Wallace explicitly denied; the denial could be found in the French translation (1891) of Wallace's book on Darwinism. Other mistakes, in chemistry, in botany, and in mathematics, made the work more undesirable for an audience of neophytes in medicine and in the licence program. A silly remark by Delage that non-Euclidean geometry could be taken seriously when it provided theorems that could be used to build an arch or a bridge left the work open for more amusing ironical comments by

Giard, perversely delighted to find such ignorance in the very faculty of Poincaré. Such were the pitfalls of thirdhand knowledge.

Delage's scientific work ranged from his early research in anatomy and taxonomy to his quasi-philosophical manifesto on heredity in 1895. This work, marking his conversion to a belief in general biology, implied a casting off of the old infertile dogmas that required all researchers to produce works in anatomy and embryogenesis. The old research rituals of histological studies and physiological experiments, no matter how well performed, usually produced few factual conclusions. To ignore the new possibilities offered by the progress achieved in knowledge of the mechanisms of life and to continue in the well-trodden path of tradition would be to fall into a "métier de dupes." What better subject to study than the deep structure of protoplasm, obviously containing in it the mechanical explanation of all the phenomena of which it is the center and holding the explanation of life. Delage piously considered it his duty to tell young people that the English and the Germans were already on this route ("la recherche des conditions et des causes des grandes manifestations de la vie dans la cellule, dans l'individu et dans l'espèce"). Because other French biologists disagreed with the new direction or kept silent about it, Delage felt obligated to preach the gospel of general biology in a book for the intelligent public as well as for naturalists. Delage's attempt to base much of philosophy on biology met with at least one striking success: In his famous book on creative evolution, Bergson often referred to *La Structure du protoplasma*. Delage completed his contribution to the paradigm by doing studies on merogeny and experimental partheno-genesis.[54]

Like most French biologists by the 1890s Delage accepted the transform-ist hypothesis as the only scientific one. In spite of Giard's jibes concerning Delage's alleged misunderstanding of German biological thought, the leading German Darwinist, August Weismann, wrote to Delage praising him for understanding and analyzing his thought better than anyone else. The botanist Carl Näegli's insistence on the need for an inner development principle was related to the category of a novel by Delage, who, preferring the organicism of Wilhelm Roux, opposed the preformation idea of ontogenesis and adopted epigenesis. Delage gave natural selection only a negative role in evolution. Le Dantec rejected Weismann on methodological and epistemological grounds but, unlike Delage, said little about Mendel, de Vries, and mutationism. Delage accepted the theory of Mendelian heredity, for, after all, it led to Weismann's theory – "the most complete of systems." Delage relegated Lamarckism to the status of a point of view, a tendency touching all biological issues; it was not a school or a theory.[55]

What puzzled Giard most was Delage's strange philosophy of science,

124

which led him to make his notorious declaration, so comforting to antievolutionists: "I am absolutely convinced that one is not an evolutionist for reasons based on natural history but because of one's philosophical opinions." Although Rütimeyer has said the same thing, he had logically concluded that such discussions should be avoided because of their sterility. Giard thought that few scientists would approve of a method based on philosophical opinions and propped up by a priori concepts. One must choose between scientific research and faith, Giard declared, and if Delage accepted evolution on faith, why try to convince others by so many complex arguments? Giard believed that research should not begin with the question of whether it is in agreement with a philosophical system; that question should not be a point of departure for science but rather a final group of scientific generalizations. How different was Darwin's state of mind, unconcerned with philosophical systems during a long career filled with fertile research. A hypothesis becomes scientific only when it is experimentally verified. To punish Delage for denying this dogma of the typical nineteenth-century scientist, Giard relegated him to the limbo of metaphysically inspired *Naturphilosophie*.

The reaction to Delage's *Structure du protoplasma* in the exalted columns of the *Revue scientifique* was enthusiastic and friendly. An unsigned reviewer marveled that such a work, attempting to rouse the French biologists from their conceptual stupor by sounding a foreign tocsin, could come from inside the Sorbonne, long an anti-Darwinist citadel. Certainly the most interesting and longest part of the review was the devastating critique made of the system of education typical in the zoological community for twenty-five years. The chief deficiency of the system was the total absence of ideas, an intellectual bankruptcy evident in and accentuated by the exclusive concentration of the zoologists on petty details of animal form and function. For French zoology to get out of its bog a change of orientation was prescribed. Enough big books on small facts! Enough subtleties and distinctions without any real interest! The French should emulate the youthful ardor of biologists in the United States, Germany, and England who were concerned with the experimental study of the big questions in biology. Compared to these foreign works, the small monographs on morphology were meager fare. And the intellectual renewal came directly from the great work of Darwin. Here is what Delage prescribed: Throw off the yoke of Cuvier and, inspired by Darwin, study really important problems – the cell, regeneration, grafts, generation in its many modalities, race, variation, and heredity. To make this more palatable to the French, the reviewer slyly noted that Delage was calling for a return to the traditions of the great French school of Geoffroy Saint-Hilaire, whose *Histoire naturelle générale des règnes organiques* had put

125

forward forty years before a research program in biology, including its aim, its methods, and its subjects. Then France would have a school of biologists in reality as well as in name.[56]

In the course of his review of Delage's book on general biology Giard returned to his criticism of Delage's monograph *Sur la circulation des crustacés édriophthalmes*, emphasizing his agreement with Delage that the work had not expanded or modified our knowledge of crustaceans or the respiratory function. But the fault was not in the subject, which Delage had presumably worked on the following orders from his *patron*. Shortly after the publication of Delage's thesis, E. L. Bouvier and Aimé Schneider discovered important facts invalidating Delage's hasty conclusions and expanding in a striking fashion classic ideas on the comparative anatomy of the circulatory system of crustaceans. Inspired by pure science, without ulterior motives, these two zoologists made the decisive observation that Delage had not even dreamed about any more than he thought of dealing with the interesting problem of the double circulation of *Lernaeidae* (Copepoda). Giard restated his view of the problem: Delage's haste to get his doctorate and, at the same time, a prize of the Academy of Sciences, which was more important in preparing for a professional future in French science than a serious scientific discovery was.

The doctoral system itself was being questioned, and Giard rushed to its defense. For Delage, a conscientious study, using a technique based on that of the best masters, nearly always gave new, exact, and positive results, although the conclusions were usually of minor importance, incapable of modifying accepted ideas on the general questions related to the subject. Giard bluntly rejected this argument: A monograph is only as good as the person who writes it. If Darwin had not written his monograph on barnacles, he would not have been able to write the *Origin of Species*. Ghiselin has noted the "unexplained mystery" of Darwin's spending eight years (1846–54) in studying barnacles when he had already written a second draft of the *Origin*. This was no mystery to Giard, although clearly the importance of the monograph was not understood at all by the zoologists who discussed Darwin's candidature in 1870 for corresponding membership in the Academy of Sciences. Giard seems to have understood that the "work was nothing less than a rigorous and sweeping critical test for a comprehensive theory of evolution."[57] So it would be disastrous for young scientists to follow Delage's advice to abandon the type of specialized study that leads to nothing on the "big questions." To leave the field of observation and "decisive experimentation" would mean dependence on the sterile vanities of the philosophy of science. Even good zoologists like Savigny and Weismann, who developed vision problems, showed the limitations of their

126

theoretical deductions when they could not control them by experiment. (After 1904 Delage had to stop his work in experimental science because of a detached retina.) Giard was alarmed that an insidious tendency to invoke the support of a priori had invaded recent doctoral dissertations. How foolish was the reviewer in the *Revue scientifique* who declared that during the previous twenty-five years doctoral theses had been cast in the same mould, making a large monograph collection of depressing reading. Without citing any of the work done under him at Lille, Giard showed how unfair such a statement was by simply referring to the theses of Chabry, of Henneguy, of Bataillon, and of Louis Léger.

If anyone was asleep, as the article in the *Revue scientifique* claimed, it was only in the temple where, Giard sneered, the last representatives of Cuvier's school were wandering like ghosts, and in the clique of the "eminent pontiff" (Lacaze-Duthiers) in which Delage was for so long a choirboy. Suddenly they had noticed that the "sun of Darwinism had risen on the horizon of science." At least they were now awake!

Giard thought that Delage's sudden hatred of monographs was based on an erroneously conceived history of science: It is absolutely false to argue that zoology passed through an evolution of four successive periods – those of taxonomy, anatomy, marine zoology, histology – after which it would ascend into philosophy. In all periods were there not minds that rebelled against general conceptions and were there not powerful generalizing intelligences? Giard also accepted the view of the mathematician (Jules Tannery) who argued in favor of the coexistence of continuity and discontinuity in science and in the mind.[58] The distinguishing characteristic of the man of science is not system but method; with the same tools different minds do different work.

> Das ist ja was den Menschen zieret
> Und dazu ward ihm der Verstand
> Dasz in innern Herzen spüret
> Was er erschafft mit seiner Hand.

What better way to clinch the argument than to quote German poetry, especially if it had been previously implied than the enemy, also poet and littérateur, had difficulty in accurately interpreting German scientific texts? Giard might also have cited the great lines of Mephistopheles:

> Grau, teurer Freund, ist alle Theorie
> Und grün des Lebens goldner Baum.

Old methods could still be the basis of excellent work. Using the methods of Réaumur, for example, Perris and Fabre had made more progress in entymology than a number of anatomists dissecting insects according to the rules of Cuvier, whose school still exercised a dying tyranny. Perhaps because he had been a scientific overlord and certainly because Delage had found something good to say about his monographs, poor old Cuvier was Giard's symbol of what was wrong with French science. Later Houssay specified the exact problem: Long dead in philosophical zoology, Cuvier lived on in didactic works and treatises intended for education.[59] With the resources of comparative anatomy, Bouvier had made more progress in the study of molluscs than many biologists armed with the best microtomes. Van Baer had no maritime laboratory when he created embryogeny. Delage seemed ignorant of the importance of contemporary research in taxonomy. This alone might have been enough to damn Delage in the opinion of Giard, who proudly advertised himself as a morphologist. Delage thought that the perfection of methods of study might have weakened the scientist's intuitive sense of natural affinities in the same way that the microscope spoiled his eyes. For Giard this was a double inaccuracy showing that Delage was totally ignorant of the brilliant systematic works inspired by the Darwinist viewpoint. After all, if the progress of science were closely tied to the perfection of its instrumentation, as Giard thought physics was – a very French view that fitted in well with the common nineteeth-century delusion that science was "complete" and perfection meant more precise measure- ment – then one ought to be able to divide up the history of physics, following the logic of Delage's history of zoology, according to the different tools of the discipline: periods of the balance, of the cathetometer, of photography, and so on. Reductio ad absurdum again. Delage's history was based on a false principle.

Even by 1907 it seems that Giard was still smarting from Delage's attack on the value of monographs, for he gave another defense of its function in science. Well-known facts can suddenly take on new significance, acquire an unforeseen value, or unhoped-for applications with the perfecting of instrumentation and the progress of observations. Intelligent observation of even minute detail can surpass anatomical values in its general morphological significance: The researcher brings his stone to the eternally unfinished edifice of human knowledge. But for a monograph to embody these virtues it has to be impregnated with a guiding idea. So Giard could gloat over the "fact" that monographs done without such an idea – Bojanus on the tortoise, Lyonnet on the caterpillar of the cossus, Straus-Durckheim on the cock- chafer – works constantly upheld as inimitable models by the last repre- sentatives of the Cuvier school, could only be henceforth of limited utility and very minor interest. The basic problem in the production of mono- graphs, according to Giard's litany, was that the great light of Darwinism illuminating the sky of biology was kept out of certain laboratories. And this in spite of the "conversion" of Delage?[60]

Giard seems to have had a bad press among the Parisian biologists. No doubt part of this was due to his break with Lacaze-Duthiers, but the seemingly trivial nature of the topics on which he encouraged research also made the men who liked the big-question approach rather suspicious of Giard's scientific prowess. Giard took a preverse delight in reporting the puzzlement and ironic reaction of his colleagues in seeing him push students into research on galls. "Encore un que vous lancez sur les galles! Prenez garde, ils vont se manger!" Le Dantec also noted Giard's obsession with morphology, which could be interpreted as the equivalent in biology of the graphic method in the physical sciences, but this was a quite respectable research neurosis for a good nineteenth-century biologist. It is certain that Giard's personal charm and willingness to devote a great deal of his time to helping researchers led to the formation of a contingent of loyal followers. Those who had contact with him usually experienced an excitement and enthusiasm difficult for outsiders to understand. Caullery was once asked what he saw in Giard: "Mais enfin, votre Giard, qu'a-t-il donc fait de si extraordinaire pour vous *emballer* tous comme ça?"[61]

For Félix Le Dantec the significance of Giard in the development of biology in France included his influence on teaching and research. "At a time when the teaching of the natural sciences in France was limited to questions of anatomy, to questions catalogued in programs [catechistic guides for examinations], Giard, inspired by the *Origin of Species*, understood all the importance attached to the exact verification of the *small facts* neglected by the older naturalists." For "in these *small facts* the trained mind sometimes finds the idea of a general law." Giard possessed a healthy scepticism toward the celebrated dictum of Claude Bernard that a good technique sometimes renders more service to science than the improvement of high theoretical speculations – such as Darwin's, no doubt.[62] Even if anyone had noticed that Darwin was no slouch as a practitioner of technique, it could hardly have been expected that much excitement would have been generated by this in the land where the traditional predominance of anatomy and physiology had long established a key role for technique in the natural sciences, with a powerful influence on the type of research done throughout the nineteenth century. The clinical staff of Asklepios no longer cast a long shadow, but zoology emerged only slowly from the tributary role in which Bernard had cast it, in spite of the efforts of Lacaze-Duthiers and Giard.

Giard's attack must be seen in perspective. It was a reaction against the abuses of technique, not an attack on technique per se. What worried Giard was the contemporary tendency to confuse procedure with science and, what was probably related, the gross neglect of observation of the living animal. Presumably Giard did not mean the type of experimentation for which Claude Bernard was famous. In his own work Giard made a great effort to understand the curious ethnological relations between parasites and hosts because he believed that to cut up into thin slices the objects deposited by fishermen on his laboratory table would be an inadequate method of knowing well the organization and behavior of marine life. This led to the study of animals in a biological equilibrium. It seems that Giard's philosophy evolved in a period of about twenty years from a belief in mechanical materialism to the conviction that biology could not be reduced to contemporary physicochemistry.

Maurice Caullery claimed that Giard introduced a really new chapter in general biology. Under the rubric of parasitic castration – one of the favorite areas of research of Giard's school of biology – he grouped "all morphological or physiological phenomena of an order that entailed in the organization of a living being the presence of a parasite that acts on the genital function of the host." In drawing attention to and in analyzing these widespread and varied phenomena Giard left his mark on general biology.

Gohau has boldly identified Giard and his fellow transformists as a group with a research program different from that of their adversaries. Without neglecting anatomy, Giard gave great importance to the study of embryos. The emphasis put on the way in which an organ is formed made little sense to Lacaze-Duthiers, for whom the anatomical determination of its presence was more important than its origin, an issue shrouded in the mists of Teutonic philosophy, impenetrable to the light of empirical science. But Giard rejected Weismann's neo-Darwinism because of its religious baggage. He could no more accept the idea of chance than the idea of an immanent directing agent. Had he lived longer, Giard might have ended up in the same camp as Delage, for before his death Giard lavishly praised Mendelism as a theory of hybridity as fertile in biology as atomic theory in chemistry. Le Dantec explained his master's apparent gyrations as changes of course necessary in Giard's remaining faithful to the great morphological god whose worship made it a pleasant duty to be seduced by this combination of forms.[63] Giard was on the verge of seeing that the work done on heredity was creating a new biology. Death spared him the joyful agony of paradigm switch.

What was the situation in French zoology at the end of the nineteenth century? A charming and vicious description was given in one of the leading French scientific journals, the *Revue scientifique*, when it hailed the appearance of Delage's diatribe against his own discipline, the introduction to his *Structure du Protoplasma*.[64]

130

It is easy to sum up the present situation of zoology. Let us trace the career of a young man who believes in the deceptive appearance of zoology as a discipline in which one can have a career with a position that is stable and, above all, official – the ideal of a majority of Frenchmen. So he enters a faculty of sciences to take his *licence ès sciences naturelles* and then his doctorate. He follows the required courses and is assiduous in the laboratories, where he is initiated into the mysteries of small creatures – big ones have been out of style for a quarter of a century. From the laboratory he goes to the library and then finishes his pilgrimage with a bout of hard work at a marine research laboratory. For each "question" he delves into the best authors, summarizes and analyzes them; his wife is English, the *Zoological Record*, and his mistress German, the *Zoologischer Anzeiger*. After two years he gets his licence. The first stage is passed. He has done some preparation for the second in noting that some group or other in the world of *petites bêtes* has been badly studied or neglected. Or perhaps the works on it are old – scientific work ages fast – or the circulatory system of some family is unclear, or there is something odd about the nervous system of another not agreeing with the way it is described in the literature. He is ready to consume from two to five years of his life in doing a doctoral thesis. The longer it takes, the more unfortunate is the whole affair.

The researcher begins. Armed with orthodox methods gained in his preparation for the licence, knowing how to use instruments and reagents, skilled in the difficult and deadly art of bibliography, he accumulates index cards, makes innumerable sections, reagents, and drawings. The chosen animal is searched down to its most intimate folds, from mouth to rectum, from cutaneous appendices to the epithelium of the digestive canal. Discoveries follow one upon the other. One day he notes that the heart has five arteries, and Cuvier had missed the fifth! A month later he exposes digestive appendices that Milne Edwards never saw. Then a great discovery: He finds striate fibers in a muscle the manuals classify as smooth. Science is turned topsy-turvy. He finds nervous networks that Delle Chiaje didn't know about; he suspects the existence of new sensitive organs – an admirable vein to exploit – and to top everything – a glory reserved to a few – he discovers a new parasite, *intus* or *extra*, which he names after his professor-*patron* or a close friend. Nothing could be more flattering to a professor than to have a tapeworm named after him or to become the godfather of an insect hitherto unclassified. The system has the advantage of immortalizing the trio of creature, *maître*, and happy discoverer. This is more than enough for a thesis, which eventually appears, is approved, signed with the proper flourish, dedicated, and adorned with generally good drawings that are appropriate, learned, and conscientiously done.

131

The rest is simple. Read the conclusions carefully: there one will find from five to twenty detailed facts, either new or renewed, modified or presented in a new fashion. The Gargantuan efforts of one researcher led him to forty-nine conclusions. But no ideas, just little differences, little corrections, or small novelties. Following this ritual, the zoologists built up over twenty-five years a mountain of insignificant facts, stored in theses written according to the same rigid rules. Only a change of direction based on a shakeup of general ideas could change the situation. And salvation had to come from outside France, from foreigners imbued with the Darwinist gospel, so fertile in new research programs.

The same brilliant, ironic critique could probably be applied to biology in any country about the turn of the century. Only the French were likely to tolerate it in a leading general scientific journal; the fact alone indicates that the discipline was really in good condition, or at least had a healthy attitude toward itself.

A generation later, biology at the Sorbonne was still an easy target for criticism from insiders. One of France's best biologists delivered a critique in 1925 echoing a combination of Giardian and Delagean themes. When Maurice Caullery gave his first lecture as the new professor in the chair of the Evolution des êtres organisés, he took the opportunity to expose with unusual frankness the sorry state of zoological education at the University of Paris. The pedagogical curse on zoology was most evident in the organization of the *certificats d'études supérieures*, which drove students to approach zoology from the viewpoint of the examinations for the certificate. Not necessarily a bad thing; it would depend on the examinations, which were, unfortunately, in this case, based on what was taught by the holders of the *chaires magistrales*, a clearly decadent lot who gave examinations based on neither the subject matter of contemporary science nor on the logical requirements of its study, but on their own science of yesteryear. So in the years preceding Caullery's ascent to the chair in the evolution of organic life, practically none of the candidates for the certificate in zoology took the course on evolution or did the related laboratory work. Because this course dealt with the issues regarded as most important for scientific progress and for basic problems in zoology it was a tragic flaw that the bad organization of the certificate system effectively ensured that students would acquire an out-of-date zoological education. In due time many of the unenlightened would pass on their ancient knowledge to unsuspecting and unwilling lycée students.[65] The efficacy of ignorance should not be underestimated.

What were the big issues in biology in 1925 that were being so badly covered in French zoology, leaving in the background not only classical zoology but even "the most recent revolutionary Darwinian zoology"? The

132

most fertile research was being done in the increasingly experimental area of general biology. One of the master components of this discipline was heredity, the very subject of Caullery's course in 1925-6. A revival of Mendelism, based on scientific observation and precisely controlled experimentation, had led to the development of the new science of genetics. Embryogeny, formerly descriptive, preoccupied with phylogenetic data, had also become mostly experimental. This science was so new that the French had no name for it except Brachet's curious *embryogénie causale*. The clever Germans had long enjoyed the silken syllables of *Entwicklungsmechanik*. Caullery had given a course in 1923-4 on the part dealing with regeneration. Out of studies on regeneration came the technique of the culture of tissues *in vitro*, part of cellular biology, then largely under the mechanistic spell of Jacques Loeb. Work on the egg was a key aspect of zoology's legitimate obsession with understanding sexuality. This topic was also treated in one of Caullery's courses (1924-5). Caullery's courses seemed like a disturbing cult of modernity in the Sorbonnard temple of scientific orthodoxy.

Caullery recognized that not all French zoologists had been content to ossify in the pre-1890 morphological and systematic paradigm of zoology. Had not Delage undergone a conversion from the morphological approach, the heritage of Lacaze-Duthiers and his ancestors, to recognition of the need for a new approach in general biology?[66] And Delage had been wrong in ignoring the pioneering modern approach long advocated by Giard, Caullery's old professor and first holder of the notorious chair of the Evolution des êtres organisés. What distinguished the new post-1890 approach was an increasingly intimate association of morphology and physiology coupled with the growing dominance of the fabulous experimental method. It was in the more lively biological journals published chiefly in the United States, in England, and even in Germany that the end of the era of simple observation and conjecture could be clearly seen. Those who wanted the new gospel in France had a few opportunities, as in Caullery's courses. There were few courses in general biology in French faculties of science; even the program at the Sorbonne had few students because the courses led to a special certificate of no value for the *agrégation* or the doctorate. Institutionally, not much had changed since the heyday of Delage and Giard, but because many of the bright candidates for doctorates went to Giard and later to Caullery, the impact of the backward programs was more on pedagogy than on research. France was hardly the only country in which the educational system was remarkably immune to new ideas.

4

THE INDUSTRIAL
CONNECTION OF
UNIVERSITY SCIENCE

Technical progress is a function of bourgeois money.
– Jacques Ellul, *The Technological Society*, p. 54

T HE functions of the faculties of science in the later nineteenth century were teaching, research, and service to agriculture, industry, and government at municipal, departmental, and national levels. An extra duty of faculties, resulting from the organization of a unified system of education and the historical connection between lycée and faculty, was their time-consuming responsibility for the baccalaureate examinations, a particularly heavy burden in large towns. The examination figures for the University of Paris in 1893–4 show the problem.[1]

	Number examined
Baccalauréats	
ès sciences	2,926
classique (lettres mathématiques)	929
moderne (lettres mathématiques)	94
moderne (lettres sciences)	103
de l'enseignement (lre partie rhétorique)	2,207
secondaire classique (2e partie philosophique)	1,466
partie scientifique	
Licence ès sciences	313
Doctorat	38
Concours pour les bourses de licence	32
Total	8,108

Fortunately a light teaching load gave the French university scientist some time for research.

Probably the most striking feature of the provincial faculties of science was the development of a system of institutes of applied science, each of which was usually headed by a leading scientist interested in regional industry and agriculture. The careers of Pasteur in Lille, Schützenberger in Mulhouse and in Paris, Haller in Nancy, Sabatier in Toulouse, and the Berthelots and Le Chatelier in Paris provide a paradigm of the activity of the academic scientist whose research was intimately connected with the economic life of the region and the nation.

134

In developing institutes of applied science and technology the University saw itself as fulfilling a vital social function. The Institut polytechnique (Brenier) symbolized for Grenoble close ties between the University and the Dauphinoise population resulting from its service to regional economic interests, its concern for public health, and the opening of its doors to everyone who wanted to get acquainted with the methods and discoveries of science.[2] Generally the faculties had to serve social needs of the period by developing programs directly related to the economic activities of their region, for all the universities could not develop a wide range of professional schools.[3] In about a generation the faculties underwent one of their most remarkable developments since the middle ages, with a striking collaboration of public and private powers in financing their growth.

Stimulation: industrial societies

The interface between science and industry that characterized the institutes was not a new arrival in the late nineteenth century; there had been some fruitful partnerships producing remarkable results since the ancien régime. Early in the nineteenth century scientists and industrialists began to organize for the infusion of science into industry.[4] A number of industrial societies helped to create a favorable environment in which the institutes could flourish. The Société d'encouragement pour l'industrie nationale, dating back to 1801, was the focus of a strong attempt by scientists, encouraged by the Third Republic, to bring to industry the innovations and improvements that science alone could provide. An election of members of the administrative council in December 1873 consecrated the elite of the French scientific community in their role of doctors to French industry: Dumas, president, Baron A. Séguier and Balard, vice presidents, Baron Thenard, associate vice president, E. Peligot, secretary, E. Becquerel, censor, along with Troost, Jamin, A. Brongniart, Chatin, and Fremy (honorary). In 1875 the society offered 108,000 francs in prizes, distributed as follows:

Arts mécaniques	12,000
Arts chimiques	42,000
Arts économiques	9,000
Beaux-arts appliqués à l'industrie	4,000

The largest prize in this category was 6,000 francs for a theory of steel manufacture based on exact experiments. There were two grand prizes of 12,000 francs each.[5]

Between 1826 and 1870 the most admired industrial society in France was that of Mulhouse, "the wealthiest of all provincial societies," supported by the big manufacturers – Dollfus, Koechlin, Schlumberger – well known for its schools of chemistry and design and its museums, and envied for its large prizes.[6] The explicit aim of the society was to advance and spread industry by a multiple effort: centralization of the teaching of a large number of subjects; communication of important discoveries and facts, as well as the observations to which they give rise; and any means suggested by members. Its library and reading room carried the best foreign and French books and journals in the sciences and in techniques; models, plans, and manufactured products were not neglected. A monthly bulletin emphasized facts especially interesting for departmental industry. Prizes were awarded for the invention and perfection of machines and procedures useful in the technical arts, manufacturing, agriculture, and domestic economy. The society carried out experiments to determine the merit of new inventions; it also promoted scientific research of potential use to industry. Although the society began with only two committees in mechanics and in chemistry and physics, it professed a strong, paternalistic interest in the condition of the working class from the beginning: Workers must be encouraged to love work, economy, and education. Forty-four regular members were joined by a few corresponding members, mostly professors in faculties of sciences and technical institutions. By 1870 the society had 375 regular and 32 honorary members with 102 correspondents. By 1870 its annual income was over 50,000 francs, with expenses just over 38,000 francs; the Ecole de dessin had a separate income of 6,848 francs, with expenses of over 5,400 francs for its 380 students. With the annexation of Mulhouse to Germany after the Franco-Prussian war, France lost an important bit of its "commitment to technical and commercial improvement."[7] But a considerable immigration of Alsatian scientists to France helped to offset this great scientific and technological loss.

In the founding of the Société industrielle d'Angers et du département de Maine-et-Loire (1830), the successes of Mulhouse served as a beacon. Especially attractive to the Angevins was what they construed as a combination of scientists developing theory and artists and technicians applying it. Not absent from their motives was jealousy of the commercial success of England, a theme to be repeated more than once over the century. France was, of course, developing its industries, aware of the fact that "industry supplies the strength of states." Angevins harped on the idea that speculative knowledge contributed to the advancement of the technical arts. The sixty-seven-page prospectus of the society carried this idea to a great excess: If industry and science were developed fully, then agriculture would be no more than the application of "plant and animal physics"; manufacturing would be the practice of chemistry and mechanics, and commerce a

136

consequence of the history of natural and industrial production. An ambitious beginning included a bulletin, prizes, a library, and four committees (agriculture, mechanics, industrial economy and statistics, and physics and chemistry). Members included sugar manufacturers, spinners, smelters, pharmacists, printers, and one surveyor. Although the committee for physics and chemistry had as its only scientist the professor of mathematics of the Collège royal, Chevreul was a corresponding member. A strong interest in agricultural industry – a term of the period – was evidence in the bulletin: Articles on the conservation of grain and on fertilizer, with reprints of other journal articles on agriculture dominated its pages. By 1873, when the society had 414 members, of whom 163 were correspondents and 225 regular members, a modest attempt was made to start an agriculture laboratory. In those days in Angers that meant 200 francs of funding.[8]

In 1872 the Société scientifique industrielle de Marseille consecrated the union of science and industry in its very name. The title would indicate a "natural association that must exist between them for the common good." Original members were mostly engineers, with some architects and manufacturers for company. It is not surprising that this group believed that in order to become useful, all scientific advances need a practical means of regulation. But they also believed that industry and business need scientific knowledge – praised as "les lumières" – to solve problems beyond the capacity of the industrialist or businessman. By 1875 seven professors of the faculty of sciences were members, whose total number was 150 plus thirty-six associated and fourteen foreign members. Income was nearly 16,000 francs, with a 3,000-franc surplus for 1875. There was a scientific and technical library. High-quality technical articles on boilers, dynamite, locomotives, steam engines, and the laying of underwater telegraphic cables delighted the readers of the bulletin. One of the more interesting of societies because of the scientific and technical renown of its members, this society also had a different sectional setup: mechanics, industrial chemistry, mines, physics, and architecture and public works. Lyon bears some comparison: in 1893 there was a merger of the Société des sciences industrielles and the Société d'agriculture, histoire naturelle et arts utiles de Lyon. Members included Jules Raulin, director of the Ecole de chimie et de physique appliquées, professor at the faculty of sciences; Saturnin Arloing, director of the Ecole vétérinaire, professor in the faculty of sciences; P. Cazeneuve, faculty of medicine; and Léo Vignon, *maître de conférences* in the faculty of sciences and associate director of the Ecole de chimie industrielle.[9] Neither society was a typical provincial *société savante*!

The Société industrielle du Nord de la France began in 1872, with medical doctors, engineers, industrialists, professors, and other professionals.

Many of the well-known professors of the state and Catholic universities were luminaries and officers of the society: Th. Barrois, Buisine, Calmette, Gosselet, Lenoble, Paillot, Pascal, Petot, Stoffaes, Swyngedauw, and Witz. Faculty scientists played a key role in the Société chimique du Nord de la France, which, beginning in 1891 with over a hundred members, aimed at bringing together all the chemists of the Nord. The traditional hostility between the Catholic and the state faculties did not prevent chemists from both faculties from associating on professional grounds. In 1891 Charles Laurent (*directeur général technique des manufactures de produits chimiques du Nord*) was president; Lescoeur (professor at the state faculty of medicine) and Schmitt (professor at the Catholic faculty) were vice presidents. The participation of technical and industrial personnel was striking: Van Ackère (*constructeur d'instruments de précision*) was secretary and Duvillier (*sous-chef des travaux pratiques de chimie, à l'Institut industriel du Nord*) was treasurer. In 1892 Lenoble of the Catholic faculty became president. Willm was among the state faculty professors who were members. The basic aim of the society was to establish friendly relations between members in order to facilitate the entry of young chemists into industry. The society met weekly and had annual dues of twelve francs. It subscribed to leading scientific publications to encourage its members to keep abreast of developments in chemistry in France and abroad, with the ultimate aim of unifying and perfecting, where possible, methods of analysis for the needs of agriculture, industry, and commerce.[10] No other organization could better exemplify the interaction of science, the economy, and the universities.

By the twentieth century a new stage was reached in the evolution of the relations between science and industry. Instead of an easy relationship between scientist and industrialist in clubby societies or a typical situation in which scientists were called on as problem solvers, we find a substantial and powerful segment of the scientific community aggressively calling for the invasion of industry by science, as the means of making French industry competitive, as well as denouncing industrialists who did not welcome the laboratory in industry. At the head of this vociferous scientific group was Henry Le Chatelier, who had explored this topic in many publications and particularly in a series of articles in the *Revue générale des sciences*. Le Chatelier, scientist and industrialist, was a tireless if sometimes tedious propagandist for Taylorism in French management. The Société de l'Ecole polytechnique was the distributor of large sums of money given by Michelin to the *grandes écoles* for teaching Taylor's ruthless gospel of efficiency. In 1902 Charles Camichel lamented the skill of industrialists in establishing an antagonism between the laboratory and the factory and their zeal in singing the praises of empiricism. The curious thing was that many industrialists knew better, even to the extent of acknowledging the need for engineers with an extensive and deep scientific culture. Industrialists seemed to confuse the practical empiricism of the worker with the scientific empiricism of the engineer, who knew the industrial difficulty of using the results issuing from scientific laboratories.

It is also true that there was the danger of scientific fools rushing into factories where engineering angels feared to tread. Superficial analysis could be dangerous in the case of a scientific explanation of industrial phenomena. A classic case was Boutigny's explanation of boiler explosions: A set of experiments by Flechter in Manchester on real boilers showed that Boutigny was wrong. Witz then did a set of laboratory experiments showing the limitations of Boutigny's experiments – the Boutigny effect. The complexity of facts in dealing with steam was shown by Chréstien's work on the long-distance transportation of steam. Experiments done at the Conservatoire and in the open hangar of the municipal warehouse of La Villette, on an entirely different scale from the typical scientific experiment of the turn of the century, were a sort of direct industrial experimenting. Camichel concluded that it is difficult to reproduce in the laboratory an industrial phenomenon in all its detail. In fact, there is no reason to do so. It is also difficult, sometimes impossible, to apply to industrial phenomena the experimental and completely determined results of the laboratory. The more specialized the industry, the easier it is to use scientific data; the empiricism of the engineer diminishes in such a case. A natural rapprochement takes place between theory and its applications and perfections.[11]

In the search to find an explanation of supposed French reluctance to admit science to the factory, it was inevitable that someone blame the venerable Academy of Sciences. Louis Houllevigue, professor at Caen, not in imminent danger of election to the Institut, compared the academy unfavorably with the Royal Society. The French academy, ignoring the educated public, disdained any form of popularization, while "a large number of the high intellectual aristocracy of England used public lectures to show their interest in the spread as well as the advance of science." Davy, Faraday, and Tyndall were squarely in this tradition. As was so often the case in France, salvation came from the public, appropriately guided by an elite in agreement with it. Louis Liard urged scientists to spread the scientific gospel to those who could be impregnated not so much by science but the scientific spirit itself. The public was ready, because by the last quarter of the nineteenth century the applications of science had pushed into the center of human existence. The public was interested in asking questions about the thing in whose name the earth was being transformed. Acting in the common interest, "science" replied. Scientists soon followed those converted to their own gospel. Science became such a glorious entertainment that even discoveries like radium, with no immediate practical application, were followed with avid interest by a curious public. The press was of key significance in the popularization of science, while scientists flattered the egos of journalists by taking them into their confidence. But it was not until after World War I that the Academy of Sciences got around to establishing a section concerned with the applications of science. By then some of the believers in the gospel of industrial science were in the academy, and many sceptics had been frightened into a reluctant belief by the threat of annihilation by the Germans, who were reputed to be the world's greatest exploiters of *Wissenschaft*, or at least as good as the Americans. The lesson was plain: Imitate, or fade as a world power.[12]

The late nineteenth-century transformation of the moribund faculties of science into respectable bastions of learning and research came about largely as a result of three factors: the appearance of genuine students, an unprecedented emphasis on research, and an attempt to serve industry and agriculture by orienting teaching and research in the direction of their needs. This service function of the faculties was fulfilled in the network of institutes of applied sciences, attached to the faculties, offering a wide range of technical courses and degrees. The development of the institutes corresponded to France's great advances in the coal, iron, steel, chemicals, and hydroelectric industries of the generation before the First World War, especially in Lille, Nancy, Lyon, Toulouse, Grenoble, and Paris. Lille was probably the pioneer, in spite of an interesting private failure in Paris earlier in the century.

In Paris during the July Monarchy there existed right in the Latin Quarter an Ecole de chimie pratique, directed by Charles Gerhardt, who, along with Laurent, taught chemistry and mineralogy. Among Gerhardt's translations from German and English scientific works were Liebig's works on organic chemistry applied to plant physiology, to agriculture, and to animal physiology. Liebig's work on *Die Chemie in ihrer Anwendung auf Agrikultur und Physiologie* appeared in 1840. Physics was taught by J.-T. Silbermann of the Conservatoire des arts et métiers, rather than by a faculty man, and industrial applications of science were handled by E. Kunemann. The aim of the school was to teach laboratory techniques to young people who intended to go into the sciences, medicine, pharmacy, or industry. The theoretical instruction given in official and private establishments was complemented by exercises and manipulations that would let students make applications of their knowledge in physics and chemistry, even then viewed as "indispensable in nearly all the professions." For advanced students there was the possibility of doing serious research that could be published in a special bulletin. Academic ritual was reduced to a minimum to permit students to register at any time during the year. Laboratories were open every day from ten to five. A library housed books and journals needed for research. This school was a brief experiment whose guiding idea came to fruition in the University institutes half a century later.

The Second Empire: the model of Lille

During the 1840s, in some cities like Lille and Besançon, attempts were made to move the faculties in the direction of industrial research.[13] By the 1850s connections between faculties and local industries had become highly developed. In Lille the faculty of science proclaimed an overtly utilitarian ideology. Its practical orientation was demonstrated from the start in the work of Pasteur. As the faculty's first dean, from 1854 to 1857, Pasteur's teaching as well as his research dealt with the problems of industry and agriculture. By the time he left Lille for the Ecole normale supérieure in 1857, he was famous for his experiments on fermentation, and he had made the industrialists happy with the utility and profits of scientific research.[14] In the first semester of 1856–7 Pasteur gave a weekly lecture on chemistry applied to the industries of the Nord, especially the manufacture of alcohol and sugar from beets. It was Bigo-Daniel, a student in Pasteur's chemistry class, who got Pasteur interested in the study of fermentation. Bigo père, who owned one of the leading factories (at Esquermes) for producing alcohol from sugar beets, wanted to find out why anormal fermentation in some of his vats resulted in lower alcohol production. Lectures were followed by visits to factories. At the end of the academic year Pasteur arranged trips to the chief chemical and metallurgical establishments of Briache-St-Vaast, Aniche, Valenciennes, Denain, and Anzin. In 1857–8, the year after Pasteur's departure, attendance for the course in theoretical and applied chemistry was up to 152.

Compared with his successors, Pasteur, the patron saint of the faculty during the nineteenth century, was a paragon of moderation in his justification of the interaction between science and industry. When Pasteur left the three-year-old faculty in 1857, the professor of mechanics, Gabriel-Alcippe Mahistre, became acting dean. During the assembly beginning the new year, 1857–8, Mahistre remarked that few innovations were tried in industry without the advice of someone in the faculty. This benefited science because industry raised not only problems that could be solved by experiment, but also a range of questions related to the high-level theory appropriate to a university. Industry, in turn, profited in gradually giving up its empirical methods. A living illustration of this mutual benefit, Mahistre was himself engaged in studies of the efficiency of steam-engines, the resistance of materials, and the theory of water-wheels. His successor as dean, Jean Girardin, was also totally committed to industrial science. Girardin declared that it was by fulfilling local needs and supporting the big industries of the Nord that the faculty would grow and respond best to the government's interest in applying theoretical science to technical problems. Even Pasteur had never gone so far in proclaiming the utilitarian mission of the faculty, which often seemed to believe Blake's proverb of hell: "The road of excess leads to the palace of wisdom."

Both the teaching and the research of the professors were directly conditioned by the local demands at Lille. In physics, Auguste Lamy, working on the applications of heat in industry, was giving a lecture every two weeks in the mid-1850s on industrial physics – the hydraulic press, distillation, and explosions. Lamy's physics course was one of the most popular in the faculty. Lamy, having lived a long time among the Lillois, knew perfectly well their needs, inclinations, and aptitudes – "le bon sens pratique de la population flamande" (Girardin). The courses in chemistry were especially important for the commerce and industry of Lille. Charles Viollette taught applied chemistry at a high level. It was Violette who in 1862 determined the cause of the gas explosion that destroyed part of Dunkerque. In 1858 the faculty began a course in bleaching, dyeing, and printing. By the 1860s Viollette was giving a course in the chemistry of plants. Fermentation was naturally of great interest to the big brewing and distilling industries. Girardin was well known for his work on fertilizers, sugar beets, and the industrial use of water in Roubaix and Tourcoing. In 1861–2 he studied industrial pollution and came to the comforting conclusion that, although the industrialist had no financial interest in producing combustion without smoke, he could reduce pollution cheaply with a properly constructed ordinary furnace. He also spent much time in his courses dealing with steam engines, especially the Woolf engine, widely used in the Nord. Girardin gave a course the next year on the chemistry of the different fuels used in industry and households. In 1872 Duvillier, Violette's former student and a faculty assistant in chemistry, succeeded in making chromic acid cheap enough for industrial use by perfecting the long-abandoned method of Vauquelin. The utility and profitability of science could hardly be made any plainer.

Even the appointment of a new mathematician had advantages for industrially related courses. Applied mathematics was already taught in the 1850s in a way "appropriate to the needs of an essentially industrial city." The courses in applied science usually started off with a good many people, most of whom dropped out, leaving a dozen serious students, once mathematics was introduced. Pasteur emphasized the need of adding to applied mechanics some special lectures to establish basic principles through experiments and descriptions of industrial machines. In the second semester of mechanics, students visited industrial plants making tin (Charlet), linen-weaving frames (Vennin), heating units (Baudon), and other exciting items. Provincial realities sometimes imposed compromises with the higher intellectual aims of science.

The technological emphasis and perambulatory pedagogy had the active interest and backing of the town council, whose members were keen to utilize the latest advances in science and technology. In 1863–4 professors from the faculty served on a municipal commission studying the distribution of potable water, a severe problem in a city of 140,000 that had 343 industrial plants. In the 1860s, even the biologist Camille Dareste was working on purely regional problems. His "Recherches sur la végétation de la betterave à sucre" (1866) was undertaken because the Comice agricole de Lille asked for a study of abnormalities in beet growth. Dareste was also the secretary of the municipal commission studying means of ensuring the sale of pork in sanitary conditions; his own work on trichinosis was inspired by the prevalence of this disease in Lille. Biology was becoming as useful as chemistry. The faculties of science had been integrated into the social structure of the country on the most fundamental of levels, especially in economics and in health care.

Although its roots were in practical regional problems, much of the work done in response to local demands won national recognition in the 1870s. The Société centrale d'agriculture de France awarded Viollette a gold medal for his work on sugar beets. Other work, not directly relevant to the region, was undertaken and rewarded. For his acoustical research Alfred Terquem won a silver medal in the annual *concours* organized by the Congrès des sociétés savantes; Jules Gosselet won a gold medal for his work in geology. By the 1870s the annual report on the personal research of the professoriate required a page or more just to list the work of each productive scientist. By the mid-1870s, with the role of research in the life of the university vastly increased, a new era had arrived in the history of the faculties of science.

The Third Republic: the faculties and the expansion of industry

By the time of the First World War, most French faculties of science were actively engaged in technologically related teaching and research. In contrast with the isolated developments during the Second Empire, the industrial

Table 1. *Number of students who took technical courses in faculties of science including institutes of applied science, 1897–1907*

1897–8	1898–9	1899–1900	1900–1	1901–2	1902–3	1903–4	1904–5	1905–6	1906–7
249	374	549	642	764	742	941	1005	1074	1263

Table 2. *Distribution of students taking technical courses in 1906–7*

Aix–Marseille	Besançon	Bordeaux	Caen	Clermont	Dijon	Grenoble
81	27	45	8	9	41	131

Lille	Lyon	Montpellier	Nancy	Paris	Poitiers	Rennes	Toulouse
104	153	15	460	82	17	85	5

emphasis in university education had spread throughout the system, with the areas of real strength corresponding to the main centers of industry. Whereas in the Second Empire scientists occasionally pursued industrial interests in addition to their teaching duties, in the Third Republic specialized technical institutes came to dominate the faculties. This was an important change of direction. The leading technical institutes developed, in the two decades before the First World War, at Grenoble, Nancy, and Toulouse, with Lille and Lyon following close behind (see Tables 1, 2, and 3).

With 502 registered students, Grenoble had the third largest provincial faculty of sciences by 1912–13, after Nancy (945) and Toulouse (857). At Grenoble 356 of the students were in the University's Institut polytechnique, spread over seven divisions: préparatoire (59), élémentaire (54), pédagogique (5), supérieure (lre année) (75), supérieure (2e année) (55), spéciale des ingénieurs (70), and the Ecole de papeterie (38). An additional hundred people attended the public course on industrial electricity. The following degrees were granted at the end of the year:

Diplômes d'ingénieur électricien		98
	d'études électrotechniques	13
Certificats	de physique industrielle	7
	d'électrochimie	8
Diplômes d'ingénieur papetier		10

The number of technical diplomas granted by Nancy and Grenoble was about the same. With 382 students attenting courses, Lyon granted the following technical degrees:

Ecole de chimie (non compris les étudiants déjà comptés aux certificats de licence)		79
Brevet de chimie industrielle		10
Brevet d'électrotechnique		8
	d'études agronomiques	2
	de mathématiques générales	1
Diplômes		
	d'études supérieures	3
	d'ingénieur chimiste	10

Nancy averaged about the same over the three-year period 1911–14 (these are total figures for the three years):

Institut chimique	107
Institut électrotechnique	127
Institut de mécanique	57
Ecole de brasserie	43

Table 3. *Distribution of students in French faculties of science*

	1890		1900		
Faculty	Total registered students	Licence	Total registered students	Licence	PCN
Paris	583	174	1,288	658	395
Besançon	67	49	82	59	24
Bordeaux	98	59	198	104	122
Caen	55	47	42	43	10
Clermont	41	35	86	38	60
Dijon	28	26	83	55	33
Grenoble	71	58	75	40	26
Lille	119	82	155	74	66
Lyon	91	45	330	233	148
Marseille	45	33	158	71	61
Montpellier	73	44	178	79	161
Nancy	114	49	230	140	79
Poitiers	66	37	77	49	37
Rennes	70	63	121	96	72
Toulouse	77	51	198	124	153
Algiers	49	32	26	14	21
Total	1,647	884	3,327	1,877	1,468

Note: All figures must be taken as close approximations, for official publications differ; in particular, there is a notorious lack of agreement between data published by the Ministry of Public Instruction and data published by the faculties in their inflated annual reports.

Sources: Bulletin administratif du ministère de l'instruction publique, 131 vols. (Paris, 1864–1932), vols. 87 (1913), 93 (1915); *Enquêtes et documents relatifs à l'enseignement supérieur*, 124 vols., (Paris, 1883–1929), vols. 35 (1890), 76 (1901), 101 (1911), 108 (1914).

1910				1913			
Total registered students	Licence	PCN	Institutes, special schools (chiefly industrial)[a]	Total registered students	Licence	PCN	Institutes, special schools (chiefly industrial)[a]
1,845	1,117	469	99	1,624	846	492	96
88	65	32	—	82	65	30	—
219	109	117	38	253	138	132	30
62	57	28	—	63	61	24	—
104	58	38	29	100	53	41	17
105	59	41	—	80	53	29	—
323	95	19	209	468	149	44	356
260	164	47	67	276	183	56	102
345	280	122	60	357	258	103	95
280	97	76	—	235	78	70	10
245	132	129	—	240	128	153	—
699	248	89	511	805	277	82	626
89	84	31		94	77	28	—
163	168	73	—	195	218	76	—
601	143	88	404	772	140	122	499
95	55	58	—	94	44	58	—
5,523	2,931	1,457	1,417	5,738	2,768	1,540	1,831

[a]The figures on the institutes indicate approximately the number of bona-fide students. Derived mostly from the better-known sources cited below, they do not cover universities where there were no formally constituted institutes. Of course, as Table 2 shows, some students were engaged in technical studies in all faculties, so that the total number of students so engaged was in reality much higher than the totals given for the institutes alone. Most students were in studies related to electricity, chemistry, and applied mechanics. Out of the 511 listed for Nancy in 1910, for example, 123 were in the Institut chimique, 363 in the Institut électrotechnique et de mécanique appliquée, and only 25 in the Ecole de brasserie. Out of the 499 listed for Toulouse in 1913, 422 were in the Institut électrotechnique and 77 in the Institut de chimie.

The only other comparable university operations in France were the Institut de chimie appliquée in Paris, with over a hundred students, and the vigorous Institut électrotechnique in Toulouse, with 422 students in 1913. The growth of these institutes was an important factor in a modest success in the decentralization of faculties. University students in Paris fell from 55 percent in 1876 to 43 percent of the French total in 1914, in spite of a rise from 13 percent to 46 percent in letters for the same period. Pharmacy, medicine, and science showed a clear shift to the provinces. Parisian science students dropped from about 40 percent of university science students in 1876 to 23 percent in 1914, when Lyon, Grenoble, Nancy, and Toulouse had 40 percent of the science students.[15] Geographically, the distribution of institutes was excellent, corresponding to the large population centers (see Table 4).

The proliferation of diplomas in the University was made possible by the decree of July 21, 1897, which allowed the universities to grant special diplomas that were not equivalent to and which did not confer the privileges of the state degrees. A clear distinction was made between the *baccalauréat*, the *licence*, and the *doctorat d'Etat*, all carefully regulated by the state educational bureaucracy, and the university diplomas, including the *doctorat d'université*, which were controlled by each institution and approved by the Conseil supérieur de l'instruction publique. The separation was broadly between degrees that granted the right to teach in the University and degrees that prepared for industry, commerce, or agriculture. Most of the diplomas were related to chemistry and electricity, two of the key sources of modernization and a new life-style in the late nineteenth and early twentieth centuries. In 1902 Nancy upgraded its diploma for chemists to one for *ingénieurs-chimistes* with a normal three-year period of study – except for graduates of the Ecole normale, Polytechnique, Mines, or Centrale – although the period could be shortened to two years by a special decision of the faculty. In 1904 Lille introduced a diploma for students of general and applied chemistry at the Institut de chimie of the faculty of sciences. The aim was to give a serious theoretical and practical chemical tool to people going into industry and agriculture. The three-year program included a detailed study of chemical, agricultural, and textile industries and metallurgy, as well as other key activities of the Nord. Practical work was included. In 1906 Toulouse added a diploma of *ingénieur-chimiste* that was specifically designed for young men entering local industry. Lyon followed in 1907 with a *brevet d'études techniques de chimie industrielle*; Lyon already had an Ecole de chimie industrielle.[16] The general trend toward using the University as a vehicle for introducing scientific and technological knowledge into economic life was firmly established well before the First World War.

148

Table 4. *Institutes of applied science in provincial universities, c. 1910*

University	Institute
Besançon	Service Chronométrique de l'Observatoire Enseignement Supérieur d'Horlogerie Laboratoire de Mécanique Laboratoire de Métallographie
Bordeaux	École de Chimie Appliquée à l'Industrie et à l'Agriculture Laboratoire de Chimie Appliquée à l'Industrie des Résines Enseignement sur les Fermentations et un Cours sur les Fraudes
Caen	Institut Technique de Normandie (électrotechnique et mécanique, chimie)
Clermont	Institut Industriel et Commercial du Centre de la France (chimie industrielle)
Grenoble	Institut Polytechnique-Brenier (électrotechnique, électromécanique, électrochimie, électrométallurgie) Station d'Essais Industriels d'Électrochimie et d'Électrométallurgie Laboratoire d'Électrochimie
Lille	Institut Électrotechnique Institut de Chimie Chaire de Paléontologie Houillère
Lyon	École de Chimie Industrielle École Française de Tannerie
Marseille	École Pratique de Chimie
Montpellier	Institut de Physique Institut de Chimie
Nancy	Institut Chimique Institut Électrotechnique et de Mécanique Appliquée École de Brasserie Institut Aérodynamique
Toulouse	Institut de Chimie Institut Électrotechnique Institut Agricole

Note: Dijon, Poitiers, and Rennes had courses and chairs in industrial physics and chemistry and in agricultural chemistry.

All the institutes were proud that no graduate had any difficulty in finding employment. With a special part of the Lille curriculum devoted to the intensive study of the chief industries of the Nord, it is not surprising that nearly all graduates in applied chemistry at Lille went into regional industry. Students were often directly connected with industry through their families, as in Nancy, where all the students at the Ecole de brassserie were the sons of brewers, managers, and foremen. The social origin of students in the

149

institutes varied according to the nature of the program. In the 1890s most of the students in applied chemistry in Paris were the sons of industrialists or directors of manufacture in chemical industries, although a few sons of civil servants and of businessmen outside the chemical industry were also recruited. Many of the Paris graduates went into their parents' factories and into chemical industries where they already had a family connection. A patronage committee made up of academics and businessmen took responsibility for placing other graduates. Buisine, the head of the Institut de chimie in Lille, noted in 1905 that his students were the sons of industrialists or young men preparing for jobs as factory chemists.[17] The situation was different in Lyon, where the school of industrial chemistry took most of its students from the Ecole la Martinière, a private professional school at the higher primary level. Fourteen or fifteen years old, without a baccalaureate, these working-class or lower-middle-class students spent two or three years acquiring the rudimentary scientific knowledge that would prepare them for foremen's jobs in the Lyonnais dye factories.[18] Nancy combined elements of the situations at both Paris and Nancy, as Toulouse was to do later. Most institutes slowly developed courses of high quality, especially after 1900.

Many of the students at the institutes came from the schools of *arts et métiers*; in 1928 at Grenoble there were 450 iron-scrapers (*gratteurs de fer* was the academic snob's old term for students of *arts et métiers*). If one believes that the best engineering and science students came out of a classical secondary education – high Latin and Greek culture with mathematics added at the end – then the students in *arts et métiers* had a real disadvantage in their training; most of them had not gone through the secondary track leading to the *grandes écoles* and to the faculties. The institute had a special value for students from *arts et métiers* that it did not have for the others.[19] In addition to acquiring a saleable specialization, they were exposed, however inadequately and briefly, to the fringes of a different social and academic world. It was not a disadvantage, although its virtues would be hard to define.

The success of the institutes rested squarely on their ability to produce trained personnel for business and industry, a function they fulfilled adequately. By 1913 about 300 students had graduated from the Institut de chimie appliquée in Paris in the first fourteen classes. Of these, 176 went into French factories, fourteen to factories elsewhere in Europe, and eight to factories outside Europe. A survey of the Institut chimique in Nancy, which had 129 students, twenty-three of whom were chemical engineers with degrees, showed that out of 405 graduates in industry, 324 were working in France. The Institut électrotechnique in Grenoble, with 358 regular students, found that industry absorbed 150 engineers in 1912. The Institut électrotechnique in Toulouse placed 164 electrical engineers in industry during its first four years; by 1914 the institute had 335 students, including 150 foreigners; this meant a total increase of 197 students in two years. By 1912 fifty-two engineers, eighteen of whom were in industry, had been graduated from the Institut de chimie in Toulouse, founded in 1909. Between 1900 and 1914 Nancy trained 480 chemical engineers, 375 electrical engineers, and 101 mechanical engineers; Grenoble turned out 361 engineers of all sorts. By 1911 the faculties granted more industrially oriented degrees (425) than licences (377). Paris remained the giant licence factory of the system. By 1913 nearly 80 percent of the nearly 500 of industrially oriented degrees were granted by the faculties in Grenoble, Nancy, and Toulouse. Most of the *licenciés* went into teaching or entered a professional school. Out of forty-nine *licenciés-ès-sciences naturelles* who graduated from Lille between 1872 and 1892, ten went into secondary teaching, ten into higher education, and ten into medical sciences.[20]

Financial support for the institutes of applied science came from several sources. State, departmental, municipal, University, and private resources were necessary to build the giant complexes in Nancy, Grenoble, and Paris. In Nancy, where there was a long tradition of teaching applied science – and a range of technical courses comparable with that of the Conservatoire des arts et métiers – the foundation of the Institut chimique for chemical engineering, consciously emulating the German example, dates from an agreement between the ministry of public instruction, the city of Nancy, and the departments of Meurthe-et-Moselle and the Vosges. This agreement, concluded in 1887, two years before the opening of the institute, engaged each local authority for the sum of 500,000 francs. The prince of patrons for Nancy was Ernest Solvay: By 1907 his Société Solvay had donated 320,000 francs. Industrial companies like his organized the funding of such giant projects as the laboratories of electrochemistry and of physical chemistry, which cost 400,000 francs. Most active were officers of the Société industrielle de l'Est, the Société technique de l'industrie du gaz, as well as the Comité des forges et des mines de fer Meurthe-et-Moselle. Financial flexibility was enhanced in the faculty by an annual departmental subsidy of 10,000 francs, paid for fifteen years after 1898.[21] Similar patterns of financing existed in Lyon, Grenoble, Paris, Lille, and Toulouse, and on a smaller scale in Bordeaux, Besançon, and Clermont-Ferrand.

For their model of an ideal faculty of science, all provincial universities turned enviously toward Lorraine.[22] Whatever their degree of success in emulating Nancy, the universities were forced to adopt a similar financial structure for the applied sciences. The Institut d'électricité industrielle in Grenoble was financed by 10,000 francs from the state, a loan of 65,000 francs from the Crédit Foncier, and a subsidy of 4,000 francs from the city. The remarkable growth of the institute would not have been possible without Casimir Brenier's gift of land in 1907; the land, valued at 600,000 francs, was intended for the building of the new Institut électrotechnique. In order to complete this ambitious project, parliament finally voted a special subsidy of 360,000 francs – the state being already committed to an annual subsidy of 55,424 francs. By 1913, 625,000 francs had been spent or encumbered.[23] In Paris, applied chemistry was endowed with a chair in 1909 as a result of a gift of 128,000 francs by Solvay and the Chambre syndicale des produits chimiques. The Solvay company celebrated its fiftieth anniversary in 1912 by giving half a million francs for the completion of the Institut de chimie. Perhaps the greatest year for Paris was 1909, when Henri Deutsch gave half a million francs for the construction of an Institut aérotechnique, with an additional annual subsidy of 15,000 francs, and Basil Zaharoff donated 700,000 francs for the establishment of a chair of aviation in the faculty.[24] Even the beer industry did not stay aloof from the process of collaboration between government, industry, and academic science: The Ecole de brasserie in Nancy was established with the support of regional brewers and of the ministries of public instruction and of commerce, along with the patronage of Nancy's chamber of commerce.

A clear illustration of the prosperity that applied science could bring was the growth of the *polytechnique* complex at Grenoble. Between 1905 and 1907 the number of science students rose from 197 to 252, including eleven foreigners working for the *doctorat d'université* in chemistry. In the same period the number of students in the Institut électrotechnique jumped from fifty-two to ninety-eight, exclusive of thirteen in industrial physics. A year later the total rose to 144. In addition to all the usual implications of growth – more professors, assistants, laboratories, buildings – the income of the faculty of science from fees increased from 24,000 francs in 1898 to nearly 72,000 francs in 1907. L'Institut polytechnique de Grenoble was above all a group of engineering schools: l'Institut électrotechnique, l'Ecole française de papeterie, l'Ecole des conducteurs-électriciens, l'Ecole des ingénieurs-hydrauliciens. By the mid-1930s, about 500 students enjoyed the facilities of the Laboratoire d'essais et d'étalonnage électriques (1900), the Laboratoire de l'Ecole française de papeterie (1907), the Laboratoire des essais mécaniques et physiques des métaux, chaux et ciments (1921), and the Laboratoire d'essais hydrauliques (1934).[25] The generosity of private benefactors like Solvay to Nancy and Brenier to Grenoble was the direct result of the development of applied science, for the prosperity thesis was a corollary of the fairy tale of science as the goose that lays the golden egg.

The institutes were the product of a healthy economy based on industrial growth. By the 1880s French success in industrial technology was evident in the production of locomotives, glass, ships, chemicals, cast iron, and, of greatest importance, steel. France played a large role in the evolution of scientific metallurgy, especially microscopic metallurgy with a theory of the composition of steel. George Charpy, Pierre Chévenard, Henry and André Le Chatelier, Léon Guillet, and Pierre Weiss were among the many engineers and scientists well known for their work in faculties, *grandes écoles*, and industrial laboratories.[26] In the chemical industry, the Solvay method of alkali manufacture spread rapidly in the last quarter of the nineteenth century, until, by 1905, nearly all alkali was made by the ammonia process. But the French lagged in introducing electrolytic methods of preparing chlorine and caustics, and the organic chemical industry did not fulfill its early promise because of a lack of capital following the fiscal disasters of the 1860s.[27] The specialties of the institutes depended on the leading economic activity of their regions, and the level of instruction varied with the quality of the technology in regional industries.

In nearly all faculties there was a strong interest in chemistry and in electricity – fields which, as Jacob Schmookler has pointed out, "owed their start to scientific discoveries," although the intensity of their cultivation was "proportional to their prospective contributions . . . towards the satisfaction of private and public wants."[28] It was possible to obtain a *brevet* and qualify as an electrician almost anywhere, but the best students who wanted to become electrical engineers chose Grenoble, Lille, Nancy, or Toulouse. In the beginning, several programs – watchmaking at Besançon, tanning at Lyon, and brewing at Nancy, for example – were little more than apprenticeship training, but as the programs developed there was a conscious effort to help make French industry competitive in the international market by raising the level of instruction; it was intended that graduates should at least be able to keep up with the newest technology. In addition to the large general programs in chemistry and electricity, there were often smaller specializations explicitly geared to regional activities, such as the wine and pine products industries at Bordeaux, paper at Grenoble, and aerodynamics at Nancy and Paris.[29] The institutes thus made up a complex set of technical schools with a wide variety of programs ranging from the most elementary to the highly advanced. An increasing emphasis was put on instruction at a high academic level, and standards rose markedly in the two or three decades before the First World War. This was due in part to the greater sophistication of industrial technology but also to the determination of the institutes themselves to bring science into industry through their programs.

The intimate connection with specific industries led to a conscious decision at Lille to provide a sort of research and development operation for industry. Since Lille and the Nord had so many special schools – the Institut industriel du Nord de la France, the Ecole des arts industriels at Roubaix, the Ecole des industries agricoles at Douai, the Ecole d'agriculture at Wagonville, to say nothing of the Catholic schools – there was no need for a school of industrial chemistry within the Lille faculty. Instead of duplicating facilities, the faculty decided to push specialization to its logical extreme and prepare young people for direct entry into the various industries in which they were already interested. The creation of special laboratories for this purpose also made possible industrial research of a certain type – the testing of products and the study of new processes – which could not be done in industry because of the absence of adequate laboratories. The Institut chimique of the faculty of Nancy adopted a similar policy. Special laboratories were opened for the chief regional industries, especially for the dye industry, which profited from a course subsidized by 3,000 francs from the municipal government, and for the breweries, which enjoyed the resources of an experimental brewery based on the best German model. Georges Arth, Haller's successor at Nancy, continued to develop the Institut chimique. In 1908 the institute opened a laboratory for industrial testing, run by a scientist who left an industrial metallurgical laboratory for the University. The new laboratory was also a boon for pure science: With its analytical section, it could counteract the baleful effects of the general neglect of analytical chemistry in the faculties, in contrast to its good status in the Ecole des mines and in the Ecole centrale.[30] If they are judged by their faculties of science, Lille and Nancy had genuine regional universities.

By the first decade of the twentieth century Lille and Nancy could boast that they had created a technical faculty, a sort of *technische Hochschule*, within the University. The claim is justified, although some programs continued to produce foremen for such industries as leatherworks and breweries. Through programs accepting young people who had not taken the baccalaureate, the faculties of science provided a limited means of upward mobility for a small number, giving them an introduction to university education. But secondary and university education maintained class segregation and may even have increased social divisions.[31] The main function of the technical institutes soon came to be seen by the faculties as that of harnessing high-level scientific work to industry. Scientists of the caliber of Albin Haller at Nancy, Le Chatelier at Paris, and Paul Sabatier at Toulouse ensured the success of ambitious programs.

The pursuit of excellence resulted in constant upgrading of programs. In 1902–3, Grenoble created an elementary division of the institute, which awarded a *brevet de conducteur-électricien* to students who successfully finished one year of study. A higher engineering section was created for those with some background in technical studies or industrial experience. By 1900 the high standing of the institute was indicated by the fact that it was receiving the same number of artillery officers from the Ministry of War as the Ecole supérieure d'électricité in Paris. The same process occurred at Lille, where, by 1912, a growing number of students earned the *licence-ès-sciences* at the same time as a *diplôme d'ingénieur-chimiste*. Similarly, at the Institut électro-technique in Toulouse, which had 422 students in 1912–13, there were 370 students studying for the *diplôme d'ingénieur-électricien* and only fifty-two for the *brevet de conducteur*; out of the fifty-nine graduates who took engineering degrees, two were graduates of the Ecole polytechnique.[32] Even Rennes found by 1899 that, in order for its certificate of applied chemistry to be useful for regional industry, the level of knowledge in theoretical chemistry had to be high. Students therefore had to obtain a certificate in general chemistry before undertaking the two years required for the certificate in applied chemistry. Rennes recognized that the knowledge needed by a chemist could only be acquired by studying for three certificates – in applied chemistry, applied physics, and general chemistry – which would entitle the successful candidate to a *diplôme de licencié chimiste*.

The combination of special scientific knowledge available only in the faculties with the practical work done in the institutes distinguished the latter from such institutions as the *écoles d'arts et métiers*. But in the lower echelons of the institutes the levels of knowledge and of attainment were by no means higher.[33] Faculties were aware that the institutes had to be saved from becoming imperfect copies of technical schools and that the only way this could be done was through permanent contact with the "pure science" solidly ensconced in the faculties.[34] Successful institutes needed good science. Of course, the democratic potential of university-related technical studies was restricted by the difficulty of high-level scientific work and by limited facilities. Restricting recruitment through entrance examinations became a badge of distinction. An extreme example: The Ecole supérieure de métallurgie de l'industrie des mines (Nancy), which had seven *polytechniciens* in its second-year group, boasted of admitting only 34 of 330 applicants in the *concours* of 1922. But this problem may have resulted less from French elitism than from the difficulty of reconciling professional standards with democratic ideology. This was hardly a unique French problem.

The growing prestige of industry-related studies was evident in the transformation of courses into professorships. Few faculties were not infected with the industrial virus. In 1889 Bordeaux created a faculty chair for its course in industrial chemistry. Strongly supporting the industrial trend, municipal councils backed their rhetoric by funding chairs. This source of support became increasingly important as state financing for faculty expansion diminished. In 1907 the municipal council of Toulouse voted funds for the creation of a chair of industrial electricity in the faculty of sciences and also guaranteed the salary and promotion increments for twenty years. So long as no one attempted to replace a chair in basic science with one directly related to industry, the educational bureaucracy gave its approval to the technological invasion.[35]

The drift toward technology also affected traditional teaching, even in mathematics. Toulouse justified its request for a fourth mathematical chair in 1907 on the basis of the yearly increase of science students, many of whom were interested in the applications of science. This category of students required a substantial scientific foundation in analysis, geometry, algebra, and mechanics, but not at so high a level as the students specializing in pure mathematics. Instruction for such students had to be practical in orientation, despite gaps in their secondary scientific education that were the result of not taking special mathematics in the lycée. The high utility of general mathematics had been shown by experience. It was also necessary to compete with other faculties. Toulouse wanted a chair corresponding to the chair in applied mathematics at Nancy. A clinching argument was that the chair would demonstrate the orientation of the faculty toward the teaching of the applied sciences, which was necessarily the source of its prosperity as well as of public favor.[36]

The influence of industry on the structure of the university and the direction of its teaching can be seen in the different diplomas created. In 1908 Grenoble instituted a *diplôme d'ingénieur-papetier*, granted after two years of study. In the first year the subjects were physics, chemistry, industrial drawing, mathematics, electricity, and industrial mechanics; in the second year, industrial and commercial law and finance were supplemented by courses in paper manufacture. There was an admission requirement to the program: An applicant who did not have a baccalaureate had to take an admission examination based on the syllabuses of the mathematics class of the second cycle of the lycées. In the preceding year the University of Grenoble and the Union des fabricants de papier de France had agreed on the creation of an Ecole française de papeterie, annexed to the Institut électrotechnique of Grenoble. This pattern of the growth of industrial schools within the University, which granted the degree, was also followed by Nancy's Ecole de brasserie, which had between thirty and thirty-five students by 1904; here the qualification awarded was a *diplôme d'ingénieur-brasseur*. By 1900 it was no longer adequate for industry to profit from the industrial research of the professors of the faculties of science. Industry needed its own institutions, but their narrowly defined purposes did not prevent them from exploiting the knowledge of the faculties of sciences and the prestige of university diplomas.

With the invention of the dynamo by Gramme in 1869, the modern electrical industry began its short gestation period. The presentation of the first machine of Gramme to the Academy of Sciences by Jamin in 1871 achieved a degree of publicity that Pacinotti, who had really established the principle of the dynamo ten years before, had not managed to get by publishing in the small journal *Il nuovo cimento*. France kept active on the electrical research front. Marcel Deprez began his work on the theory of the dynamo in 1884.[37] In 1885 he carried out experiments on the transmission of current between Vizille and Grenoble and in 1886 between Creil and Paris (La Chapelle). Industrial current proper was really "invented" between 1870 and 1890. Deprez first carried out a set of experiments in the shops of the Chemins de fer du Nord. Wishing to use the many powerful waterfalls around Grenoble, the city Council asked Desprez to repeat the experiments there.[38] By 1900, when the Institut polytechnique de Grenoble saw the installation of the Laboratoire d'essais et d'étalonnage électriques in the old quarters of the Institut électrotechnique, Grenoble had become one of the world's leading centers of electrical studies.

Grenoble was accepted as a model for other schools wishing to produce electrical engineers; its Institut électrotechnique was closer to the German model than most French institutions. Not that there were many basic differences between schools on opposite banks of the Rhine. Higher education in electricity was an important component in the *Hochschulen* in Aachen, Berlin, Charlottenburg, Brunswick, Dantzig, Dresden, Hanover, Stuttgart, Munich, Darmstadt, and Karlsruhe. It was only in the last two cities that there were autonomous electrotechnical institutes; in the others electrotechnology coexisted with another science to form a special teaching section. Electrical studies in the *Hochschulen* corresponded to those in the French electrotechnical institutes that were special sections of the faculties of science in cities of the rank of Grenoble, Nancy, and Toulouse. At a lower level in Germany the *Technica* (technical schools) put the emphasis on practical rather than scientific aspects of electricity. As least from the viewpoint of electrical studies, the *Technica* had their counterpart in the French *écoles d'arts et métiers*. There were also schools that taught only electricity. These *écoles d'électricité*, nearly in the same class as the *Technica*, gave theoretical instruction at a higher level than the *Technica* and not far below the *Hochschulen* and the faculty institutes. Both governments and industrialists had a high opinion of the *écoles d'électricité*.[39]

German educational programs were usually more practical than the French ones, with the mathematical program being separated as much as possible from any theory not leading directly to practical results. There was not the same obsession with the establishment of formulas as in France. German instruction put much more emphasis on calculations applying the theoretical formulas than on their establishment and foundation. So perhaps French studies needed to take a more practical direction. If French industrialists had supported the laboratories, as they did in Germany, the French change of direction would have been inevitable. And the French would have established better and more specialized libraries. In Grenoble the electro-technical institute was strongly supported by the leading manufacturer of hydraulic turbines. The example needed copying by many more factories and schools.

The rapid growth of the electrical industry was a particularly powerful stimulus to the creation of university-related institutes and courses. Grenoble's *diplôme d'ingénieur-électricien* (1902) could be obtained after one year by engineers with diplomas from the *grandes écoles* or foreign equivalents, or even by students with a high level of technical knowledge as determined by admission examinations. Examinations included an electrical installation and personal laboratory work as well as written and oral sections. In 1903 Besançon started giving a diploma in applied electricity; the examining jury was made up of professors of industrial electricity and general physics, and a third member who could be an engineer in the region. The diploma provided an opportunity for young people who could not pursue traditional science degrees but who could profit from two years of study in the faculty and then get jobs in industrial electricity. In 1905 Poitiers created a *brevet d'études électriques*, conferring the weighty title of electrician of the University of Poitiers. This two-year course had no admission requirements. An electrical engineer could sit on the jury. Poitiers also gave a *certificat d'électricité industrielle*. In Nancy electrical studies were under the same roof as applied mechanics in the Institut électrotechnique et de mécanique appliquée. Nancy created a higher *diplôme d'ingénieur-mécanicien* in 1905 for those completing three years of study in the mechanics section of the institute.[40] The quality and range of studies available depended on local and regional factors, but the system as a whole provided a respectable offering of courses in most of the latest technological miracles.

By 1913 the faculties of Paris, Lyon, and Lille had about a hundred students each in the applied sciences, mostly in chemistry; Nancy, Toulouse, and Grenoble had a total of nearly 1,500 in chemistry, electricity, and mechanics. The faculties of science had found a national role relating directly to industrial growth. This role was the most successful part of the plan to regionalize the universities. It was not unreasonable to claim that at least some programs also achieved a successful integration of science and technology, but this was contested by some scientists – and not only by the Solons of theory – who thought that the technological saturation point had been reached in the faculties of science by the early twentieth century. Money agreed with theory: State funding for science faculties dropped by between 9 and 21 percent in the period 1901–14; generally, up to 1908 local governments and industries made up the difference, but between 1909 and 1914 French industrial funding dropped by 30 percent.[41]

In 1905 Nancy created a *diplôme de hautes études commerciales*. Students studied economics, law, geography, commercial accounting, and modern languages (English and German) for two years. In 1912 the Section permanente of the Conseil supérieur turned down a proposal by the University of Nancy to transform, for the benefit of foreign students only, the diploma of higher studies in agriculture into a *diplôme d'ingénieur-agricole*. The grounds for rejection were that it would also have to be done for French students; this was impossible, for the Ministry of Agriculture wished to reserve that title for students coming out of its schools. Acting on the recommendation of a report by Appell, the section also turned down Nancy's attempt to upgrade a diploma of higher commercial studies into a *diplôme d'ingénieur-commercial*; the old title corresponded exactly to the program organized in 1905 and there was no justification for creating a new title.[42] Title inflation had reached a temporary peak.

The most serious problem resulting from the rapid expansion of the large institutes was finding qualified teachers. A very thorny issue was the status of engineers and technicians whose primary job was teaching, but who had a secondary occupation outside the institutes. The arrangement gave a more stable personnel than hiring regional technicians whose chief activity was outside work. But few of the technical teachers had a doctorate, or if they did it was likely to be the University rather than the state doctorate. So in spite of their age, experience, and expertise, it was not possible to integrate them into a faculty of sciences or to give them the rank they deserved as employees of the institutes. The problem would have been greater if doctoral theses in technical areas like electrotechnology and in metallurgy, rather than in pure physics and chemistry, had not been acceptable as subjects for a state doctorate. The problem was most acute at Grenoble, which, having a very small science faculty, had to hire a big technical staff to teach courses. Another possible solution to the problem of how to integrate professors without doctorates was the ingenious but not very fruitful idea of letting engineers take a type of doctorate (title of *ingénieur-docteur*) that did not require the candidate to navigate the pedagogical labyrinth of the licence. A barrier was erected by the state doctorate, and it was hard to find a substitute that would come up to its level.[43]

The aim of the new degree invented for engineers in 1923 was really to open faculty laboratories to the top graduates of the *grandes écoles* in order for them to learn how to do scientific research. Although governmental decrees in 1922 had given graduates of the Ecole centrale an automatic scientific certificate, in addition to one in general mathematics given on entrance, few engineers showed interest in this scheme. Those who did wanted to keep their jobs and spend as little time as possible in the laboratory. And unconsulted faculties were afraid of the danger of weakening degrees by granting concessions to privileged graduates of the Centrale. The degree of *ingénieur-docteur* was supposed to overcome these difficulties, but it was not terribly different from the University doctorate that had come on the scene in the late nineteenth century without much more success.[44]

By 1930 it was clear that something had to be done to improve the quality of engineering education in France, at least in the schools under the Ministry of Education. It was more difficult to get a degree in the United States and in many European countries than in France. In order to avoid the possibility that French engineering degrees might become the bargains of the Western educational world, reformers pushed for an improvement in quality. An upgrading of standards was certain to make things more difficult for foreigners; so there was a fear that any remedial measures would result in a smaller number of French-educated foreigners, who historically had an important propagandistic function in France's *mission civilisatrice*. Besides, foreigners often brought a valuable quality to the school: better *scientific* training than that of the average French engineering student. As in the medical profession, there was the fear that foreigners would be eligible for French jobs once they had a state degree. Italy had passed a law requiring state employees to have a national diploma. France was not ready to imitate Fascists in 1930, but the economic and social value of the state degree was being protected in many countries, and France was not immune to the protectionist virus.[45]

Raising standards would also mean fewer foreigners in the institutes, although most foreign students were in the more elementary programs. Few foreigners could penetrate the *écoles d'état*, perhaps not so much because of the *concours* – although they were formidable, the idea that they were an insuperable barrier is largely a French intellectual conceit – but because these schools, like the notorious *doctorat d'état*, gave access to governmental jobs. The schools had – and have – a *caractère fonctionnariste*, opening up the royal road to the plums of bureaucracy. So the faculty institutes were some of the best producers of propagandists for French civilization. To cut off the flow of foreigners would threaten a return to the days when all that foreigners knew about France was, on one level, Notre-Dame, the Eiffel tower, and the dubious delights of tourism, and, on another level, Maxim's and the influential Ecole des beaux-arts.

Improving the quality of engineering meant more mathematics and more science. In an already heavily loaded program this posed serious problems for both professors and students, especially when, as in the case of physics, the basics of the science itself were changing rapidly and radically. By the 1930s it was publicly admitted that the teaching of physics in schools of engineering – the elevated status now reached by several of the old institutes of applied science – was in trouble. First, there was the enormous development of science since technical education had begun in the universities; a vast body of new data had to be assimilated, often within the severe limits of a student's embryonic structure of scientific knowledge. Second, the material had to be taught in a short time. The situation was not improved by an annual rise in the complexity of required mathematics, which was directly related to the increasing difficulty of basic science. About 1900 an engineer knowing second-order differential equations possessed high-level theoretical knowledge *for an engineer*, but in a generation the level rose to Bessel's functions, Lagrange's series, and other recondite but powerful mathematical weapons used by science in its conquests. Comparable high levels were imposed on the engineer in areas like chemistry, mechanics, and metallurgy. Worst of all, a new physics had appeared.[46]

Aging professors were distressed to witness the ruin of the old physicomathematical structure – assumed to correspond to reality itself – with which they had fallen in love in their youth and loved ever since with rigorous fidelity but less passion. But it would have been an intellectual crime not to teach to the eager youth in the institutes the subject that delivered new discoveries daily. The problem was that the structure of studies for degrees did not easily permit the intrusion of new material on the nature of matter, the kinetic theory of gases, or relativity. What Louis Barbillion proposed was an improvement in the courses in classical physics, eliminating some parts and placing increased emphasis on generalities or on applications as required for contemporary engineering: General physics, optics, acoustics, and thermodynamics would be on the engineer's menu. In the first year of engineering a course in the new physics would introduce students to the work of contemporary physicists. X-rays, radioactivity, the disintegration of matter, and the behavior of light would be among the subjects introduced, for they impinged on the work of the engineer more than did some other new topics.[47] The new physics had become an essential ingredient of the engineer's intellectual and practical repertoire because new theory opened up the possibility of more technical conquests.

Mathematics, with its dazzling advances, was a serious challenge to both physicist and engineer. What was not directly relevant was often condemned as an intellectual amusement of the aesthetic class. Sometimes the degree of abstruseness goaded the physicist into protest. And the social and academic prestige of mathematics aggravated the situation. Henri Bouasse, a whimsical physicist at Toulouse, enemy of quantum physics and relativity, counteracted with a peevish outburst in the preface to his book on the resistance of material: Mathematics was useless for the formation of the mind – a serious charge in France.[48] The engineer had to contend with both physicists and mathematicians. Although the engineer needed only a limited mathematical baggage even in handling the most complex public works, the fact of the matter was, as Barbillion regretfully admitted in 1936, that in France the criterion for the classification of engineers was their degree of mathematical knowledge. Higher, if not transcendental, mathematics was necessary for recognition as a true engineer.[49] This was a large portion of the cognitive component of the prestige and power bestowed by the *grandes écoles*.

What parts of mathematics did the engineer need to have an absolute mastery of? A reasonable answer just before the Second World War would begin with an emphasis on all elementary mathematics – algebra, trigonometry, geometry, and related studies. Certain areas, such as the analytical and graphic procedures for the construction of curves, could not be overemphasized. And here was a gap in the teaching of general mathematics in the faculties, at least from the viewpoint of the engineer's ideal education. The engineer could do with a lot less of the endless discussion on conic properties, for, like all other curves, they could be studied algebraically. Pure geometry, a French delight, had no exceptional interest for the engineer; it was an occupation for dilettantes. However beautiful the theory of elasticity, it could not take the engineer very far unless there was substituted for it the half-empirical subject of the resistance of materials. New areas of investigation could be kept out of the engineer's program until basic science and its practical applications became clear. New theories dealing with the mechanics of fluids and of viscosity fell into this category in the late 1930s.

It was true that the engineer should not be drowned in a flood of advanced mathematics, but neither should he be given too much of the science relevant to his engineering specialization. This second error became a serious flaw in the education of some schools, none more egregiously so than in the schools of chemistry, which taught so much chemistry that they forgot about engineering while trying to produce chemical engineers. So the graduates were more chemists than engineers, not knowing simple engineering construction procedures. Because this mistake had a sorry result in the job market, with graduates of the Centrale, for example, getting jobs that would ordinarily have gone to the faculty's chemical engineers, the Syndicat des ingénieurs-chimistes de France and other professional groups got universities to make changes in and additions to educational programs. It had to be made clear to employers that an engineer knew his engineering as well as his science.

The reasons for the technological emphasis in the faculties

In the last two decades of the nineteenth century, provincial faculties of science were encouraged to shift the source of their funding from money supplied by the central government to subsidies given by local authorities and regional industry. The bureaucracy of higher education worked hard to make universities financially dependent on their regions and to stimulate regional economic development by promoting cooperation between professors and industry. There were definite problems in convincing some businessmen that investment in faculties of science could be profitable, but the new programs generally worked well. Of course, it must be emphasized that in some cases, as at Lille and Nancy, the government was carrying coals to Newcastle, for the interaction between the faculties and industry sprang from local mutual interests dating back to the origins of the faculties. Between 1885 and 1900 municipal and departmental generosity made even conservative Lyon the "most generously endowed of the fourteen provincial institutions," with about 300,000 francs coming from private sources.[50] It is difficult to say to what extent the encouragement from Paris helped, for the movement of local support for provincial faculties of science was complex, with many variations. Perhaps in the end nothing was more of a stimulus than the provincial academic perception of a grim reality: Faculty growth would not be funded by Paris beyond a certain minimum needed to produce personnel for the secondary and higher educational systems, which would mean retrenchment for many faculties.

The growth of institutes of applied science was sometimes the result of overtures to industry by the faculty, such as that by Haller in Nancy, sometimes the result of calls for help on the part of industry, as in the case of the brewing industry, or sometimes the result of a fruitful interaction between industry and science – the typical situation in Lille after Kuhlmann's arrival there in 1823. In these cases interaction between industry and the faculties of science was easily achieved because of the existence of well-developed local industries and because of dynamic imperialists in the faculties. By the end of the nineteenth century any good faculty of science had to create industrially and agriculturally related institutes. If an industry did not exist, it might be necessary to invent one, or if industry were dying, it could be resuscitated in order to ensure the growth of the faculty of sciences. Science could be the white knight rescuing the damsel in distress, as it was in Besançon in 1897, when a course in industrial and agricultural chemistry was established. The lecturer, Pierre Genvresse, encouraged by an annual subsidy of 4,000 francs from the departmental and municipal councils, concluded that it was the function of the faculty to halt the decline of industry, in particular the chemical industry, in the Franche-Comté. Factories needed chemists in order to survive and, *a fortiori*, in order to grow and improve in a brutally competitive market. Since the Francs-Comtois possessed the scientific spirit and the faculty was at the disposal of industrialists, all that was needed was the use of Genvresse's course to open up the mutual flow of benefits between industries and faculty. He started his course with an audience of forty.[51]

By the 1880s it was clear that the historic function of producing lycée teachers offered little opportunity for faculties of science to expand. Faculties had produced so many *licenciés* that it was difficult for them to find jobs. Between 1880 and 1885 Paris produced 210 *licenciés* and 151 *agrégés*, of whom 191 went into secondary education and 12 into the faculties. When the number of candidates passing licence and *agrégation* examinations fell in Marseille in the late 1880s, the dean was pleased, for an equilibrium had to be established between graduates and jobs available. A high failure rate – 50 percent for two examination periods in Marseille (1887–8) – was the brutally effective way of adjusting supply to demand. Lyon came to the same conclusion.[52] But the faculties did not abandon their expansionist ideology. A new demand for students had to be created. Fortunately, the Goblet reforms of 1885 permitted regional and local agencies to fund the faculties for the production of students trained to work in industry, agriculture, and business. For Nancy, Lille, Grenoble, and Toulouse, and, to a lesser extent, for the other science faculties, the result was an explosive growth that would have been inconceivable had students been directed exclusively to posts in the educational system.

The pot-pourri of arguments used to justify the new practical direction of the science faculties was neatly brought together in the Darboux report on the controversial *certificat d'études physiques, chimiques et naturelles* (PCN – dubbed "petit cochon noir" by wicked students), introduced in 1893. The general scientific PCN education for medical students could be useful for others: In addition to students preparing for special schools like the Centrale and the Institut agronomique, there was a larger number of students destined for industrial and agricultural careers who needed the appropriate practical education. Lyon and Nancy had successful programs. Experience showed the success in industry of those who took a general practical education. In certain centers the PCN program could become the point of departure for a technical education useful to national industries. The faculties of science, with only 1,900 students preparing for licences, *agrégations*, and doctorates, could easily handle this new program. The enlargement of the scope of the faculties would also join the higher to the lower parts of education, thus delighting lovers of unity. The general justification was the increasing role of science in society. Contemporary developments in the chemical and electrical industries, especially in Germany, could be attributed in large part to university teaching or institutes run by university professors. The conclusion of the report ended on a patriotic note: The faculties were ready to repay the debts they owed to their country by teaching science to physicians – even if they did not want it – and by making industrialists and farmers aware of scientific methods.

There is a great deal of fantasy in these arguments for the PCN, whose main purpose was to provide science students for the faculties the Republic had created. It was believed that the survival of France as a great power depended on the implementation of rather Baconian schemes in the new houses of Solomon. If students did not see their salvation in science, they should be forced to find their new freedom in the program designed by the republican scientific mandarins. There was no necessary connection between the programs of the institutes harnessing science to industry and, given the direction already clear at Lille, Lyon, and Nancy, probably no need for the superfluous stimulus of the PCN. The immediate results of the new program verged on disaster. Faculties soon complained that enrollments in the PCN courses varied with fluctuations in the medical faculties. The worst result of the program was an inundation of science faculties that had just become or were becoming adequate for other categories of students. Bordeaux, Grenoble, Lille, and Dijon enjoyed groaning about the overcrowding of lecture halls and laboratories. The first notable application of the new financial regime made by the sciences faculty council of the Sorbonne was to borrow money for a new building to handle PCN students, who made up 25 percent of registered science students in 1914. As late as 1906–7 PCN laboratory work at Bordeaux was being done in the corridors.[53] The attempt to inject science into medicine through this particular program was of dubious value and also added unnecessary fuel to already blazing quarrels in medical education.

In the smaller universities the practical orientation of science was directly linked to the idea of the survival of the faculty. In 1902–3 Caen was down to thirty-seven students, including seventeen for the PCN.[54] People being most interested in immediate practical results, the faculty had to develop applied science courses in the Station agronomique to deal with agricultural problems and to insert industrial issues into the course in industrial chemistry. Guinchant's course in industrial electricity attracted over 120, including some electrical workers. Professors in Clermont-Ferrand were alarmed when the number of science students had dropped to sixty-eight, including thirty-four for the PCN; there was even a rumor that the faculty would close. The remedy for this anemia was to drop general courses in favor of local and regional studies in harmony with the tastes and traditions of the Auvergnat and connected with its material prosperity. Chemistry was soon catering to the needs of the three flourishing industries of the region: rubber, paper, and the dyeing of thread, wool, and silk for carpets. By 1912–13 Clermont had 111 students in the sciences, including sixteen for the diploma in industrial chemistry and seven for the *brevet* in industrial electricity. The revival also produced spacious new buildings for chemistry, natural history, and mathematics. Facilities for industrial chemistry were provided in the Institut industriel et commercial du Centre de la France, financed by the Clermont chamber of commerce. Industrial chemistry also had an annual subsidy of 120,000 francs, most of this sum coming from the chamber of commerce.[55] At Besançon the faculty, frankly recognizing its limitations, coordinated its efforts with the Ecole d'horlogerie and the local observatory in the interests of the horological industry; "tout pour l'horlogerie" was a slogan taken quite seriously.[56]

Even the institution of the PCN as a requirement for medical students did not eliminate the threat of extinction. Cries of alarm were still heard in Lille in 1906: In 1904–5 the number of science students dropped from 213 to 186, with the PCN enrollment shrinking from fifty-five to thirty-eight. The rector warned that provincial faculties of science could not expect an increase of students in the pure sciences. Just as the geologist Charles Barrois was elected to the Academy of Sciences, the number of students in geology was dropping. The contrast in the applied sciences was evident, with a steady growth in the number of students in the Institut électrotechnique and in the Institut de chimie. An expansion of the Ecole de chimie, as it was popularly known, from sixteen to forty students was planned to capitalize on the demand for a new diploma of *chimiste de l'Université de Lille*.[57] The gift of a chair of industrial physics to Lille by the Conseil général du Nord was typical of the general interaction of industry, government, and education in the late nineteenth century. Other disciplines were not slow to learn from the experience of chemistry and physics. Geology was "saved" at Lille through the creation of an institute catering to the needs of the coal industry. The departmental councils of the Nord and the Pas-de-Calais gave subsidies, and the Chambre des Houillères of the two departments gave the money to establish courses on coal paleontology. Barrois also found enough money to open a useful coal museum, and in 1906–7 he organized a permanent center of technical information for the region's mining engineers.[58] Discoveries of coal in the Longwy and Briey basins of the Lorraine created a similar favorable situation for geology at Nancy, where a chair of geology was created in 1907 for René Nicklès, famous for his work on coal deposits in the Lorraine.[59] A discipline that in most universities could have been only of minor interest on academic grounds became quite significant in areas where mining was of major economic importance.

An eternal theme of the propagandists for the institutes of applied science was the need to emulate the industrial and commercial successes of the chemical paradise in Germany. A. Léger, writing in the new publication *Lyon scientifique et industriel* (1882), made the classic case in pointing out the solid profits that other countries made from "scientific speculation." In Germany and Switzerland, where first-class scientists were harnessed to laboratories whose research was directly related to industry, factories employed legions of chemists who proliferated marketable specialties. The value of dyes derived from coal tar was only 3 million francs for France, compared with 7 million francs for Switzerland and 65 million francs for Germany. Although French production had remained at the same level in the six years since 1876, that of Germany had doubled in the same period. As a result, Lyon was dependent on Germany for many of the chemical products required by its industry. The lesson was clear: To defend itself against this new invasion, France had to look to the chemists for salvation. Lyon could make a good beginning by supporting general chemistry in as lavish a fashion as it supported chemical studies in the new medical school. The weakness of the French chemical industry in World War I led the government to organize research in some key sectors, especially the dye industry, which was supplying only one-ninth of the country's needs in 1913. The final result of reorganization, the Compagnie nationale des matières colorantes (1917), was taken over by Kuhlmann, which also ingested the Société des produits chimiques et colorants français de Saint-Denis. The need for munitions led to a great surge in the demand for organic chemicals, stimulating the French into increases in the quality and quantity of chemicals produced as well as in efficiency of production.[60]

Practically the same lament issued from Besançon in 1898. Genvresse argued that one of the industries that could be started in the Franche-Comté was the manufacture of synthetic dyes, for, with the exception of the Poirrier firm at St. Denis and the Société chimique des usines du Rhône, most of the dye business in France was in the hands of the branches of foreign firms.

These included Badische Anilin- und Soda-Fabrik at Neuville-sur-Saône, the Fabwerke of Hoechst near Compiègne, Bayer of Elberfeld at Flers, and the Société lyonnaise des matières colorantes, a branch of the Casella firm at Mainhur. But it was not clear that the imbalance between France and Germany could be corrected through the production of chemical engineers by the institutes, even over many generations. In fact, if one took into account such factors as population, industrial strength, and the scientific-technological strength of the educational system in Germany, parity seemed impossible. The director of the Ecole de chimie industrielle de Lyon pointed out pessimistically that France had only about 4 percent of the number of chemical engineers active in Germany.[61]

A good deal of support for the institutes came from scientists and progressive industrialists who claimed that there were grave deficiencies in France's system of technical education because it ignored information on the latest scientific and technological innovations important for industry. Both the Ecole centrale and the Ecole polytechnique were criticized.[62] For some years the *Revue générale des sciences* campaigned for the establishment of laboratories for the experimental study of mechanics, an area in which France lagged badly; in 1899 it welcomed the creation of Gabriel Koenig's laboratory at the Sorbonne for the contributions it would make to the study of friction and the functioning of articulated systems used in industry.[63] This particular research lag, as well as many others, was eliminated by the establishment of mechanics laboratories in the institutes of Nancy and Lille. In an economy in which too few industrialists believed in the utility of introducing science into the factory, the institutes provided one of the few places where science and industry could meet. At the Institut électro-technique in Lille, the professor of applied mechanics devoted his courses in 1905 to the theoretical study of the automobile. The most successful French industries, such as the hydroelectric and metallurgical industries, were directly linked to institutes or had their own laboratories for pursuing the industrial application of scientific research. So the institutes were an important symbol in the campaign of scientists like Le Chatelier and Appell for the reform of higher technical education in France.[64]

It was clear that the successful factory would be dependent on the faculty, with the implication that the faculties would find their most important social role in training young people for scientific employment in regional industry. Their traditional functions – to train a few people to become professors or teachers and to fashion generally cultivated minds – could appeal only to a small élite. Expansion demanded greater numbers of students. For this purpose those interested in industry were an exploitable source; and the very nature of modern industry made the reorientation of the faculties intellectually respectable and socially justifiable.

The institutes and the transformation of University science

A strong implication of a report in 1909 by the Société de l'enseignement supérieur was that what France really needed was a system of *technische Hochschulen*. French technical schools – the Polytechnique, the Centrale, Mines, and Ponts et chaussées – received a small homogenized body of students and gave them an exclusively scientific education followed by an encyclopedic technical education dealing with a large number of different industries. More emphasis was given to lectures than to laboratories. What did the French admire in the *technische Hochschulen*? First, they were open to all who had the minimum preparation needed to follow the courses; technical education was not restricted to a minority. Diffusion of learning, development of individual initiative, and personality formation were favored in the German system. Second, the German schools permitted specialization according to the divisions of industry itself; so the young engineer was at home in the factory from the day he arrived. Third, the German schools, like the American, devoted most of their time to practical work and gave much more importance to laboratory time than to attendance at lectures. The similarity between the laboratory and the German factory gave the student a vital preparation for his role in industry. This was the technical educational system which, allied with German industry, explained the dominance in France of German dyes, chemical products, lighting fixtures, and porcelain; it further explained why many French metallurgical plants were constructed by Germans and why German refractory materials were used to construct the ovens of Parisian gas plants. Models of the *technische Hochschule* did exist in France: the industrial institutes of the University of Nancy, and the Ecole municipale de physique et de chimie industrielles and the Ecole supérieure d'électricité, both in Paris. This type of technical education had strong roots in many other French universities; so what was needed was a perfecting of existing institutes and a concerted governmental effort to introduce industrial science courses into universities where they did not exist. But pure science and original research should not be allowed to suffer.[65] The Société de l'enseignement supérieur had given its pedagogical blessing to the trend that had been stimulating the growth of the institutes for a generation.

The growth of high and low technology provided an opportunity for a genuine regionalization of part of the university system through the exploitation of applied science related to local economic activities. Given the prevailing view of science and technology as universal activities untainted by national or regional factors, this development was not without a certain irony. When Raulin established courses in industrial chemistry at Lyon, he tailored them to existing industries in the region, with the aim of showing industry how to increase profits.[66] Implicit in the new trend was a revolt against the old model of successful teaching – interesting lectures followed by the applause of auditors charmed by Isocratic rhetoric. This was now regarded as sterile; instead, what counted was the preparation of students capable of carrying into local industries the habits of order, accuracy, and rigor that had become the basis of the best industrial practice.

It would be naïve to insist on a procrustean division between "pure" and "applied" research. There are, of course, some clear-cut cases, but most research falls into both categories, directly or indirectly, immediately or in the future. Few scientists anywhere spend most of their time plumbing the mysteries of the universe, however intellectually exhilarating that may be. More important, much of the "pure" research that did take place was made possible by the existence of large, well-equipped faculties of science that could never have developed had they depended solely on funds from Paris and on the limited laboratory facilities appropriate to the training of lycée and faculty personnel. Uniting Apollonian knowledge with Vulcan skill, the institutes offered good research opportunities that could not otherwise have existed. Sabatier attracted to Toulouse students who could go on from the technical program in chemical engineering to research for the *doctorat d'université* – a degree just as useful for industry as the *doctorat d'état*, whose more complex requirements were necessary only for a University career.[67] The best institutes also attracted a teaching staff of high quality. Nancy and Grenoble had notable success in winning prizes. The year 1911–12 produced a great vintage in Nancy: Victor Grignard shared the Nobel prize in chemistry; Nicklès won the Jospeh Labbé prize and Lucien Cuénot the Cuvier prize, both awarded by the Academy of Sciences; and André Wahl won the academy's Berthelot prize.[68] On two occasions Haller won the Jecker prize: in 1887 with Arnaud and again, in his own right, in 1897. Sabatier of Toulouse won the Lacaze prize in 1897, the Jecker prize in 1905, and shared the Nobel prize with Grignard in 1912. Clearly there was a fruitful interaction between fundamental research and industrial research. It is true that "a utilitarian bent characterizes Sabatier in his methodological approach to chemical theory," but it would be misleading to ignore the theory in order to emphasize the pragmatism in any consideration of his science.[69] The same can be said of many other scientists in the institutes, especially in Nancy, Grenoble, Toulouse, Lille, and Paris. The interaction of academic science and industrial science can be seen also in the careers of leading *normalien* scientists: Paul Janet, Coryphaeus of industrial electricity at Grenoble from 1886 to 1894, and then head of the Ecole nationale supérieure d'électricité in Paris, is probably the best known. Charles Friedel, "an incomparable master of the laboratory," known for his work in atomic theory and chemical nomenclature, was the driving force behind the creation of the Institut de chimie appliquée in Paris.

It may be that the ease with which the faculties of science spawned institutes of applied science was a direct consequence of the strength of the French experimental tradition, especially in chemistry and physics. Pierre Duhem explicitly justified the interest of scientists in practical questions with the historical argument that rational mechanics was born out of the need to understand and perfect mechanisms, especially in dealing with clocks, naval architecture, hydraulics, acoustics, artillery, and the resistance of materials. The desire to improve industry was intimately connected with progress in theoretical and experimental science. Victor Regnault's revision of the experimental laws of heat, especially those related to the study of the steam engine, led to the possibility of important improvements in heat engines and was at the same time a "true revolution" in experimental physics.[70] If the new direction of the faculties had a historical foundation in French scientific practice, it did not really represent the departure from tradition suggested by alarmists like Emile Picard. The hoary distinction between science and technology may be totally unenlightening theoretically because it obscures an interaction that was actually characteristic of the development of science in nineteenth-century France, and which was recognized by the nation's leading theoretical scientists.[71]

Duhem knew that industrialists generally welcomed contact with researchers in pure science. In 1899 he was president of the jury examining Albert Turpain's doctoral thesis in the physical sciences, an experimental study of electromagnetic waves. Turpain's research had not been done in what officialdom called one of the "vast laboratories" of the Bordeaux faculty of sciences, but in a factory outbuilding lent to him by a sympathetic industrialist. For Duhem this was a comment on the deplorable conditions that prevailed in the faculty. The basic problem was that the administrators of the nation's universities were intent on creating monuments to science that were totally unsuited for experimental work. The snobs of science were erecting richly ornamented stone palaces for science all over the country, whereas science required temporary, adaptable buildings, with laboratories capable of modification in response to changing patterns of research. It was all too easy for a physicist to find himself in the absurd position of being given yesterday's laboratory for tomorrow's work. Shifting research interests in physics were dictating totally new priorities in laboratory construction to an unprecedented degree in 1900. There was an evolution from concern with "freedom from the disturbing influence of iron upon magnetic measurements" to "concern for freedom from vibration," coupled with a growing "attention ... to the facilities for the production and distribution of electric current." In such circumstances, "Factories, not monuments!" was not a bad slogan.[72]

The changing character of the universities raised an obvious question: how were they to be distinguished from the *grandes écoles*?[73] Jules Gosselet recognized that the answer depended on the programs that were possible, the students available, and the official recognition given to their work. The hold of the *grandes écoles* on the most desirable posts, and their consequent ability to attract large numbers of the ablest students, were recurring sources of contention. *Polytechniciens* monopolized the highest positions in the state engineering corps, most notably in Mines and Ponts et chaussées, and graduates of the Institut agronomique were no less dominant in the Ecole forestière. There was also the informal monopoly existing through courses and diplomas. Professorships of agriculture and directorships of agricultural stations were the preserve of the Institut agronomique. Gosselet's solution to this problem was the only possible one: Examinations and degrees granted by the institutes should be given equality with degrees in pure science; an institute's courses should carry credit for the licence (a basic requirement for teachers), and should qualify for the same dispensation from military service as was granted to licence students. But such appeals had little effect. The barriers of snobbery and monopoly did not fall, and industry remained the chief outlet for students in the applied sciences.

Offering various independent specialties, the applied science programs appealed to several categories of students. It was certainly not a drawback that one category was composed of rejects of the *grandes écoles*, men recognized as capable, hard-working students, whose minds, unsuited for the canned programs of the *écoles*, were probably better adapted to the more flexible University programs. It was clear, too, that some engineering graduates from the Ecole centrale or the Ecole industrielle de Lille, for example, could profit from a year in institutes that specialized in electricity or in brewing. Even a graduate of the Institut agronomique could take the special courses in regional agriculture available in the faculties. There was also the potential for some democratization of higher education through the admission of able students who had not completed secondary education but who could profit from this type of higher education. The special courses in applied science did not demand the general knowledge given in secondary schools. In industrial chemistry, for example, it was not necessary to have taken extra courses in mathematics or even the baccalaureate. Even though applied mechanics and industrial physics required more mathematics, they could still be open to those who had not completed the full, regular secondary course.

If the institutes of applied science in the faculties came off second best in the competition for students, and if their graduates were not ranked first in the job market, they did possess a major advantage over the *grandes écoles* in the flexibility of their programs. As professions became more specialized the uniform programs of the *grandes écoles* became less appropriate; the Centrale, for example, had to divide its program into five sections, and the Ecole industrielle de Lille divided its program into three sections. But these divisions masked a basic uniformity. The *grandes écoles* still concentrated on giving a broad preparation for the engineering profession, whereas the university graduates were trained in a particular area of specialization. A student who did applied electricity at a faculty was a better *electrical* engineer than was a graduate of the Centrale, although he lacked the general qualifications for other professions. In a time of increasing specialization this was an advantage, provided the "mafiosi" of the *grandes écoles* let the free market prevail. The complaints made about *polytechnicien* incompetence in business and industry, as in the state-owned telephone system, was an expression of resentment against their privileges if not evidence of their incompetence.

Gosselet, like Duhem later, emphasized that the programs of universities could not imitate the *grandes écoles* because of a more fundamental reason. The spirit of scientific research inspired the University and gave it its distinctive character. The curricula of the *grandes écoles*, by contrast, were too encyclopedic to give the time and freedom required by effective research. In contrast with the universal nature of the *école* programs, universities could accommodate regional interests. The initiative of professors in creating applied science courses was possible because these courses were free of the bureaucratic regulations governing the programs dreamed up in Paris. It looked like the best of all possible worlds.

In the beginning these institutes, created under the authority of the faculties, admitted students on the basis of reduced or no examinations and eliminated the weak ones in the courses of study over the years. This was dangerously close to the education described in an Oxford broadsheet about 1914: "He gets degrees in making jam / At Liverpool and Birmingham." Gabriel Hanotaux summed up the system of institutes: "Young students, limited periods of study, and practical knowledge that is immediately usable."[74] Many people's ideal modern American university. Here at last, it was vainly imagined, was an antidote to the poison of the *concours* of the *grandes écoles*, whose rejects, brains burned out by two or three years of arid coaching in mathematics, formed a vast reservoir of young psychological wrecks, youth full of hatred and envy, resenting their wasted talents. Between 1,200 and 1,800 students, who had done little between their seventeenth and eighteenth years except prepare for entrance examinations, were rejected by the top schools. One of the advantages women derived from the discrimination practiced against them was that they escaped this atrocity. But the institutes were not content to serve humbly as technical universities or higher technical schools. By the 1920s the institutes wanted to adorn themselves with some of the glamor of the *grandes écoles* through selective recruitment. Visions of the Centrale, of Mines, of the Polytechnique danced in the heads of ambitious professors and administrators.

The new students recruited by the institutes gave general satisfaction when they went to work. Not all were engineer-heroes of the type admired by the public in the fanciful novels of Georges Ohnet. A model of the sentimental engineer's journey: graduation from the Ecole centrale, work hard, save a failing business, marry the owner's grateful daughter, who had been on the verge of suicide. One play had a female engineer, a graduate of Grenoble's institute, fall in love with hydroelectric power. Socialist realism may be bourgeois sentimentality. Top state and big company jobs, including the railroads, were the preserve of Mines, Ponts et chaussées, Tabac, and the military schools. But new institutions had appeared to provide technical personnel for new enterprises. University institutes and new types of economic activity were part of the same social development. Areas in which the *grandes écoles* were weak became the outlet for the institutes; electricity, hydraulics, metallurgy, electro-metallurgy, chemistry applied to papermaking and to brewing provided some of the most striking new opportunities. So tempting were these opportunities that the *grandes écoles* loosened their academic straitjackets to develop enough specialization to permit their graduates to compete with the institutes in the new industries. What the *grandes écoles* lacked in industrial expertise they more than made up for in general culture, social connections, and snob appeal – an unbeatable combination in any society.

This whole development was the result of the accelerating growth of technical complexity in industry. Louis Barbillion pointed out that at the turn of the century a few weeks of courses on alternate current and a few experiments would give a student a marketable skill in industrial electricity. This was the end of the period of "heroic inventors," many of whom became rich and supported the installation of a great many expensive and useless machines in faculties and schools. In the 1920s specialization in electricity was not a simple matter. To play with 150,000 or 220,000 volts required

174

some scientific baggage. General scientific education became important as a prelude to training in mechanics, which took up about 80 percent of the electrotechnician's training. So the institutes were forced to evolve, developing their specializations far beyond any potential competition from the *grandes écoles*, condemned by their very grandeur to remain rather general. Some territories might have been surrendered by the institutes, but others were firmly subjugated and a few more conquered.

The striking development of the institutes of applied science was inseparable from the great industrial progress that France made from the 1880s to 1914 – first in the metallurgical and chemical industries, and later in the hydroelectric industry and in the production of automobiles and aircraft. From 1890 to 1904 France was the world's leading producer of automobiles; from 1905 to 1914, a period when the United States conquered first place, France remained the biggest producer in Europe. The French educational system used to take much of the blame for a supposedly poor economic performance. But we now have to reassess the role of that system in the light of Maurice Lévy-Leboyer's thesis of the intelligent exploitation of the growth potential by big business over a long time-span, chiefly in the period of rapid growth after 1905, which was conditioned by a widespread entrepreneurial ability in the business community. "The success of the 1900–30 period can be taken as a result of the employers' earlier interest in technical education."[75] As even the harshest of critics, Henry Le Chatelier, recognized, the institutes at Nancy, Grenoble, and Lille had remarkable success in joining science to industry, and they showed excellent organizational skills. Although technical education had been integrated into the faculties, the faculties themselves, especially at Nancy and Grenoble, had been largely transformed into technical schools comparable with the *technische Hochschulen*. The transformation was less evident at Lyon and even Lille, and in Paris the Institut de chimie appliquée was an adjunct to the faculty. Each system had its virtues and vices. Yet the achievement and its results were still noteworthy, a tribute to one of the Third Republic's great creations, the provincial university system.

The *gadzarts* (graduates of the *écoles d'arts et métiers*) enjoyed a dominant position in metallurgy, mechanical construction, and especially in the railways. Shinn thinks that "one of the major obstacles to industrialization . . . and to the development of a modern industrial elite" in France has been the scorn of the powerful Corps de l'état for industrial technology. So new institutions had to be created to train personnel for modern industries. It was the growth after 1880 of relatively high-tech sectors of industry – rubber, chemicals, synthetics, steel, aluminum, and electricity – that created the need for a new type of engineer. From 1883 to 1909 new engineering schools were established: the municipal Ecole (supérieure) de physique et de chimie industrielles, the private Ecole (supérieure) d'électricité, and the faculty institutes. From 1910 to 1914 these schools produced 120 engineers annually; after 1918 the annual number of graduates rose to 200. By 1895 the Corps d'ingénieurs de l'état and the Polytechnique, worried over the possible threat to their monopolies from the competition of the new schools, opposed their development. But the schools fought back vigorously with the help of their graduates. This fruitful squabble was over by the interwar period, when it was clear that there were more than enough jobs in France for all engineering groups.[76]

There is some truth in the claim, made by Gosselet, that the institutionalization of the teaching of the applied sciences in the faculties began a revolution in the educational system by breaking down the dyadic structure that formally reserved pure science for the faculties and applied science for the *grandes écoles*. In developing institutes of applied science and technology, the universities believed that they were fulfilling important social and political functions. An institution like the Institut Brenier in Grenoble could strengthen its ties with the local community by serving regional economic interests and by promoting public health measures. On the political level the democratic implications of an open-door policy for anyone who wanted to learn the methods and discoveries of science in order to enter the scientific and industrial professions seemed entirely in accord with the ideology of the Third Republic. In more limited terms, we can say that the institutes developed because the *grandes écoles* did not fulfill the needs of industry for trained personnel, especially at the lower levels, and because they did not produce or teach the specialized scientific and technological knowledge essential to a dynamic industrial state. These were the functions in which the faculties of science excelled.

The success of the institutes in the decade before the Great War made them the target of harsh criticism, much of it unreasonable, some of it justified. Criticism ranged from the absurd – the institutes had not enabled France to match Germany's industrial growth – to the plausible – applied science was gobbling up the resources of pure science, a view that assumed that the money going into applied science would go into the empty coffers of pure science if there were less applied science. The history of the provincial faculties provides no evidence that this would have been the case. Most of the other criticism of the institutes centered on the paucity of creative scientific research useful for industry; indeed, the institutes seemed to reflect industry. A move to separate the institutes from the faculties, clearly evident by 1909, peaked with Senator Joseph Goy's bill of 1916 to create faculties of technology. The bill failed, fading into a general postwar campaign to improve the conditions of basic science.[77]

It is tempting to argue that the greatest failure of the institutes was in not placing a primary emphasis on scientific research and persisting in teaching standard science courses and in analyzing manufacturing processes. Such a passive role precluded any effort to penetrate French industry with the fructifying tool of industrial research science. But even condemnations and pessimistic studies show the existence of good basic scientific research in the faculties of science that were famous for their institutes. And French industry did have a considerable amount of industrial research being done in its own research departments. In 1925 Henry Le Chatelier fulminated once more his brutal indictments of French industry and agriculture for their failure to keep up with science, or, in the case of small industry, total neglect of it. His book *Science et industrie* points out that there should have been a hundred or more research laboratories in French industry but that there were only a small number: for mining, the laboratory of the Comité des Houillères; for metallurgy, the laboratories of Montluçon, of Creusot, of Saint-Chamond, and of Imphy; for mechanical construction, Dion de Bouton; for chemical products, Saint-Gobain, Alais et la Carmargue, the Société d'électrochimie, Lumière (Lyon), Michelin, the Compagnie parisienne du Gaz; for cement, Teil and Boulogne-sur-Mer. Le Chatelier could have added the refrigeration industry to his catalog; by 1910 it was much more developed in England and Germany, although Louis Marchis's lectures promoting it were very popular in Paris. Horrible technological lags certainly existed in France. The most notorious was in the telephone system – or lack of it – a joke for engineers and a source of exasperation for clients, the most archaic and inefficient system in Europe. In an acerbic analysis written after a fire destroyed the Hôtel des téléphones in Paris, the physicist Albert Turpain, who taught a course in industrial electricity at Poitiers, blamed all of the problems on the incompetent Polytechnique graduates running the system. Even if Le Chatelier was right about the Taylorian efficiency of their learning process, permitting the school's students to learn in two years what it took six years to master in the faculties, whatever science the polytechnicians knew had little to do with modern communications.[78] In a recent study of 34 companies between 1880 and 1940, Terry Shinn concluded that less than a quarter of them developed industrial research departments before well into the twentieth century. The Société des produits chimiques, Air liquide, the Société française radio-électrique, the Société d'électrochimie, and Héroult did have remarkable success in integrating industrial research and the development of new products. Not much changed until the 1930s, when companies like Pechiney, Saint-Gobain, and the Cie générale d'électricité organized their own industrial research rather than continue depending chiefly on outside work and patents, however profitable that had been in the past. It was also in the 1930s that the government finally concocted a new scientific policy for basic and applied research.[79]

By the 1930s the technical institutes of the science faculties had fallen from their earlier high status to become a burden on the faculties. Jean Delsarte went so far as to classify them as quasi-useless because of difficulties of recruiting students, even mediocre ones.[80] Since most of these students had not gone through the licence program, Delsarte did not even consider them to be real students. This condescension was not merely a matter of the snobbery of a mathematician but of a mathematician who was a professor in the faculty of sciences at Nancy, one of the most successful of the technical institutes. This failure of the institutes may be considered part of a more general problem of the inferiority of technological research in France. Indeed, Gustave Ribaud, physicist, expert on high temperature and its uses, professor at the Sorbonne, simply referred to the abandonment of technical research in France.[81]

Making the situation more alarming for the French was the fact that they detected no such decline in Germany, where the number and quality of publications showed the strength and prestige of technical research in numerous laboratories of applied research in the technical institutes for the production of specialized engineers and in a well-endowed institute of applied technical *research*. The specialized laboratories of the Kaiser Wilhelm institutes, such as the Kaiser-Wilhelm-Institut für Eisenforschung, complemented the more general laboratories. Ribaud thought that the only institution in France providing its students with *research* laboratory facilities beyond ordinary requirements was the Ecole de physique et chimie industrielles, then under the stimulating directorship of Langevin. French institutes, even that at Grenoble, had not kept up with the research capabilities developed in German institutions. In Paris the Ecole des mines did not even have the physical facilities or material means to train the graduates of the Polytechnique who wanted to go in the direction of technical research. One of the important aims of the new Service de la recherche scientifique would be to recommend ways of adapting the leading French institutes to their role of pioneers in technical research. Some institutions, like the institute of applied chemistry in Paris, known for its applied as well as pure research, could play an important part in a general movement of coordinating the research efforts of all the institutes to avoid duplication and to profit from the advantages of regional or even national specialization. After the Second World War the organizations conceived in the late 1930s, the CNRSA, followed by the CNRS, would be able to implement many of the plans and programs of the prewar critics. The results were eventually even better than predicted.

5

SCIENCE IN AGRICULTURE
AN INCREASING ROLE IN THE NEW LAND
OF PLENTY

In France Cybèle has more worshippers than Christ.
                                    – François Mauriac

Higher agricultural education

A perusal of nineteenth-century scientific literature, especially the journals, soon reveals the large role that science was capable of playing in agriculture and the degree to which farmers and all sorts of politicians were coming to rely on scientists to tell them how to increase production. The patron saint of French agricultural education is Mathieu de Dombasle, who opened near Nancy the first serious agricultural school in 1824. This noble failure encouraged the engineers Polonceau and Auguste Bella to found in 1828, with royal support, the more scientifically oriented Institution royale agronomique de Grignon, near Paris. In 1848 the Second Republic created a national organization of agricultural education: seventy departmental farm-schools to produce good farm workers; regional schools of agriculture; and, capstone of the system, the Institut national agronomique de Versailles. In line with policies in most western countries, the Third Republic promoted scientific agriculture on a scale unprecedented in French history. An autonomous Ministry of Agriculture emerged in 1881 from the clutches of the ministries in which it had been held captive for most of the century. Nineteen different ministers of agriculture served the forty-two governments that ruled France from 1881 to 1914. Given the composition of the Chamber of Deputies, it is not surprising that the ministry fell victim to a near monopoly by lawyers and doctors. The republic reestablished in Paris the Institut national agronomique, which had been suppressed in 1852; the law of 1875 maintained the existence of the *fermes-écoles*, although the number dropped to five by 1926, authorized the creation of *écoles pratiques de l'agriculture*, and made improvements in practical and scientific instruction by establishing laboratories and agricultural stations and putting agricultural teachers in each department. The Ecole nationale d'horticulture was established at Versailles in 1873. Perhaps as some sort of compensation for losing its faculty of letters to Lille, Douai gained the Ecole nationale des industries agricoles in 1892. Some *écoles nationales de laiterie et de fromagerie* also appeared. It was a period of birth for a series of new organisms.

Eugen Weber's *Peasants into Frenchmen* celebrates the requiem of nineteenth-century religion on an altar erected to chemical fertilizer, which replaced the *ersatz* fertility rites and manure of the old days.[1] The new authorities were the curés of science, with a few high priests, and their churches were agricultural laboratories and schools. In 1908 graduates of the national schools were given hieratic status with the title of *ingénieur-agronome*. In 1912 Grignon started processing women in a program of *enseignement ménager agricole*, although the production of "grignonnaises" ended in 1923, when an Ecole nationale d'agriculture pour jeunes filles was created at Rennes to supply France with "maîtresses ménagères." A national structure of scientific agriculture, with all sorts of support systems and a few spinoffs for the lower classes, was firmly rooted in France by 1914.

Some of France's leading scientists were active members of agricultural organizations; many more showed up at meetings organized under an agricultural umbrella. Marcelin Berthelot, professor at the Collège de France, a perpetual secretary of the Academy of Sciences, was also director of the Station de chimie végétale de Meudon. Dehérain was a member of the Academy of Sciences. And Pasteur was one of the leading researchers in matters of vital interest to agriculture. When the Société nationale d'encouragement à l'agriculture held its general assembly of 1881, the president of the republic, Jules Grévy, was on hand to tell the meeting that politicians believed that how to increase agricultural production was "la grande question." Only science could show the way to a big increase, was the smug, if grim, opinion of L. Grandeau, head of the Station agronomique de l'Est and dean of the faculty of sciences in Nancy. Agricultural stations were the agents of this program because they were the "natural intermediary between science and the practical." France spent only about 150,000 francs on these stations, whereas the United States spent the equivalent of more than 3 million francs. Grandeau calculated that if the budget were raised 200,000 francs, the investment would be a potential gold mine. A modest permanent increase of one hectolitre of wheat for the average French yield in 1880 would produce 6,850,000 hectolitres, worth twenty-five francs a hectolitre, for an extra 1.71 million francs. This would mean a return of a thousand to one on the money spent for the agricultural stations.[2] Science could pay, and scientists were not reluctant to encourage speculation.

The strong interest in the use of science in agriculture existed not only in departments with faculties of science. The departmental agricultural laboratory of Nîmes, approved by the Conseil général du Gard in 1884, was organized by a departmental commission in concert with the prefectoral administration. Subsidies came from the state, the city of Nîmes, the Société d'agriculture du Gard, the Comité central du phylloxéra, the Comice agricole du Vigan, and the Comités phylloxériques du Vigan et d'Uzès. Phylloxera did more for agricultural science than a gaggle of savants would have persuaded the various levels of government to do in a generation. The aim of the laboratory, like that of the agricultural station, was to improve agriculture in the Gard by spreading a knowledge of chemistry, "the base of all agricultural progress," by informing the farmer exactly what type of soil he had, and the fertilizers necessary for maximum yield. Laboratory analysis could also solve one of the farmer's expensive problems: There was at least as much fraud in the selling of the new fertilizers as of the old religion; only testing by a laboratory technician could identify the purveyors of worthless fertilizer. Water could be also analyzed, an important matter in heavily irrigated areas, where the farmer often had daily questions. The most frequently adulterated commercial product since antiquity may be wine, but in the nineteenth century it became possible to be reasonably accurate about the worst kinds of fraud, some of which were made possible by science; the laboratories and stations could give useful advice on vinification. Scientists did not always win in law. In a set of famous lawsuits on wine fraud the experiments of Pasteur, Balard, and Wurtz were used as evidence, but a court in Montpellier rejected Pasteur's conclusions in one case.[3] In the latter part of the nineteenth century the organization of agricultural syndicates coincided with the passing of a law against fraud in fertilizer. With the help of testing offices in places convenient for the farmer to reach quickly, it was possible to avoid the numerous, long-established crooks in the fertilizer business. Prices were low for laboratory work: three francs for each sample tested in Nîmes. These institutions also gave farmers advice on what action to take in practical matters like the purchase of fertilizers and the treatment of wine diseases. This long process of the diffusion of practical science in the countryside was one of the most remarkable achievements of the century. The process was fraught with many serious long-range questions with which we are now wrestling, but the happy beneficiary of science in the 1880s was as ignorant of them as the scientist was.

Although the structure of agricultural education reveals a surprising complexity, there was necessarily only a small number of institutions doing high-level research. Among the few at the top were certainly the Ecole de Grignon, which by 1910 had a full range of laboratories in areas from botany to zootechny and ran three agricultural stations. Its brilliant period was from 1870 to 1910, with professors of world-wide reputation: Dehérain and Maquenne in chemistry, Stanislas Meunier in geology, and

André Sanson in zootechny are some of the better-known names, one or two of whom are still known today to a few historians of science.

The Queuille law of 1920 gave to agricultural schools funds amounting to a rakeoff of 1 percent from the state's pari-mutuel profits. Grignon benefited enormously, especially in expanding the Centre national d'expérimentation (station expérimentale de grande culture). Grignon's professors were prolific publishers, especially of tomes in the *Annales agronomiques*. There were entrance and graduating examinations for the two and one-half years of study leading to a diploma for a good student and a *certificat d'étude* for a weaker one. Courses existed in general and agricultural chemistry, rural economy, legislation, rural engineering, botany, technology, silviculture, animal husbandry, meteorology, geology, accounting, hygiene, entomology, and arboriculture. Most of its graduates went into various types of agricultural exploitations or into administration, but a few became agronomists, professors, and teachers. Many of the leaders in French agriculture were graduates of Grignon.[4]

When the Institut national agronomique was resurrected in 1876, it too was located in Paris, first in the Conservatoire national des arts et métiers, and, after attracting 140 students in 1883, at two successive, bigger locations near the University of Paris. The old institute had been out in Versailles, too far for its illustrious professors to travel, especially when they had good jobs in the heart of Paris. The law of 1848 stipulated that the institute would be the Ecole normale of agriculture, but it tried to be an *école polytechnique*. Many of its students went from two years there to more specialized schools or studies: Ecole des eaux et forêts, Ecole supérieure du génie rural, Ecole des Haras, Institut supérieur d'agronomie coloniale, and many special training sections and agricultural laboratories of ministries. After the 1870s its graduates got the best jobs in agriculture. For the practical support of its mission of giving a high scientific education oriented toward agricultural production, the Institut had a large model farm at Noisy-le-Roi, just outside Paris.[5]

The agricultural programs of these two essentially Parisian institutions were complemented by two famous chairs at the Conservatoire, flatteringly referred to sometimes as the Collège de France of industrial science. Boussingault, who occupied the chair of agricultural chemistry from 1836 to 1873, was followed by the Schloesings, father and son. The chair of agriculture (including, after 1922, agricultural production in its relations with industry) was occupied by L. Moll, Lecouteux, Grandeau, Blaringhem, and F. Heim de Balsac. Moll set the level in production of scholarship by scribbling thirteen volumes on everything in agronomy. What was said in the courses at the Conservatoire had to be said with wit and elegance because the lectures were one of the serious amusements of property owners who passed the winter in Paris and "la belle saison" on their estates. A good deal of research was published on colonial and tropical products, including cotton, sugar, rubber, cocoa, and coffee. Two journals were creatures of the Conservatoire: *Riz et riziculture* and *Coton et culture cottonnière*. Agriculture was an important area of research in one of France's *grandes écoles*.[6]

Agricultural research was of high quality when it was done either close to or in a scientific institution. Like the Institut agronomique, the schools at Montpellier and Rennes were initially both victims of ignorance of the rule that the quest for bucolic novelty is unlikely to succeed if it is more than walking distance from a faculty of science or a *grande école*. The origins of the Ecole nationale d'agriculture de Rennes went back to 1841, when the Institut agricole de l'Ouest was organized at Grand-Jouan (Loire-Inférieure). It was elevated to the status of Ecole régionale in 1849, on a level with Grignon and La Saulsaie, near Montpellier. Not able to attract enough good students, it was moved to Rennes in 1895. Rennes was chosen over Nantes because Rennes had a university, with five faculties, even if the sciences were rather weak there. Cooperation between state, department, and city ensured a speedly and fiscally sound move. Beginning in 1907 the school published the chief works of professors in its own *Annales*. In 1908 it began to grant the degree of *ingénieur-agricole*, which by 1921 entitled its holder to take courses for the licence in the faculty of sciences. Its nine chairholders directed their scrutiny to all aspects of the region's cereal and fodder crops. A special effort was made to use films in teaching. Nearly a quarter of the graduates went into teaching and administration. The Ecole de Rennes could take credit for extensive and happy changes in the rural economy of Brittany.[7]

Like Grignon and Rennes, the Ecole nationale d'agriculture de Montpellier was the result of private initiative. Founded in 1842, it became a regional school in 1848, moving to a national classification as its area of student recruitment grew from four departments (Ain, Loire, Rhône, and Saône-et-Loire) to thirty-one. Yet the school languished. A cure was suggested by the Société centrale d'agriculture de l'Hérault: Move the school to Montpellier, where it could find both students and science in abundance because of the town's active university. It cost the department and the city nearly 350,000 francs to set up a new school in 1871–2. Viticulture was the most famous specialty of the school; in a sense, oenology was created in Montpellier by Bouffard, Foëx, Viala, Ravaz, Maillot, and Lambert. Viala's research trip to the United States was critical in the selection of the proper vines for grafting to French vines without going through the agonies of trial and error during the phylloxera epidemic. The number of students rose from

13 in 1874 to 176 in 1900. Nearly 70 percent of its graduates went into agriculture. Montpellier's claim to fame was its scientific role in reconstituting French vineyards in the late nineteenth century. The school became first-rate, with its greatest strength in science relevant to producing regional products, because it was able to draw on the scientific resources of the university.[8]

A great many pressures resulted in the introduction of courses related to agriculture into the faculties of sciences as early as the Second Empire. Local societies passed motions and forwarded them to the minister of agriculture, who in turn suggested the introduction of courses into higher education and the lycées. Napoleon III used his own funds to promote the movement starting the introduction of courses in agricultural chemistry into higher education. True, up to 1870 agricultural education was not very impressive, but before 1860 there had been little agricultural education at all in France. Most of the credit should go to Dumas. As minister of agriculture and commerce (1850), he made it possible for these courses to be given in towns where they would succeed. Malagutti at Rennes, Isidore Pierre at Caen, and Girardin at Rouen gave public courses in rural economy. Dumas gave grants, the towns provided the places for the courses, the research expenses, and the money for publication of the results. The professors themselves did quite well academically as a result of this program, even to the extent of getting encouragement from the Institute. Soon Baudrimont in Bordeaux, Ladrey in Dijon, Filhiol in Toulouse, and Bonet in the Jura created courses suited to the special needs of local production. The value of these courses greatly increased after the commercial treaty with England in 1860, which also fitted into the Emperor's plans for agricultural reform. As Dumas said, it was not cotton and coal that France sent to England but rather vegetables, fruit, and wine – all products of French agriculture. Agricultural expositions showed a great improvement in products; regional competition and the introduction of science into agriculture played a large role in that improvement.[9]

Several universities devoted a considerable amount of attention to agriculture, offering courses of regional interest and benefit, with a few faculties of sciences actually developing an *institut agricole*. Nancy pioneered in organizing a comprehensive program within the Institut agricole de Nancy, founded in 1901. Regional interests predominated in the institute's five sections of agriculture, milk production, the economy, colonies, and forestry. The two or possibly three years of study were open to qualified candidates, but there was no formal requirement like the baccalaureate. Many agronomists thought that the program at Nancy was too theoretical. Viewed from Toulouse, it was, and therein lay its strength. Competing within the faculty of sciences and the region for scarce funds, the institute was financially weak; Nancy's money went into industrially related institutes.

The most successful *institut agricole*, founded at Toulouse in 1909, was in large part the achievement of the dean of the faculty of sciences, Paul Sabatier, who had the generous financial support of the departmental councils and agricultural societies of the Sud-Ouest. The city of Toulouse gave the 100,000 francs needed to buy the land. Another 400,000 francs got the institute built. The polyculture (cereals, wine, maize, fruit, cattle) of the vast and largely agricultural Sud-Ouest received a substantial infusion of science into its agriculture from the dynamic institute. Admission requirements for students were reasonably high: a baccalaureate or suitable diploma, although the mathematics program in the institute was kept to a bare minimum during the three years of study. Science was taught by professors from the faculty of sciences. The general aim of the institute was not to produce bureaucrats but young people who would run their own enterprises. This made the application of what they had learned more likely than if they worked for someone else. In 1925 the institute became an Institut d'université, with the same autonomy over its budget as a faculty had. There were 115 students, many of whom were foreign, studying in the institute during 1925–6. A few of the graduates went on to do theses for the university doctorate. The institute of Toulouse was one of the most important agricultural research operations in French industry.[10]

Many of the provincial faculties were keenly aware of the need for research related to agriculture. This was true of Lille from the beginning; and when Catholic universities were established in Lille and Angers in 1875, agricultural research was not neglected. In 1902 the University of Dijon at long last instituted a *brevet d'oenologie* and a higher diploma of oenological studies. Candidates had to register in the faculty of sciences and follow one semester of courses for the *brevet* and two semesters for the diploma. In the same year Rennes decided to change its *maîtrise de conférences* in applied botany into a chair of botany applied to agriculture. Out of the total salary of 6,000 francs for the professor, 1,250 francs was a subsidy voted for a period of thirty years by the municipal council of Rennes. Poitiers decided in 1905 to introduce a diploma in agricultural studies which gave the recipient the title of agricultural chemist of the University of Poitiers; no university degree was required for admission to the one-year course. Also in 1905 Besançon abolished its chair of industrial and agricultural chemistry, a relic of a nonspecialized age, to replace it with a *maîtrise de conférences* in applied chemistry and a chair of agricultural botany. The general council of Doubs and the city of Besançon agreed to pay 2,000 francs for each of the positions. Institute programs concentrating on agriculture could not compare in achievement with those devoted to industry. Although by no means stagnant, French agriculture was far less innovative in its structure and methods than industry, and it is not surprising that the institutes connected with growth industries were the most successful in both research and increases in numbers of students.

The voice of modernity: Louis Grandeau

The French seem particularly addicted to diagnosing France's quite ordinary problems as diseases, with the diagnosis usually done by a doctor whose cure works only when the government pays for the treatment. Louis Grandeau (1834–1911), one of the great preachers in the little chapel believing that the cure for "le mal agricole" lay in the commercialization of French agriculture, argued incessantly for an increase in yield through intensive fertilization and use of genetically selected seeds. His knowledge of foreign agriculture convinced him that the trouble with France was that its farmers did not know that discoveries of the laboratories are of vital importance to a healthy agriculture. It was therefore the duty of the state to introduce agronomy to the bourgeoisie and the peasantry, who seemed, particularly about mid-century, to lack confidence in commercial fertilizers and unwilling to risk capital in a systematic improvement of their land. It was the scientist's job to remind the state of its duty, which it could only fulfill by endowing the scientist with the means of research and providing him with an efficient and fruitful means of dissemination of the teachings of agronomy. In a report to parliament on the agricultural section of the universal exposition of 1889, Grandeau reminded politicians of the justification for spending a few millions to improve French agriculture: Its capital was 100 billion francs, its gross production was worth 13 to 14 billion, farmers paid annually half a billion in income tax, and the state collected 73 million in duties on cereals. According to the enthusiastic pleas of the scientific beneficiaries of spending, increased funds for agriculture could only swell the coffers of the Ministry of Finance.

While Grandeau was an assistant in chemistry at the Ecole secondaire de médecine in Nancy a memoir he wrote attracted the attention of Henri Sainte-Claire Deville and got him into Deville's laboratory at the Ecole normale. After ten years in the intermittent company of Deville, Bernard, and stars of lesser magnitude, Grandeau returned to Nancy's faculty of sciences in 1868 as a *chargé de cours* in chemistry and agricultural physiology. While in Paris Grandeau had acquired doctorates in science and in medicine as well as the title of first-class pharmacist; this not sufficing to exhaust his enormous energies, he also taught at the Association philotechnique, which gave workers free education. A founder of *Le Temps* in 1860, he remained a contributor and editor of the agricultural section for over forty years. The course in agricultural chemistry attracted "un public d'élite" from the agricultural sector. Grandeau had spent some time looking at farming in England and in Germany. This experience, combined with his excellent laboratory, made him popular with proprietors and cultivators of the region. By 1870 his program was advertised under the rubric of Station agronomique de l'Est (Laboratoire de chimie), although Brongniart, the inspector, noted in his report that the program was outside the regular programs of the faculty. Then began a steady stream of publications both on his specialized research and on general agricultural issues.[11]

In 1860 there were only three serious European establishments doing scientific research in agriculture. In England there was the organization of J. Bennet Lawes and Henry Gilbert at Rothamsted, founded thirteen years after the French one, and the establishment founded by a prince of Saxony, Crusius de Sahlis, in Moeckern, directed by E. Wolff. The French organization was the private farm of J. B. Boussingault at Bechelbronn in Alsace. Typical of many German agricultural stations, that of Moeckern and its branches gave farmers information on improving soils, on chemical fertilizers, and on different agricultural products. Under the domination of Liebig's theories the institute of Moeckern aimed more at disseminating existing knowledge than creating new.

Boussingault can easily be compared in stature with Grandeau as one of France's leading agronomists, although the earlier period in which he worked made it impossible for him to have the same influence on agriculture as Grandeau, whose basis of influence and power lay in his wide institutional connections and his influence in publishing and journalism. After a period as an officer in Bolivar's army, Boussingault became a professor of chemistry in the faculty of sciences in Lyon. A graduate of the Ecole des mines (Saint-Etienne), he recorded valuable observations during his short military career. In 1820 Boussingault went to Paris as professor of agricultural chemistry in the Conservatoire des arts et métiers, where he stayed for the rest of his life, although he went into politics during the Second Republic. A period as a moderate republican deputy for the Bas-Rhin and then election to the Conseil d'état did not lead him to continue in politics after Louis Napoleon's coup d'état. He was the rare example of a scientist turned politician who returns to creative scientific research, although some nineteenth-century French scientists seemed to have done quite well in the curious game of combining research and real politics.[12]

Like Grandeau's, Boussingault's agronomy was firmly anchored in mainline science: In 1821 Boussingault began a lifelong collaboration with the *Annales de physique et de chimie*, of which he became one of the chief editors. "Boussingault's experiments on nitrogen fixation from 1834 to 1854 and his work on nitrification from 1855 to 1876 brought the study of plant nitrogen essentially to the threshold of its modern microbiological formulation."[13] When he set out to bring the scientific gospel to French farmers, Grandeau certainly had an influential precursor in Boussingault.

Grandeau's Station agronomique de l'Est was also a private organization; minor subsidies from the Ministry of Agriculture supplemented the budget. In spite of the limitations stemming from scanty resources, the station became a model for other departments and for foreign countries: Italy (1870), Belgium (1872), Spain (1876), and Sweden (1877) were the best known. Blaringhem became particularly interested in the Swedish station of Svålof, founded in 1886 with funds from big property owners; famous for its research in the selection of seeds, it became responsible for distributing seeds in most of Sweden and gained a reputation abroad for top-quality plants at high prices.[14] At least in Sweden Grandeau's dream of better agriculture through seed genetics was a reality. If only France could follow the way of its own prophet! In 1881 the Ministry of Agriculture made Grandeau the "Inspecteur général des stations agronomiques et des laboratoires agricoles," a title valuable in his gaining some control over the use of applied chemistry on culture and breeding. Nancy's well-known school of forestry gave him an appointment which culminated in his collaboration with Fliche and Henry in a series of works in forestry, especially on pine and chestnut trees. Grandeau's activities in Nancy embraced a wide range of theoretical and practical research. His applied work reached a culmination perhaps in 1882 with the founding near Nancy of the Ecole pratique d'agriculture Mathieu de Dombasle. In the Château de Tromblaine he created model vegetations for the comparative and analytical study of the physical and chemical effects of different soils on plant growth. At the height of his career at Nancy, Grandeau also became dean of the faculty of sciences and stayed at this not excessively burdensome post between 1878 and 1889. In his twenty years at Nancy Grandeau gave French agricultural research considerable international stature.

A polemicist of considerable power, Grandeau carried on his campaign for the spread of science in agriculture in a wide range of publications. His research and his arguments were showered on the readers of *Le Temps*, the *Bulletin de la Société d'économie sociale*, and the *Annales de la science agronomique française et étrangère*, of which he was a founder in 1884, and the *Journal d'agriculture pratique*, which he edited after 1893. Among his long list of publications a few survived the rest: His *Études agronomiques*, four volumes of articles from *Le Temps*, appeared first in 1886–9, expanded to seven volumes, and enjoyed many reprintings and a sixth edition in 1892; his *Traité d'analyse des matières agricoles* of 1877, part of the *Bibliothèque des stations agronomiques*, expanded to two volumes in a third edition of 1897; and no work was more popular than that on fertilizer.[15] Translated into German in 1884 and into Italian in 1888, the *Traité* became an important basis for governmental regulations. Grandeau also translated two works of Friedrich Wöhler: *Eléments de chimie inorganique et organique* (1858) and the *Traité pratique d'analyse chimique* (1865).[16] His pen rarely rested.

In an age not yet bloated with academic meetings, Grandeau encouraged gatherings of specialists. His talents as an organizer of congresses and expositions became clear in his assembling European agronomists in the *Congrès agricole libre de Nancy* in 1869. In 1881 and 1891 he took the lead in organizing the international congresses of the heads of agricultural stations and laboratories. In the international exhibitions of 1878, 1889, and 1900, Grandeau was responsible for writing reports to the Ministère du commerce, de l'industrie, des postes et des télégraphes. For the exhibition of 1900 Grandeau orchestrated a collection of four volumes of documents, paintings, and photographs on agriculture in France and its colonies. As a result of this type of activity he was inevitably condemned to collecting and publishing statistics on world agriculture.[17] But Grandeau enjoyed showing the progress of world agricultural production, for it was, after all, irrefutable evidence that he was right in his prescriptions for increasing yields. Condorcet, not Malthus, was a suitable hero for this age of ebullience and optimism.

Because of his close relations with the Ministry of Agriculture, his increasing collaboration with *Le Temps*, and a set of experiments he was carrying out on the scientific feeding of horses for the *Compagnie générale des voitures*, Grandeau left his vast educational enterprise in Nancy to go to Paris. Thus the Station agronomique de l'Est ended up in Paris. The *Jardin d'essais du parc des princes* replaced the Ecole Mathieu de Dombasle as an experimental station, and the *Annales* of the station became *the* national organ with the title *Annales de la science agronomique française et étrangère*. Nancy's loss was France's gain.

Six months after his arrival in Paris, a new attack of the pedagogical virus led Grandeau to take over for the ailing Lecouteux at the Conservatoire. 1889 was a year of revolution in the teaching of French agronomic doctrines. Lecouteux's cure for "le mal agricole" was different from Grandeau's. Lecouteux's experimental base was a big estate in the Sologne, which he had slowly turned into an agronomic showcase. But this "prudent and enlightened practitioner" and "remarkable economist" believed that

190

while French farmers were improving their property the state should protect them by tariffs from the competition of lower-priced foreign products.[18] In the enemy economic camp, Grandeau argued for the speedy transformation of agriculture through the use of commercial fertilizer, scientific breeding and feeding of cattle, and the general use of scientific research in all areas of animal and plant production. The result would be, and Grandeau never tired of proclaiming it, a price low enough for French agricultural products to compete on the world market, then being assaulted by cheaper food from the United States, Argentina, Australia, Denmark, and New Zealand. Protectionism, already established for French industrial goods, could only retard the improvement and commercialization of agriculture. While Grandeau was preaching free trade from Lecouteux's old protectionist pulpit, the Third Republican politicians were busy emulating the rest of the Western world in imposing tariffs and fighting tariff wars. In 1892 Jules Méline (1828–1925), minister of agriculture, succeeded in getting parliament to end the "anarchy of the millionaires" (free trade) with a tariff of 5 to 20 percent on agricultural products. By 1910 the average duty was up to 11 percent. Lecouteux, speaking with the tongue of republican protectionism, had retired on the eve of the triumph of his doctrine; Grandeau, a voice of Second Empire free trade, became a political tinkling cymbal at the height of his scientific influence in French agriculture.

A final irony: In 1893 Grandeau replaced Lecouteux as head of the *Journal d'agriculture pratique*, founded in 1837, last bastion of intellectual resistance to the slow penetration of scientific agriculture in the countryside. Leconteux would have been pleased by a drop of 13 percent in imports in 1877–86 and again in 1887–96, and over 31 percent in the next decade. Grandeau could take grim satisfaction in the price exacted for this help to the French farmer: Over a twenty-year period the value of food exports dropped 27 percent. True, this industrial country had the good fortune to remain 90 percent self-sufficient in food production, but at a heavy price, for its wheat was produced at twice the cost of that of other leading wheat growers.[19] No doubt while starving during the dreadful turnip winter of 1916–17 the Germans wished that they had been as backward as the French. At least French farmers were able to forget their problems during a period of artificial prosperity (1895–1914), which the perverse peasantry found more acceptable than the great agricultural depression of the generation before. In a fundamental sense, of course, Grandeau was right, as seen in the blithe way we historians now babble about the transformation of the peasant world in the second half of the nineteenth and in the twentieth centuries. And although migrations, newspapers, unions, and schools were as important as the spread of machinery and the increase in yields, we should forget neither

191

the scientific and technological component of the transformation nor the tireless agronomist-scientists like Grandeau.

On the research periphery: botany in Paris and in Poitiers

Botany was one of the major participants in the general nineteenth-century process of the transformation of several areas of science from an epistemo-logical foundation based on classification, description, and analysis to one consecrated to experiment. In France "Gaston Bonnier was one of the botanists responsible for the transformation."[20] The facilities required for experimentation in plant biology, especially with its applications to agriculture, could hardly have been established in the Latin Quarter – or anywhere in Paris, notorious among cities for its lack of green space. Liard's new Sorbonne, like the old one, did not have facilities in botany commensurate with Bonnier's ambitions for his subject and himself. In 1890 a sandy clearing of three and a half hectares in the woods of Fontainebleau, three hundred meters from the railroad station, provided enough arable land to support large-scale comparative studies in plant physiology. After a very modest beginning, this new laboratory of the faculty of sciences expanded considerably with funds supplied by the University of Paris.[21]

Botanists had to be rescued from the shame of being practitioners of a low-status science, or, what was worse, being the subject of yawns and jibes. Bonnier summed up the situation with a definition from the littérateur Alphonse Karr: A botanist is a gentleman who picks the most beautiful flowers in the woods, flattens them between sheets of paper, where they are deformed and discolored, and then he insults them in Latin. It was not yet possible for botany to achieve status through mathematics, but experiment was quite adequate to the task. First it was necessary to remove the ludicrous stigma associated with the name itself. One should not take too lightly Renan's remark that in France whoever provokes laughter wins an argument. Bonnier himself told the story of a woman who asked her companion as they were walking in front of the building with its plaque advertising the Laboratoire de biologie végétale what such a subject was. She was told that Bonnier had used the term as a secret synonym for botany because botanists were the subject of ridicule in novels and vaudeville plays. Bonnier said the fellow was correct in his answer.[22] At least this made a good story at the inauguration of the new Station de biologie végétale de Mauroc (Poitiers) in 1912. And even in science it is a sign of elevated status to laugh at oneself, provided the laughter is not loud enough for the public to hear.

Bonnier's institution was a success. In addition to French students, some doing research for their doctoral degrees, Danes, Swedes, and Russians came for stays in the unique laboratory for botanical research at Fontainebleau. Foreigners trained in French research methods and work habits would, after seeing how science was cultivated and honored in France, return to their own countries knowing that the French scientist was serious, methodical, clear, and full of new ideas. Thus the foreign student was perhaps the best auxiliary of French intellectual expansion in the world. An instrument of *la mission civilisatrice*, he would serve the French propagandistic cause in arousing a love of France. In an exhuberant moment Bonnier summed up this improbable task: "partout ils font aimer la France." A connection was established with secondary schools and higher scientific education for professional-track students when the laboratory became the chief source of plants and flowers for courses in botany in lycées and in the PCN courses in the faculty of sciences. Students came from the Collège de Fontainebleau and the *écoles normales* in order to see experiments that could not be done in their institutions. Yet the chief function of the experimental station that developed at Fontainebleau was the production of high-quality scientific research worthy of publication in the proceedings of the Academy of Sciences, the bulletin of the French association for the advancement of science, and especially in the journal funded by Bonnier, the *Revue générale de botanique*. The publication of a journal in botany – the name didn't matter, for it would be read only by friends, i.e., specialists – marked a high point for both the subject and the laboratory as well as for Bonnier's ego. It was a very personal venture and not, as was usually the case for similar journals, connected with scientific institutions.

Bonnier never tired of preaching the gospel of experimental research to captive but receptive audiences. Billed as the principal speaker – "Gaston Bonnier de l'Institut" – for the opening festivities at Mauroc, he made it clear that the station had not been built for the purpose of housing herbariums, or collections, or the inevitable botanical garden, but for research on plant life. By life Bonnier meant its exclusively physical and chemical manifestations, although "its essence is still surrounded by impenetrable mysteries," which science could happily ignore. For research it would need instruments, the means of examining the interior structure of plants. "The study of the functions of living beings demands knowledge of their most intimate structure." In Poitiers, where research of any sort had hardly been a priority since the sixteenth century, millions were not available for facilities. Bonnier comforted the Poitevins: Fine research could be done with a minimum of equipment, for good workers are better than good tools. No doubt this is true, but both are better than one. Contrary to the beliefs of some local journalists, the purpose of the Mauroc station was not primarily horticultural, but to attack the highest scientific problems of the day. Researchers would add their mite to the vast edifice of biological knowledge, then so fascinating for "human intelligence." After all, the station was not only for the Poitevins. Being only the second such laboratory in France, it would cater to its share of foreigners, who would not necessarily be intrigued by the stunted trees of the region. If the situation of plant biology were ever to approach the high research status of France's eighteen laboratories of marine biology, it would have to concern itself with important issues. The need for Poitiers to rise above purely local matters took on a special importance for Bonnier. He knew the danger of the temptation for provincial researchers to sink into the slough of services to local industry and agriculture while forgetting about the general discipline.

It was the obvious popularity of botany in Poitiers that was the threat to the research orientation so ardently desired by the new botanists. In 1911 news of the Institut de biologie végétale was bruited abroad in a way that would have made Bonnier shudder if he had read it. The institute would be a station open for the observation of plants during "la belle saison" and vacations. Students needed no diplomas to enroll, just the wish to do botany, which was in great favor in Poitou, with its botanical society of more than 600 members, among whom were some well-known mycologists. But the nightmare of having hundreds of enthusiastic amateurs pursue their hobbies was never a danger to be taken literally. It was the problem of ensuring the triumph of a research ideology in the new institution that presented a challenge.

In 1912 the Station de biologie végétale de Mauroc of the University of Poitiers was formally opened in a festival of the pomp and self-congratulatory discourse so loved by the academics of the Third Republic. The greatness of their achievement allows us to forgive them their foibles. Four kilometers from town, Mauroc comprised thirty hectares of land, twenty-three of which were wooded, all taken from the church as a result of the law of separation (1905) and given to the University in 1910. Extensive repairs were done to dilapidated buildings, and the laboratory was installed for an initial expense of 30,000 francs. Things were done on a much smaller scale, also much more cheaply, in Poitiers than in industrial cities like Lille, where the Institut de physique alone cost a million francs. The entire faculty of sciences in Poitiers was opened in 1893 for 100,000 francs. Getting even the paltry 30,000 for Mauroc was a considerable accomplishment in Poitiers.

Back in 1898 Garbe, dean of the faculty, had written to the Conseil général de la Vienne asking for talks on the possible establishment of an agricultural program. After waiting over ten years for a reply, the University decided to establish the program itself. Perhaps the University could wake up the department and eventually "stimulate a spirit of initiative" that was certainly still sleeping in 1912. Some people thought that Poitiers had not changed much since Colbert had put his curse on the town for ignoring his ambitious plans: "Ce qui rend la ville de Poitiers gueuse et misérable comme elle est, c'est la mollesse et la fainéantise de ses habitants." By means of the agricultural station Poitiers could now redeem itself by emulating the other provincial university towns that had established such a fruitful interaction between science and the regional economy. Garbe held out the promise of the cure of plant diseases, better selection of seeds, methodical improvement of existing plants, creation of new agricultural varieties adapted to the region, and the acclimatization of nonnative species. Because only science could solve the problems of the farmers, they should rally around the new flag. Although even Garbe did not want to be asked to predict the weather, it was only on the grounds that forecasts were not yet scientific enough.[23] Bonnier's theoretical science certainly found a solid empirical basis in practical botany in Poitiers.

Botany had fallen on hard times by the mid-1930s. Basically, the problem may have been its historical failure to develop into a powerful discipline on the model of physics and chemistry. So when the financial crunch of the Depression led to a penny-wise, pound-foolish policy that abolished educational and research positions, botany was one of the chief sacrificial lambs in the sciences. But in the Paris faculty of sciences botany never had more than two chairs, which it had for a brief period in the twentieth century. In 1934 both a chair and a position of *chef de travaux* were abolished at the Sorbonne. Worse, two chairs were abolished at the center of botanical power, the Muséum. Beginning with two of the twelve professors in 1793, botany had acquired six of the twenty chairs available in 1931. By 1934 the number was back to four out of nineteen, nearly the same as the situation in 1894, when it was four out of eighteen. An inevitable protest of the council of the Société botanique de France warned of the danger of cutbacks: Medicine, agriculture, and the colonial economy would suffer from a reduced research establishment in botany.[24] But botany had to bear bravely what the other sciences supported sulkingly.

The wines and pines of Bordeaux

The *Guide vert* of Michelin for the Côte de l'Atlantique ends its potted history of Poitiers with a section entitled "a sleep of four centuries," lasting until after the Second World War, when a rejuvenated population finally proved Colbert wrong in a fit of dynamism that made Poitiers the regional capital of Poitou-Charentes. The culmination could not be other than the establishment of a Michelin tire factory in 1972. Over two hundred kilometers southwest, the large port and industrializing commercial city of nineteenth-century Bordeaux was certainly a world apart from Poitiers. Yet it too lacked what the *Guide vert* calls dynamism, which it finally found – no, refound – after 1945. But in the late nineteenth and early twentieth centuries Bordeaux did witness the origin of the troika of science, technology, and business characteristic of many other French towns. Until after 1945, when a new industrial city developed with the new university complex of Pessac-Talence, the most striking interaction between science and business may be told as the story of the wines and pines of Bordeaux – mostly the pines of the Landes.

The faculties of science of Paris, Lille, Nancy, and Lyon, impregnated by the industrial germ, became heavy with schools and institutes of chemistry in the late nineteenth century. This was not a total surprise in France's leading industrial centers, but Bordeaux also gave birth to an Ecole de chimie appliquée à l'agriculture et à l'industrie. Agricultural chemistry was what tied the school to the agrarian Southwest. After 1855 Bordeaux had a course in agricultural chemistry, taught for over twenty years by Alexandre-Edouard Baudrimont (1806–80). Interested in musical theory, medicine, the Basque language, atomic theory, and educational theory, he also published on fertilizers, vine diseases, wine, and applied chemistry. After 1854 he had the heady title of Vérificateur en chef des engrais for the Southwest. While in Lille Pasteur had enjoyed a similar function for the Nord. With the increasing availability of chemical fertilizers came more subtle forms of fraud, perversions of science that could be best countered by use of its own analytical powers. This regulatory function was first bestowed on the chair of chemistry, then passed to the agricultural station. In 1905 the station began to function as the Laboratoire de la répression des fraudes for the testing of food and especially wine.[25] Ulysse Gayon (1845–1929) succeeded Baudrimont in 1880 as the sole professor of chemistry in the faculty. Head of the agricultural station, Gayon reorganized the course in agricultural chemistry, a standard item in the repertoire of the sciences at Bordeaux between 1854 and 1934, turning it into a cycle of two to three years covering recent progress in all agronomy.

Research was of regional importance but not without national significance: The theory of fertilizers and their application, the composition of agricultural products, vine diseases, vinification, the agricultural role of microbes, and the scientific feeding of animals were among the subjects investigated. Although it did not have an experimental vineyard, the station was able to test crops and vinification on parcels of land lent to it by friendly property owners.[26] Gayon held up Vienna as a model for Bordeaux: "la magnifique station chimico-physiologique expérimental de viticulture et de fruiticulture de Klosterneuburg," directed by Leonhard Roesler.[27] The laboratory gradually extended its service of analyses, which were free for certain governmental agents, to departments next to the Gironde. Thousands of samples of agricultural products, wine, soil, water, and fertilizer were subjected to scientific scrutiny. The prefect of the Gironde gave his approval of a reduced charge for agriculturally related businesses. Under Gayon's forty years of directorship the station acquired an international audience as foreigners, especially Americans, Russians, and Greeks, flocked to take the courses in oenology. Like several other powerful and distinguished provincial scientists, Gayon turned down a chance to move to the University of Paris. He declined an opportunity to succeed Duclaux in the chair of biological chemistry at the Sorbonne and as director of the Institut Pasteur. Part of his reward was getting E. Dubourg appointed to the first chair of physiological (later biological) chemistry in the Bordeaux faculty. It is not surprising that one of the key subjects treated by the new professor was the group of industries based on fermentation. In 1907 the associate director of the Station agronomique et oenologique, J. Laborde, published his *Cours d'oenologie*, the latest word on the subject presented "in the modern analytical and scientific spirit." A steady stream of later publications, especially by Ribéreau-Gayon and Peynaud, has kept Bordeaux in the front rank of oenological research.[28]

Although oenophiles may be revolted by the pasteurization of wines, it was regarded by the end of the nineteenth century as one of the basic applications of science to the preservation of wine as a "boisson hygiénique." Today much of the debate over the pasteurization of wine centers on Burgundy, where Louis Latour efficiently sanitizes his products at a flash temperature of 70 degrees centigrade. Between 1865 and 1872 Pasteur's experiments on the wines of Burgundy, the East, and the Midi showed that wine was not changed in flavor and could even be improved by pasteurization.[29] Later studies by A. Bouffard and by Müller-Thurgau demonstrated that heating wine to 60 degrees centigrade – generally 55 degrees was not hot enough – prevented changes induced by malign microbes (*tourne, pousse, amertume, graisse, piqûre, fermentation manitique*, etc.) and also prevented "la maladie de la casse" by fixing the coloring matter. Would pasteurization also work for the glorious *crus* of the Gironde? Burgundy and Bordeaux are for the French gastronomic mind the equivalent of yin and yang. All scientific doubt was eliminated by Gayon in a set of experiments carried out on forty-two samples of three groups of wines, one healthy white wine and two reds, one sick and one healthy. The experiment was also designed to allay the fears of suspicious gastronomes: A commission of the chief winetasters of the Gironde did four tastings at four different periods – three and one-half months, one year, two years, and six years after pasteurization. Some wines were heated to 60 degrees and some to fifty-five degrees, but the latter group showed evidence of incomplete sterilization. The quite elaborate results showed no appreciable differences in the tasting ratings of the heated red wines and their unheated counterparts. The wines did not lose the development that comes with aging. Heated white wines did even better than the reds in the competition. Among the wines tasted were a *premier cru* Médoc 1878, a *deuxième cru* Margaux 1874, a *premier cru* Pomerol 1870, and a *premier cru* Sauternes 1878. What was good for Burgundies was good for Bordeaux.[30]

Pasteur's panacea for the prevention of wine diseases, although tolerated by a generous law, did not gain immediate wide acceptance in spite of the long experiment in its favor concocted by U. Gayon. In more exuberant public moments the Bordeaux oenologists were quick to salute Pasteur as the creator of modern oenology – a harmless if misleading piece of rhetoric – but in a more serious mood they were not loath to emphasize the grave empirical limitations of the Pasteurian approach. Nearly a century later, Ribéreau-Gayon declared that Pasteur's method of work represented in a typical way an intermediary step between empiricism and the effort of scientific research. In explaining alcoholic fermentation and alteration in wine, Pasteur did not go beyond the explanatory level of vital phenomena. He did not understand the language of his contemporaries – Liebig, Bernard, Berthelot – or the fact that vital phenomena were a mask for the chemical phenomena that needed to be discovered. Pasteur did not have the means to know of the chemical transformations carried out in wine by bacteria, as, for example, in the destruction of malic acid. The problem is not the elimination of bacteria but avoiding their harmful action and favoring their useful intervention. A large part of the scientific work underlying the emergence of a viticulture of quality in Bordeaux was the work of the Station oenologique de Bordeaux. The significance of this achievement can only be appreciated within the geographical context of the difficulty of the recent achievement of quality *crus*. The vineyard was difficult to create and maintain in regions not especially favorable to the culture of the vine and the ripening of the grape because of low temperatures and excessive humidity. By the 1960s wine was worth about 40 percent of the value of the agricultural products of the Gironde, with an annual wine production worth about 500 million francs from the vine and another 1,000 million in commerce. Exports were worth 100 million. Thirty thousand *viticulteurs* in the Gironde were supporting about 5,000 employees in wine establishments. Success has been based on scientific research on the ideal marriage of types of grapes, terrain, growing, and vinification. Pasteur's work was rather remote from the "oenologie scientifique" that became the object of worship in Bordeaux.[31]

After phylloxera devastated the wine industry, the importance of satisfying the new demands of viticulture, coupled with the growth of overseas commerce, gave a fillip to the growth of chemical and other industries in the southwest. In 1887 industrial chemistry was introduced into the faculty; in 1891 a school of chemistry was organized under Gayon and Joannis as directors with help from Dubourg. In this Ecole de chimie a course of study lasted two years, including eight hours of general chemistry, ten hours of applied chemistry, sixty-four hours of laboratory work and exercises, and two hours of elementary physics taken in the second year. The *diplôme de chimiste* was given after two years of study, but another year or two were available if a student wanted to prepare for a specific industry, or to get certificates in general chemistry, industrial chemistry, or experimental physics in order to qualify for a licence. The high research competence of the professors gave the school a good reputation. The team of Dubourg, Gayon, Gossart, Vèzes, and Vigouroux trained laboratory personnel for faculties and chemists for governmental agencies in agriculture, customs, war, colonies, and railways, as well as for industries in agriculture, gas, and metallurgy. Science had made itself indispensable to successful industrial practices; its reward was that industrial support became vital for its material well-being.[32]

Old maps of France show a sort of desert – really a savannah of sand and marsh – in the southwest. The proverb claiming that the rare bird crossing this arid area carried grain on its back for nourishment during the trip was not an exaggeration. A few fever-crazed, often hungry, rude families lived with their few sheep and drank vinegared water in the *pays landais* until the dawn of the nineteenth century. The two enemies to be conquered were sand and water. Wind produced moving dunes of sand from the beaches. Sand swept nearly into Bordeaux. Nicolas-Théodore Brémontier solved this problem by the early nineteenth century, when he finally succeeded in growing the maritime pine on the dunes, which then formed a barrier against the corrosive elements. It was in the Second Empire that water was conquered. Jules-François Chambrelent, an engineer of Ponts et chaussées, drained the plateau of the Landes; stagnant lagoons disappeared, to be replaced by the prolific pines whose enticing general designation became "la forêt landaise."[33]

Of the million hectares of forest in existence by the twentieth century, about 516,000 hectares were in the Landes, 461,000 in the Gironde, and 100,000 in Lot-et-Garonne. Most of this great national resource was little exploited until well after World War I, except as a source of wood for mine props (especially for England), for general and industrial use, and for firewood.[34] Four million tons of charcoal also came from the area. Imperial Germany was quite successful in exploiting the forests that covered a quarter of the empire. As late as 1895 agriculture and forestry provided jobs for 35.8 percent of the German population. In the early twentieth century, forests provided direct employment for more people than the German chemical industry. An absence of industry in the French Southwest meant that there was little demand for the more sophisticated products developed from resin. Germany, England, and, above all, the United States dominated this market. The price of many wood products, determined in the London commercial market, could drop up to 75 percent: The price of spirits of turpentine dropped from 1,000 to 250 francs per 100 kilos in 1920. Those supporting the development of industry in the Landes held out as an incentive to all levels of government the prevention of the financial exploitation of French products by foreigners.

And science could lead the way. In this case the specific agent of *Wissenschaft* would later be the Institut du pin, headed by Georges Dupont, professor of industrial chemistry in the Bordeaux faculty of sciences. What was needed was a transformation of the nearly exclusive use of wood in mechanical industries to a mix of mechanical and chemical industries: paper mills, distillation of wood products, extraction of resin from dead wood, cellulose extraction, and the manufacture of a cornucopia of chemicals. The faculty of sciences pointed the way to the Bordeaux of today.

It might be thought curious that the faculty of sciences in Bordeaux did not develop in the nineteenth century an institute connected with industries based on the pine. No doubt part of the answer was the simple nature of the industry itself, not requiring much technology and even less science. Chemistry professors showed little interest in resins. Auguste Laurent, doing some work on pine resin when he was on the faculty in the early 1840s, discovered pimaric acid and prepared a number of derivatives. But Laurent's interest was theoretical, of little use to Landais industry at that time; and Laurent was even less interested in staying in Bordeaux than industry was in his discoveries. No one followed up on Laurent's use of resins as a basis for discovery. Baudrimont, a fixture in the faculty from 1848 to 1880, and his successor, Ulysse Gayon, were more interested in showing wine producers the profits and glories of applied chemistry.[35] This was a quite legitimate activity for a professor of chemistry in Bordeaux, which dominated the

Gironde; meanwhile, Mont-de-Marsan and the Landes had to wait for a bigger faculty before the pine could approach the vine in the research priorities of the faculty.

From the viewpoint of later chemists, injury was added to insult when a couple of physicists did some work of benefit to the wood industry. Aignan, a lycée professor and later an *inspecteur d'Académie*, spent some years doing polarimetric studies of spirits of turpentine. Gossart, professor of experimental physics in the faculty, collaborated with Captain Croizier in using Gossart's ingenious homeotropic method to detect the adulteration of turpentine. A call from the *conseiller général* of the Gironde to the chemists attending the Bordeaux meeting of the French association for the advance of science in 1872 found no response for a generation. The fault was not totally that of the chemists. The faculty of sciences was too poor to encourage the faculty to interact with industry. A couple of scientists made up the typical provincial faculty. The professor of chemistry was placed within the straitjacket of programs and examinations set in Paris; no local or regional issues could sully the intellectual austerities of a higher science. Without assistants the professor would have had little time for turpentine, even if limited state-controlled expenses for matériel had permitted him the secret luxury of extracurricular experiments.

By the beginning of the twentieth century, after the much-touted University reforms of the 1880s and especially 1895, which encouraged a distinctive provincial physiognomy to lure regional funding, it was still only barely possible to get a minor course on resins going at Bordeaux. In his annual report of 1899 the dean of the faculty of sciences, Georges Brunel, reported that Vèzes had been able to respond to the requests of departmental councils by giving a few lectures on resins. Vèzes was congratulated by the director of higher education, for Liard was pushing one of his favorite nostrums to improve French higher education: expansion on the basis of funds from regional interests being served by the universities. If the recipient proved ungrateful, the effort must not stop. Brunel noted that the research of Millardet and Gayon in viticulture had not been properly recognized by the wine interests or the Bordeaux region. So Brunel advocated that the faculty expand its research to plums and mushrooms, as well as pushing on to serve the Périgord and taking an interest in exploiting the riches of the Pyrenees.[36] The spirit of the faculty was willing even if the flesh of the capitalists was weak.

Stimulated by the requests of an industrialist in La Teste and the departmental general council of the Landes, Vèzes created his course on resins, chiefly distillation procedures. Certain issues troubling industrialists were studied in faculty laboratories. Once launched into the resin orbit,

Vèzes became serious enough to ask for a laboratory. The Landes should venerate the icon of Hofmann, patron saint of the German coal-tar–dye industry, which earned the equivalent of more than 100 million francs a year in exports of artificial dyes for the fatherland. Laboratory research on resinous products could make the fortune of the Landais region, something duck livers could never do no matter what the price of foie gras might be. Basic requirements were scores of chemists, lots of money, and well-equipped laboratories. A modest beginning would be in a corridor with two assistants and 3,000 francs' worth of equipment. The whole operation would cost 7,000 francs, but it would have to come from outside the faculty, which, although fiscally liberated by Paris, had funds only for the courses required by the state. Departmental and communal governments, chambers of commerce, industrialists, business firms, and proprietors would have to pay up.

A plea for support went out in the periodicals and newspapers of the Landes and the Gironde. Although it was late in the year to get a new item into the budgets of local governments, the sum of 4,500 francs was collected by November 1900. More funds were promised; so the goal of 7,000 francs seemed to be a modest one. The laboratory could begin.

The basic thrust of the laboratory was practical: Laboratoire de chimie appliquée à l'industrie des résines ("une sorte d'usine d'essais"). A steam generator was the first big piece of equipment bought. Regional roots for science inevitably meant extremely useful if hardy beginning plants. Typical of most of the provincial scientific research efforts serving local economies, the resin laboratory flourished. Subsidies were unusually numerous: The departmental councils of the Gironde and the Landes, numerous communes of these two departments, several chambers of commerce, the Société d'agriculture de la Gironde, cooperatives, industrialists, and propertied exploiters of the forests assured that the laboratory had an annual budget of between four and seven thousand francs. The first request for contributions got 1,000 francs from the departmental council of the Gironde, 500 francs each from the departmental council of the Landes and the chambers of commerce of Bordeaux and Mont-de-Marsan. Forty-two communes contributed from between two hundred to five francs each. With a contribution of seventy-five francs, industrialists were the group needing enlightenment on the profits to be derived from supporting science.

A modest operation with three or four supporting personnel – it began with one – the laboratory did enough research to fill 2,000 pages in seventy-five publications between 1900 and 1918.[37] In 1908 the ministries of agriculture and of commerce and industry made the laboratory the national center for the analysis of the resinous products that had to be tested under

202

the law for the repression of frauds (August 1, 1905).[38] Fraud and adulteration led to a great expansion of low-level laboratory work – analysis of various sorts requiring some science, an average brain, and much *Sitzfleisch* – but there was some spinoff value for research, as contrasted with analysis. And laboratories that looked like purely regional enterprises had national functions.

After the First World War a new emphasis was put on the need for methodical exploitation of "la forêt landaise." Industries based on pine resin cannot develop haphazardly, for they are generally chemical industries that can develop only in close collaboration with science, that is, organic chemistry in this case. There were two evident problems in Bordeaux – and they were not unique to this region. First, most industrialists thought of laboratories as tools for analysis even if they believed laboratories were good for industry. But the great need in French industry was for laboratories of research employing selected and specialized researchers motivated by bonuses for industrial discoveries. Second, most French industries were too small to support the expense of a research laboratory. It was therefore essential that they pool their resources in order to create a large research laboratory that would be the scientific tool of all the pine industries. Only science could make a national treasure highly profitable.[39]

Such a laboratory would have certain well-defined characteristics. It would be, above all, an industrial laboratory in constant close contact with the factories it advised. At the same time it had to be based on first-class science, engaging in theoretical research in organic chemistry with the potential of opening up new avenues of industrial application. The laboratory also needed an educational function: For the resin industry this was the training of specialized engineers with a profound knowledge of chemistry as well as the professional skill and initiative coming from the habits acquired in laboratory research. Such a laboratory could only be created by a wider organization devoted to the pine industry. This was the achievement of the Institut du pin.

The Institut du pin had five components: the laboratories of research and of analysis, a school of technical instruction, an office for documentation or information, and an industrial office. The research laboratory was divided into sections specializing in the harvest and treatment of resin, in theoretical studies of spirits of turpentine along with the improvement of industrial applications, and in wood and forest by-products including paper, products from distillation, and the manufacture of tannins and coloring materials. The laboratory of analysis exercised quality control over both the raw material used by industry and the manufactured products. To provide technical instruction an Ecole d'application took students from chemistry schools or the *grandes écoles* and gave them a year's intensive training in applied chemistry, industrial and commercial administration, silviculture, and practical and industrial work. A written piece was required on a subject relevant to industry. Of considerable help to the students in the Ecole d'application was the office of documentation, charged with amassing material on everything pertinent to the pine industry, including legal and commercial as well as scientific and industrial documents from French and foreign sources.

This ambitious program was only possible with the heavy support of regional wealth and political power. An organizing committee got off to a good start in 1921 with a substantial gift from François Dupouy, a big property owner of Pontenx-les-Forges in the Landes. Once the Institut du pin started in 1922 it was directed by a committee made up of property owners and industrialists of the Gironde and the Landes. The committee included members of the chambers of commerce of Bordeaux and of the Landes, a couple of engineers, a judge of the Tribunal de commerce de Bordeaux, the president of the general council of the Gironde, and three chemists of the faculty of sciences. Two widows (proprietors) were on the committee. The budget of the Institut du pin in 1922 was heavily dependent on its political connections: grants of 10,000 francs each from the departments of the Gironde and the Landes, with another 500 francs thrown in by Lot-et-Garonne. Contributions from industrialists, payments for analyses, an annual grant from agricultural research services, and a grant from the Caisse des recherches scientifiques made up the rest of the limited resources. But by 1922–3 the Institut du pin boasted a staff of seven, including, of course, the three professors of the faculty. Two of the seven were classified as *mademoiselles*. Although the group had made only a modest improvement in housing – from corridor to basement in the building of the faculty of sciences – the future looked quite a bit brighter than in 1901.

The organization of the Laboratoire des résines as the Institut du pin, an association based on the law of July 1, 1901, was done in large measure to find far more funding for research than was possible with the modest laboratory. By the 1930s the payoff was evident. The budget for 1931–2 was over 110,000 francs. Other types of growth had also helped the operation. Beginning in 1906 a modest growth in low-level laboratory activity had resulted from the new system-wide requirement that candidates for the *agrégation de physique* had to spend two semesters in a laboratory. In 1924 a new doctoral degree in engineering (*ingénieur-docteur*) at Bordeaux required candidates to do two years of laboratory research. This supplied new researchers on resins. Other degree seekers came in respectable numbers. In the ten years or so after 1923, the laboratory provided facilities for work done to earn seventeen *diplômes d'études supérieures*, seven theses for the engineering doctorate, and twelve state doctorates. This was a good record, in large part the achievement of the directeur technique de l'Institut du pin, Georges Dupont, who was also the dean of the faculty of sciences. The connection between the institute and the faculty was institutional, scientific, and personal. And, as Dupont never tired of chanting, the Institut du pin was, above all, an industrial body that had to work closely with industrialists.

This industrial connection gave rise to an unusual problem for a French public institution. When an industrialist supported research he did not want publication of potentially profitable results. Why should he pay for research benefiting his competitors? Discoveries had to be patented. Many scientists rejected this demand without taking the trouble to find out how the problem could be solved. One ostensible solution would have been to allow royalties to accrue to a scientific research organization like the Institut du pin. This source of funds should have appealed to scientists, but it never tempted them greatly, possibly because it was illegal under the law of July 1, 1901, which prohibited such organizations from engaging in commerce. A legal maneuver was necessary if funds were to be derived from patents. So in Bordeaux a commercial and industrial Société d'études et d'applications pour le progrès de l'industrie résinière was established to exploit any discoveries of the Institut du pin that could be patented. The company was certainly a modest, if unique, operation, with a capital of 100,000 francs. But it did subsidize the institute with most of the profits from several of thirty patents taken out in the twelve years after the company's founding.

This company could also undertake projects possibly outside the moral principles and certainly beyond the resources of a university. It took the initiative in founding Les papeteries de Gascogne, a large paper mill at Mimizan in the Landes, to turn into pulp all the wood left over from other operations. With the collaboration of the Administration des eaux et forêts, it organized recently acquired property of Eaux et forêts in Pierroton (Gironde), a laboratory annex especially equipped for the study of the maritime pine, its resinous secretion, and possible improvements in harvesting the resin. All quite amazing when one considers that all this was conceived in the not overly fertile womb of the old faculty of sciences of the University of Bordeaux, but perhaps it has usually been judged too harshly and according to rigorous Sorbonnard criteria for pure science.

205

High-level success, low-level failure

Agricultural research enjoyed high status in the Third Republic, just as the big agricultural schools do today, when their status has been raised in the research hierarchy by the growth of genetic research, the challenge of plant physiology, and the support of agribusiness. Pasteur, Duclaux, and the Berthelots were among leading scientists who did agricultural research and promoted it. The Collège de France had a Station de recherches at Meudon with laboratories of plant chemistry and of plant physiology. The Station de chimie végétale was attached to Marcelin Berthelot's chair of organic chemistry. Berthelot received an extraordinary grant of 4,000 francs in 1883 for the Meudon station; the annual subsidy was 2,000 francs. In 1907 Müntz (Institut agronomique) became head of the laboratory of plant chemistry and physicist Daniel Berthelot (Ecole supérieure de pharmacie), Marcelin's son, the head of plant physiology. The annual budget was 15,000 francs, which paid for two laboratory assistants, two daily employees, and supplies worth 7,600 francs.[40] Marcelin Berthelot's presidential address to the Société nationale d'agriculture de France in 1892 was a paean to scientific agriculture. "Les temps bénis de la vieille ignorance érigée en principe sont passés." The lead article in the September 1893 issue of the *Revue scientifique* was an eloquent piece by Dehérain, a member of the Institut de France, on farm manure.[41]

In agriculture there were surviving glimmers of the continuation of a minor tradition of research by the upper classes, especially if there was some ancestral hero in the annals of science. Paul Thenard did research on phosphates and phylloxera and invented ingenious machines on his big estate in Burgundy.[42] But in the nineteenth century the overwhelming part of agricultural research was done in schools and research institutions, including the traditional elite ones. Blaringhem started research for his doctorate in the botany laboratory of the Ecole normale under J. Costantin, who became professor at the Muséum, and L. Matruchot, adjunct professor at the Sorbonne. Bonnier invited Blaringhem to use his laboratory of plant biology at Fontainebleau, where there were facilities for the study of wild plants. It was Giard's notorious course on evolution that introduced Blaringhem to research problems deriving from mutation, the laws of sexual reproduction, and parthenogenesis. Blaringhem also did work in Giard's laboratory at Wimereux, then at Coxyde (University of Brussels), and in Hjalmar Nilsson's seed laboratory (Svålof) in Sweden. But without fields for experimentation Blaringhem could not have done his work. These fields were provided near Paris by the Société d'encouragement de la culture des orges de brasserie en France, which showed a keen interest in the possibility of the discovery of laws governing the improvement of varieties of barley.[43] There could exist in agricultural research a complex interaction of laboratories, both French and foreign, and of science and business.

In sad contrast to the existence of a high-quality agricultural research network in some elite schools, there were few educational opportunities for the peasantry. On the hustings the politicians may have described the Third Republic as a peasant's republic, but this was scarcely evident in agricultural education except as a policy of benign neglect. Gordon Wright thinks that "the agrarian legislation of the Third Republic was so meager and unimaginative that it hardly added up to a policy at all." There was a high tariff policy, of course, although the Méline tariff was a footnote to industrial protectionism. As in the Soviet Union, there was also gradual electrification of rural regions after 1918. State intervention plus electrification for private property equaled French capitalism. But Wright states bluntly that the Third Republic "failed miserably" in providing "facilities for the education of young farmers and the advanced training of agronomists and technicians." The first part of the statement is true. Less than 5 percent of the sons of peasants were given any agricultural instruction. An inevitable comparison with imperial Germany: The Germans supposedly had as many agricultural schools as the French had agricultural students.[44] Between 1876 and 1891, according to Michel Boulet, less than 4,000 students followed courses in the écoles pratiques, while 300,000 persons followed adult courses. In a country of three million farmers the system ground out about 800 students annually: 250 at the level of engineer, a number more or less adequate for France's needs at this level; 350 at the middle level, and 200 from the fermes-écoles. The last two categories made up a marginal group of professionally educated farmers. The basic problem in the supply of students was that agricultural education was left under the Ministry of Agriculture, which used it as a component of farm politics. Had the programs been in the Ministry of Education, they might have become part of the mass school system taking shape in the late nineteenth century, with the aim of producing the model educated worker and citizen.[45]

Agriculture as a competitive industry

French agriculture was not stagnant. And there was a considerable amount of trickling-down from research to agricultural practice, especially in times of crisis. High production costs and foreign competition led to low profits or losses. In 1889 France produced over fifteen hectoliters of wheat per hectare, compared to ten and a half for the United States, but French wheat was not a competitive export after 1880. France became a major grain importer during the periods of poor harvests after the 1840s, and especially after the 1870s. Phylloxera wiped out a third of France's vines, with a drop of 50 percent in wine production between 1873 and 1885. But reconstruction of the vineyards was all too successful by 1900, beginning a period of high production of low-quality wine in the Midi. An unexpected result of the conquest (not annihilation) of aphids by the agronomists was peasant rioting in 1907 over the collapse of wine prices. Largely because of bad harvests, the period from 1908 to 1913 was a good one for the wine industry. From a long-range viewpoint the period from the 1850s to the 1880s saw the "structural transformation of French agriculture." Growing cities and towns increased and changed the demand for food: Productivity in cereals increased by 50 percent. It was the railways that made it all possible, including agricultural specialization, enormously increased imports, and the distribution of goods. The bottom line is that wine and meat became fare for the common man, who was now liberated by an abundance of grain from the fear of famine that had been a feature of the psychology of the old economic regime.[46]

The four big French agricultural industries at the turn of the century were sugar refining and three fermentation industries – distilling, brewing, and winemaking. Agriculture still accounted for about 35 percent of France's gross domestic product, with industry accounting for roughly the same percentage; in Germany agriculture accounted for 30 percent and industry for 40 percent of the gross domestic product. The extraction of sugar from beets was the most important agricultural industry. In the bonanza year 1901–2 this industry produced over a million tons of sugar, worth about 300 million francs. World production of sugar from beets was about seven million tons, far exceeding the nearly five million tons coming from cane. The importance of sugar beet production in France went far beyond its use in food processing and in the chaptalization of wine: It was a source of increased wheat and beef production. The pulp left after the sugar was extracted from the beets was fed to cattle, whose increased consumption produced increases in animal size and in the amount of manure. Because sugar beets required heavy fertilization the land used was improved for the production of wheat in a system of three- or four-year rotation. An increase of 18 percent in wheat production in the thirty years before 1907 made France the world's third wheat grower after the United States and Russia. Sugar and alcohol put 500 million in taxes into the state budget, one-eighth of total receipts. It is not surprising that sugar beet growers became a powerful lobby in Third Republican politics.[47]

The sugar beet was also the most important source of alcohol for French distilleries: 1,200,000 hectolitres, as compared with 700,000 from molasses (sugar cane), 350,000 hectolitres from grains, and 200,000 hectolitres from wine and cider, although the last figure excludes the *bouilleurs de cru* – that is, those who both grow and distill their own products. Unlike the Germans, the French did not produce a potato cheap enough to send to the distilleries; so industrialists in sugar and in alcohol bid against one another for the sugar beet. This industrial competition gave much pleasure and some profit to the growers. Some factories were big: A giant in the Pas-de-Calais gobbled up 1.5 million kilograms of beets daily and regurgitated 1,000 hectoliters of alcohol. But most operations were too small to achieve the economies of scale loved by economists and, it seems, even many capitalists.

Brewing was an important industry in France, especially in the North and the East. True, Germans guzzled 117 liters of beer per capita yearly, but the French also drank 36 liters. Annual English per capita consumption of beer in 1913–14 was 27.5 standard gallons, not very different from a century before. The French were not more sober than the Germans or the English, for if we add wine and cider to the French figures, annual French per capita libation rises to the sodden total of 195 liters. With a billion francs in agricultural revenues, wine remained the biggest of French agricultural industries.

What was the importance of science and technology in these agro-industries? Two major, related arguments were frequently made. No one made them more strikingly than Emile Barbet, the president of the society of civil engineers in 1908. First of all, there was the eternal litany of the imperative need to catch up with the Germans, who might not have known where they were going but who were going there fast by science and technology. German superiority dictated the second argument, namely, that France had to invest heavily in scientific and technical education with a direct pipeline into industry and agriculture in order to have the slightest hope of being in the same competitive league as the second Reich.

In the sugar beet industry Germany produced more than twice as much as France and did it more efficiently in 350 factories handling a daily average of 550 tons of beets. The French scientist's perception of German success was heavily colored by the obviously greater role research played in the agricultural segment of the German economy than in the French case. *Wissenschaft* had become a tedious *deus ex machina* in French commentary on Germany. In the end, perhaps unity and order were considered equally important, for in place of French particularism in business, the German sugar beet industry accepted technical direction from the institute of sugar manufacturing in Berlin, which was heavily subsidized by the sugar manufacturers themselves. Professor Herzfeld's well-funded studies, done in the institute's laboratory and small experimental factory, were supported by a small army of chemists and students in technology. After spending four or five years in technical school, which was usually closely connected with industry, students went into factories that were really part of one system, a sort of extended industrial family. The German sugar market was international while the fermentation industries, especially distilling, filled the needs of an essentially national market. Yet there was a splendid institute of industrial fermentations, located opposite the sugar institute. The association of the producers of industrial alcohol and that of the brewers were generous in subsidies to the institute. Generally, big French breweries copied the techniques worked out with admirable Teutonic patience and detail in their neighbor's industry. Whatever the wisdom of their choice, the French preached, and often followed, a slavish imitation of the German model of interaction between science, technology, and industry. Use of the word *choice* is, however, in itself the exercise of a later option, the illusory bourgeois luxury of our sated age. The nineteenth-century mind saw only the inevitability of imitation, unless one were to choose the unthinkable path leading to tertiary power status for the nation. And the French have never been bush-league thinkers on the status of *la grande nation*.

Barbet also warned the wine industry that it would have to follow the path of industrialization, in accordance with the same principles followed in sugar plants and distilleries. Big-name oenologues were already approving large-scale production; all that remained was to convince the producers, who, laboring under the difficulties of small holdings, often had their grapes spoiled by black rot and mildew, and suffered from a deep ignorance of scientific vinification. Barbet thought that two new wineries being built in Burgundy and in Algeria would go a long way in spreading the technological gospel. The winery would be an intermediary between the producer and big business, bringing under the iron rule of economic law an activity that would be France's biggest agricultural industry. But waiting for a French Gallo may be waiting for Godot, although Duboeuf has arrived.[48]

The input of science and technology into the sugar industry showed the extent to which an industry could be dependent on scientific and technological schools for its personnel. Other agricultural industries would not differ substantially as they advanced into large-scale efficient production. A small army of specialists in various sorts of engineering was enthusiastically listed by Barbet as vital to ensure the health of the industry: the engineer-agronomist to study the culture of the sugar beet and its progressive enrichment; the chemical engineer, a true soubrette to the industry in all stages of production from growing – analysis of the soil, the fertilizer, and the beet – to the final production of the syrup or liquid sugar; the mechanical engineer, a key figure in an industry employing so many generators, motors, and varied equipment; the electrical engineer to take care of lighting, transportation, oven lifts, and other electrical equipment; and the architectural engineer, essential to an industry needing enormous buildings and all sorts of rail transport. It is not surprising that the president of the association of civil engineers would express his admiration for the sugar industry – "la plus belle de nos industries agricoles."

Whenever scientifically soluble problems faced French agriculture, the state was able to mobilize enough talent for solutions, even though they might take a long time. The presumed needs of French agriculture – mechanization, consolidation or enclosure, and large-scale commercialization – were social and political in nature, not scientific or technological. That was the work of the Fourth and Fifth republics, although strong beginnings were made in the 1930s. Peasants perversely invested in land, not techniques. Using fertilizer depended on having the money to buy it and on a certain amount of technical knowledge. The bilious Marxist Jules Guesde said that "the only way to make the peasants fecund is rape." But in place of this Stalinist agricultural technique the French art of slow seduction was preferable and more fertile. It was also the only feasible policy in a country where politics came to the peasants generations ago. In any case, a great deal of the economist's cherished innovation did take place in the nineteenth century on properties of more than forty hectares, it is true, but they made up about 47 percent of agricultural land if one includes large private forests.[49] The spirit of Grandeau finally triumphed by 1984, when France had an agricultural export surplus of over 25 billion francs.

L'Assassin dans votre verre? science for the consumer

Modern urban growth was dependent on large increases in world agricultural production. The sale of food in cities required regulation by new laws protecting the health of consumers. It was the development of rigorous scientific techniques of testing that made the process of regulation different from past efforts. But the creation and implementation of these rules required extensive and controversial governmental action, which sometimes seemed based on flimsy science and dubious technology, weakened versions of the two instruments that have changed the nature of food and drink since the middle of the nineteenth century. In France the tremendous growth of Paris and its suburbs, along with the extension of male suffrage, made it

211

indispensable for politicians to assure that food and drink were reasonably safe for human consumption. One of the ironies of the progress of science and technology was that it became more difficult for some of the more subtle adulterations to be detected by traditional methods. A counterattack by science itself became vital.

The scientific approach to public health appeared in early nineteenth-century France, with French leadership clearly emerging by the 1840s, especially in Paris, where *hygiène publique* blossomed into a science. Except in the diagnosis and treatment of communicable disease, however, the French had fallen behind in public health by the late nineteenth century. Compared to Berlin and London, Paris was full of dirt and disease, with a far higher death rate, especially for infants. French problems in public health were aggravated by the density of the population of Paris. The Department of the Seine, with a population about the same as that of New York (3,600,000), covered an area of under 48,000 hectares, compared with about 80,000 hectares for New York; worse, over 26,000 hectares of the Seine were densely populated.[50]

In 1791 mayors had been given the right to inspect food; in 1884 the municipal police were given the uncontested right to assure the whole-someness of victuals. In Paris this meant the Préfecture de police, not the mayor. Laws passed in 1851 and in 1855 provided the legal precedents, although an unfortunate and curious omission in 1851 left out liquids, one of the chief victims of frequent crude falsification and dangerous additions.[51] Perhaps it was assumed that every Frenchman could tell when a wine had been illegally altered; experience soon showed this to be an unwarranted assumption.

Shortly after dyes were synthesized from coal tar, manufacturers and producers tried to attract the dull eye of the consumer and to lull his jaded palate into accepting artificial rather than natural colors in food and drink. Enormous profits could be threatened by public fear or, more improbable, governmental prohibition. At the request of Badische- Anilin und Soda-Fabrik, two of Germany's leading scientists, Hofmann and Virchow, allayed the fears of the public and cleared the fiscal conscience of the chemical companies by certifying the innocuity of as many artificial colors as even the most imaginative aesthetic addict could want. Acting on the principle that what is not harmful should be legal, the government of the Reich passed the law of July 5, 1887 (effective May 1, 1888). A clear separation was made between dangerous matters like antimony, arsenic, lead and mercury, and aniline and azoic colors, which were shown by experiments to have no harmful effect on the human body.[52] Where the Germans rushed in, the

French were not slow to tread, but not without much anxiety and squabbling.

In the 1880s the Parquet de Lyon carried out a campaign over a number of years against the artificial coloration of wine. Science decided that the law against it was based on bad evidence. Experiments done by Cazeneuve, Lépine, and Arloing came to the generally reassuring conclusion that if some colors derived from coal tar were harmful, all those usually found in wines could be tolerated in high doses by man and animals without any risk of accident. Opponents of the use of artificial dyes could therefore be easily subdued by the *a fortiori* argument that the small doses in food and drink could not be harmful. In the cold light of toxicology and hygiene, these colors were absolutely comparable to cochineal and other natural colors, which were never viewed as harmful. Riche and Gautier were of the same opinion. In Switzerland Graebe and Vincent agreed with Hofmann and Virchow. The European scientific community told the public that it could continue to ingest its gaily colored bonbons with the assurance that what was good for the chemical industry was good for their tummies. Most people still act on the assumption that this is not yet shown to be false.

By the 1870s the debate over edible and potable poisons, concocted by scientists and pushed by industry for commercial convenience, eye-luring aesthetic appeal, and profit, was well under way in the scientific community and on its saner fringes. (The *Revue d'hygiène et de police sanitaire* began in 1879.) In 1877 Feltz and Ritter, professors in the faculty of medicine at Nancy, stirred up a lively quarrel between chemists when they published the alarming results of their experiments on the effects of fuchsine when ingested by animals. Publications the year before by Dr. Georges Bergeron and J. Clouët had attempted to show the innocuousness of pure fuchsine.[53] True, Wöhler and Friedrichs regarded aniline as harmless, but most chemists considered it and its derivatives to be toxic. An order issued by the police in Berlin in 1863 stipulated that all colors derived from aniline had to be examined chemically to ascertain the absence of arsenic products. The experiments by Feltz and Ritter on man and dog produced striking symptoms of poisoning. Reports were also common of people being poisoned by caramels and wine loaded with arsenical fuchsine. When Feltz and Ritter recommended that this type of coloring agent be outlawed, it was objected that their fuchsine was impure. Not so, the doctors protested, for they have got their poison pure from Dr. Willm, *chef de travaux* in Wurtz's laboratory. A repeat of the experiments produced the same results.

Although debates in the Académie des sciences showed that adding fuchsine to wine had disadvantages and deserved to be repressed as a fraud,

213

the practice continued. Poisoning from colors known to be dangerous and therefore declared illegal continued because it was commercially advantageous to use a color that was four times as effective as the fuchsines obtained by nonarsenical means, as in the case of the Coupier process, which used nitrobenzene, not arsenic. The experiments done by Feltz and Ritter were not accepted as reliable. Work done by Paul Cazeneuve, professor of toxicology at the faculty of medicine in Lyon, and by S. Arloing was widely accepted as showing that the ingestion of even high doses of pure fuchsine had no effect on animal organisms. It was not surprising that the Paris prefecture of police lifted the ban on acid fuchsine in confectioners' products on December 31, 1890.[54] The debate did not end, but the advocates of the use of artificial color had won.

The 1870s saw an important change in the regulation of food and drink in Paris. On November 2, 1876, Dumas, chemist, politician, a great power in the scientific establishment, acting in his capacity of secretary of the seventh municipal commission, asked for the establishment in the Préfecture de police of an office for testing wine samples for buyers who would be willing to pay a small fee to be certain that their wine was not artificially colored. The council voted on August 1, 1878, to establish a municipal laboratory of chemistry as an annex to the Service de dégustation. Beginning in October 1878 the laboratory immediately showed its usefulness by indicating the widespread use of wine made from raisins ("piquettes de raisins secs"), of glucose for hiding the dilution of wine with water, and of various coloring materials. The devastation of French vineyards by oïdium in the 1850s and by phylloxera in the 1870s had led to an enormous business in the fabrication of artificial wines. Consumers, who were probably less discriminating in the nineteenth century than now, did not care much about wine as long as it had enough alcohol in it to produce the illusions necessary for survival.

In December 1880 the laboratory was reorganized and opened to the public. Inspectors of victuals were upgraded to the title of *inspecteurs experts attachés au laboratoire* (March 1, 1881). Between 1879 and 1883 the budget grew from 14,000 to about 200,000 francs. Fifty-five employees carried out twenty-three types of investigations, including an analysis of the gases produced in graves, in order to satisfy the morbid curiosity of the Commission des cimitères concerning the mephitic vapors emanating from their charges. A total of 10,752 analyses were done in 1881 and 1882. Results were not encouraging for commerce: 2,707 items were classified good, 2,679 passable, but the rest were bad, although the actually harmful were 1,544, with 3,822 bad but not harmful.[55] In logical logomachy the laboratory technician is not greatly inferior to the scholastic theologian.

Howls of protest came from the commercial bourgeoisie and their minions when the founder and director of the laboratory released his first report in 1883. A report dealing with water, wine, beer, cider, vinegar, liqueurs, milk, butter, meat, coffee, tea, fruit, vegetables, etc. made it clear that the French often bought bad products and, worse, that they consumed them without much protest. A reply came from a distinguished member of the faculty of sciences in Paris, H. Pellet, whose attack on the report carried on the title-page a solemn warning that he was an "Officer d'académie, chimiste-conseil." Girard's report was dismissed with a heavy dose of professional scorn for a neophyte's efforts. Pellet found the report full of errors and falsehoods about the sugar and other industries. Methods were not rigorous enough; experiments were not done in duplicate, or at least this was what Pellet judged from the large number of tests done by a limited number of people. One of the great quarrels of the day was over the advisability of feeding draff (refuse of malt after brewing is completed) to cows. Learned attacks were mounted in the *Journal de l'agriculture* and the *Revue de la brasserie*.[56] At least Girard did not have to think about our experiments of feeding the poor creatures cardboard for roughage!

In the second report (1885) Girard, in replying to his numerous critics, justified at some length the work of the laboratory. A strong attack was launched on the vicious if ingenious practices of some businesses. Girard's diatribe gave no quarter to his enemies. "Commerce today is quick to change the discoveries of science into instruments of fraud. Falsification, which was formerly based on a few clumsy formulas, has become scientific – and we cannot be successful against it if we don't attack it with weapons equal to its own." Partially replying to Pellet's criticism of the way in which tests were done, Girard noted that some of the best employees of the old Service de dégustation, although incapable of detecting the new types of frauds invented by the wicked wizards of *Wissenschaft*, were kept on to "test" the wine before it went to the good fairies posing as laboratory chemists, who countered the evil of their opposite numbers in the employ of commerce. The two results were compared. One cannot help getting the feeling that this was a harmless humanitarian gesture that would disappear when the amateurs retired from the laboratory. It is doubtful that Girard had any more confidence in "dégustation" as a means of detecting fraud than Pellet could have had in this quaint method of repeating experiments to ensure their validity. Just to be on the safe side, however, Girard emphasized that the ultimate result of the work of the laboratory would be to strengthen French commerce, chiefly by modifying the state of things that had led Richard Cobden to make the statement that the French don't know how to sell because they are not clever enough to be constantly loyal and make sure that the merchandise delivered always conforms to that promised.[57] Everyone was for the triumph of French commercial interests, but the worshipers of science split into two camps on how laboratory science could contribute to this triumph and especially on what publicity should be given to the results of experiments done on French products.

The squabble over publishing the reports was probably carried on within governmental and bureaucratic circles as well as in the general community, but the most intense debate was between business and the municipal authorities. A short report prepared for the Paris chamber of commerce in 1883 admitted that the laboratory incontestably served the interests of Parisian consumers, but it insisted that the form of the monthly bulletins especially heaped discredit on wine, the national product par excellence. In 1881 the laboratory estimated that 56 percent of French wine was drinkable; at least this modified the cruel judgment of the London *Times* that 61 percent of France's wine production was not fit to drink. Another agrument for keeping the results from the general public! The report did not detail how this scandal could be stopped, even if it should.

Dissatisfaction with the functioning of the laboratory led to a high-level political discussion of its transfer from the jurisdiction of the police to the prefecture of the Seine. To encourage the transfer, J. Bruhat, a former chemist at the laboratory, wrote a hundred-page attack in 1884 on the dictatorial management and "administrative chemistry" of Girard. Bruhat's articles in *Le XIXe siècle* led it to engage in a lawsuit against the laboratory; to its great surprise, it lost. When "higher" and "lower" sciences clash, the law does not necessarily decide in favor of the "higher," especially when the fight is over the quality of wine. In 1890 Bruhat was still campaigning for the separation of the laboratory from the Service des experts de police sanitaire: Under the prefecture of the Seine, a permanent and paid scientific commission of five scientists and one legal counsel would meet every two weeks to supervise the laboratory. Bruhat supported the argument of Jeannon that the trouble with the management and situation of the laboratory came from the fact that it was under the police. The implication of this was presumably that it was in Girard's interest to invent fraud if he did not find any.[58] When Alexandre Millerand, deputy of the Seine, attacked the laboratory and Girard, the affair reached the highest but not the most enlightened political level. Millerand accused Girard of carrying out the greatest possible number of analyses, thus artificially increasing the volume of business in order to build up his empire.[59] It is a good thing that Parkinson's law had not yet been discovered.

Most criticism of the laboratory came from its pronouncements on wine.[60] The Griffe law dealt with the obligation of informing the consumer of the nature of the product he bought under the name of wine and of the related need to prevent fraud. Technical information for the application of the law came from Girard.[61] The qualities of the best wines are easily mistaken for faults, especially when judged according to the criteria of an unsubtle science. Pasteur said that it is extremely difficult to tell if a wine is adulterated. Boussingault required eight to ten days to analyze a wine. The committee of the international congress of chemistry (1889) declared it impossible to recognize with certainty the presence in natural wine of wine made from raisins. It was not surprising that some of the results of the analyses done by harried employees of the laboratory turned out to be wrong, and sometimes, what is worse for science, amusingly wrong. Some well-known good wines sent in anonymously were pronounced bad but not harmful: so much for Gruaud-Larose (1864), "second premier cru classé du Bordelais." Then a fraudulent wine sent in by the Le XIXe siècle emerged from the laboratory unscathed as a good natural wine.[62] It is probably safe to assume that these were the inevitable exceptions, resulting from human error or sloth, to the general detective work of the laboratory; there certainly was plenty of adulteration to detect in nineteenth-century Parisian food and drink.

In one respect the laboratory showed itself distressingly modern and in tune with public demand. It was absurd enough to insist that wine should have 12 percent alcohol and twenty-four grams of dry extract. For some of the wines of the sunny south this was no problem because of the high sugar content of the grapes that had basked in the inebriating powers of the Provençal sun. But for Burgundy and Bordeaux the case was quite different. Many excellent wines came up to nine or ten degrees of alcohol and between eighteen and twenty-two grams of dry extract. Indeed, Wurtz had established that the average dry extract of the French harvest was just under nineteen grams. And of course there were abominable wines with fourteen to fifteen degrees of alcohol and twenty-five to thirty-five grams of dry extract. Readers of Le Canard enchaîné and even Le Monde have become enlightened in recent years about the adding of sugar to wine to bring up the percentage of alcohol. The French may not have invented the process, but they supplied the proper word: chaptalisation, appropriately for the chemist who crowned the process with the method of science. Governments, often inspired by sugar beet growers and their powerful lobbies, made legal what was profitable, although it is not legal yet for certain areas of the Côtes-du-Rhône, where its superfluity does not prevent growers from demanding that chaptalisation be made legal anyway. (1980, a great year for sugar beet growers – "l'année du betteravier" – showed magnificently to what extent the wine industry, especially in a poor year, had become "la belle captive" of sugar.)

Excitement was high in the wine business by 1889, and not only in the community of the master blenders of the Quai de Bercy. A list of demands was published by the *Chambre syndicale* of liquor wholesalers under the guise of a report on the working of the laboratory. The wholesalers affirmed the legality of the manipulations performed on wine and indignantly announced to the public by the laboratory's guardian saints of the consumer's conscience.[63] After protesting against the publicity given in the *Bulletin municipal* to bad wine, the report entered into a justification of the dubious practices of the wine dealers. *Vinage* (putting alcohol into wine) was legal. *Sucrage* (sugaring the wine) is legal and encouraged by the government, which wants to collect taxes on sugar used for this purpose. The legal references for this practice were carefully cited: "Jugement de la Cour d'appel de Bordeaux du 14 février dernier (1889)," supported by article 2, *loi*, July 29, 1882, and a decree of July 22, 1885, regulating the *sucrage* authorized in 1882. Although the defenders of adding colors to wine cited the legality of the practice of using coal tar derivatives in Germany, the defenders of *sucrage* did not cite the nearly universal practice of sweetening in Germany. *Mouillage* (adding water to wine) was difficult to determine. *Plâtrage* (the horrible but now happily rare clarification of wine by adding calcium sulphate was a recognized process of vinification necessary "for the good appearance of nearly all the products of our southern provinces"; no law or decree limited the dosage. The use of raisins was very difficult to recognize by tasting. The wholesalers did not want to abolish the laboratory but to confine its activities within the strict interpretation of the law and to stop it from making its findings excessively public. Its list of demands included cheaper prices for tests, use of procedures approved by the Academy of Sciences and published in a special review for all chemists, and a double analysis procedure for control. There should be three categories of judgment: *bon, malade*, and *mauvais* (for harmful substances). The bulletin of qualitative analysis released to the public should not go beyond these simple words. If this procedure were not adopted, the monthly publication of results, in all their disgusting detail, should be stopped. In any case, the laboratory should not give information to the newspapers on the falsification they *think* that they have discovered. *Caveat emptor* was far from being changed into *caveat venditor*.

The clear acceptance of a role for science in public health also led to changes and growth in research and teaching institutions. Under the directorship of Duclaux, the Institut Pasteur established in 1901 a Service d'analyse et de chimie appliquée à l'hygiène, headed by Auguste Trillat ("Expert chimiste au Tribunal civil de la Seine"), one of the leaders in the study of the application of chemistry to items of human consumption. The aim of the Service was to teach analysis in a way that would incorporate pharmacy and medicine as well as emphasize current knowledge about food consumption. Because well-known drinks like cognac were often adulterated, one of the topics dealt with was the use of chemical analysis to determine the nature of the adulteration of eaux-de-vie.[64] But it was not until 1905 that the government got around to recognizing explicitly the necessity of providing for the training of *chimistes-expertes* to be employed as government agents to implement the new law against fraud in food and agricultural products. The model recognized was the *Nahrungsmittel-Chemiker* in Germany.[65] At a lower level, an Institut de technique sanitaire et d'hygiène des industries was founded in 1934 as a state organism attached to the Conservatoire national des arts et métiers and authorized to grant a Brevet de technicien sanitaire.

Charles Girard retired from the municipal laboratory in 1911. He was succeeded by A. Kling, who had a doctorate in the physical sciences.[66] This was a sign of the end of the old tension between a group of low-level scientists, who wrapped themselves in the mantle of defender of the consumer, and the top scientists, who were scandalized by the sloppy work done in the laboratory in the early days. The new practical science was a distinct offshoot of the higher science, with its new methods, technology, and manuals. The whole movement of the detection and control of adulteration in drugs, food, and drink could never have developed as it did without nineteenth-century physics, with its precision instruments of analysis, especially in optics (microscope, refractometer, and polariscope). Similarly, the use of the new methods and instruments of chemical analysis in the organic area, along with inorganic quantitative analysis, made the detection of adulteration easier. But testing and standardization based on biology were developments of the turn of the century.[67] In 1911 the municipal laboratory of Paris made the important innovation of carrying out short but rapid analyses for merchants wishing to take immediate delivery of items such as wine at the Gare de Bercy. A quick test could tell if the wine were watered or *plâtré*, or if a tank of milk were skimmed or watered down. Here was one of the obvious reasons for the success and growth of the laboratory: It was just as useful to the merchant as the consumer. By the eve of the Great War the laboratory was carrying out between 20,000 and 30,000 analyses annually. This form of social control was not only a necessity for the continuation of large-scale city life, it was also a boon for the commercial bourgeoisie, which discovered that it could make more profits from selling good products than from selling bad ones.

The city had much more difficulty in stopping infectious disease than in controlling adulteration. About 1900 the death rate in Paris was 47 percent higher than in Berlin, 16.9 compared to 11.5 per 100,000 inhabitants. Morality and mortality intersected, as Catholics never tired of insisting. In Berlin the rate of illegitimate births was 65 percent lower than in Paris, 139 compared to 283 per 1,000 living births. The mortality of illegitimate babies was much higher than the legitimate. The low French birth rate made the issue of sanitation a matter of national concern. The striking fact was that the fecundity of married Parisian women was 50 percent lower than married Berlin women. The *Revue scientifique* and the *Revue générale des sciences* both thought these issues important enough for extensive serious coverage in 1905 and 1906. Confronted by the brutal reality of medical fact, the city of Paris bought time by fudging their own official figures. The good result of this bad news would be a big increase in the funding of medical biology.

6

SCIENCE IN THE CATHOLIC UNIVERSITIES

Emparons-nous de la science!
— Mgr. d'Hulst, *L'empoisonnement de la science* (1883)

Secret du monde, va devant! Et l'heure vienne où la barre
Nous soit enfin prise des mains! . . . . J'ai vu glisser dans l'huile sainte les
   grandes oboles ruisselantes de l'horlogerie céleste,
De grandes paumes avenantes m'ouvrent les voies du songe insatiable . . .

Ils m'ont appelé l'Obscur et j'habitais l'éclat.
                                        — Saint-John Perse, *Amers*

N 1875, after nearly half a century of battle, French Catholics broke the
tenacious monopoly of the state University in higher education.
Although still hindered by onerous state regulations, the episcopacy
founded five universities, strategically established in Paris, Lille, Angers,
Lyon, and Toulouse. All served a certain group of dioceses. Except for
Toulouse, they had at least the three faculties legally required for the
establishment of a private university. The law also required that a faculty of
science possess physics and chemistry laboratories as well as collections of
instruments and material in physics and natural history.[1] In spite of the
jealous reassertion of some of the state University's old monopolistic
prerogatives by the Ferry laws, involving, among other things, the
elimination of professors of the Catholic faculties from examination juries
and the denial of the title university to the new establishments, the Catholic
institutions survived and still exist. Ferry was minister of education from
February 4, 1879, to November 14, 1881, and from February 21 to
November 20, 1883. Catholics interpreted the law of March 18, 1880, as a
serious blow against the private faculties. Of the 1,372,000 signatures
collected against it, 150,000 came from the Nord and Pas-de-Calais. In the
shock following 1881 the Catholic faculty of sciences in Paris could not call
itself a faculty because after losing three teachers with doctorates it no longer
had the number of doctors required by law for the existence of a faculty. So
the university was rechristened as the Ecole libre des hautes études
scientifiques et littéraires, metamorphosizing later into the Institut catholique.[2]

In the new universities the sciences were given a prominent role.
Especially in Paris and in Lille big amounts of money were spent to establish

relatively modest programs and laboratories. A hundred years later, when the institutes quietly proceeded to abolish a science faculty, as was done at Angers, or to transform it into specialized schools, especially for electronics, as has been done in Paris, it is difficult to understand why Catholics devoted such colossal efforts and large sums of money to establishing and maintaining the cult of Isis, especially in view of the overwhelming competition of the ambitious program of expansion undertaken by the state soon after 1875.[3]

What was the "collective mentality" of the Catholic intellectuals of the period? In the 1850s an interesting precursor of a type of Catholic scientific education can be found in the foundation of the Ecole préparatoire des Carmes by Msgr. Sibour. Completely distinct and separate from the Ecole ecclésiastique des hautes études, it was opened for young people who wanted to do science in order to pass the entrance examinations for the Polytechnique or for the Ecole militaire de Saint-Cyr. Sibour wanted to ensure that the army and the country had a supply of officers full of faith as well as faithful to their duties.[4] But there was a radical change in the outlook of the Catholic establishment between 1875 and 1880. In 1875 the bishops associated with the founding of the University in Paris could recognize the attempts of the state to reform education as evidence of true understanding of the needs of the nation. The justification for establishing Catholic universities was that the monopoly of the state could not be an instrument of progress because, in addition to restricting the rights of the head of the family, it enslaved the spirit of initiative and deprived minds of the salutary stimulus of competition. Equally regrettable was the absence of any role for religion in the state system.[5]

An attitude similar to that of the bishops was expressed by M. de Vareille's speech at the inaugural session of the Catholic faculties of Lille on November 18, 1875. He emphasized that he did *not* accuse the state faculties of attacking religion. In the case of the state faculty of law, Catholics had been guilty of much exaggeration and of great excesses in their attacks against the state University. Nearly all the professors were scrupulously correct in their behavior and in their respect for the opinions and beliefs of others. But they performed no service for religion; no attack but no defense. Finding themselves in a very delicate situation, the professors chose neutrality, which manifested itself in a cold politeness. "En résumé, c'est le silence de l'abstention."[6] Illustrative of the same attitude was the decision of the founders of the Catholic university at Lyon not to protest *collectively* the government's decision to establish a state faculty of law in Lyon after forty years of refusal, partly on the grounds it would hurt the faculty at Grenoble. Although the aim of the government may have been to fulfill the wishes of the enemies of Catholic higher education, there was no point in creating an atmosphere of irritation and hostility.[7] Fearful of the state's great creative and destructive powers, Catholics interested in building Catholic institutions were careful not to antagonize it unnecessarily.

With the new universities recruiting heavily from the state system, Catholic administrators were anxious to avoid any clash with it. Valson, dean of the Catholic faculty of sciences and professor of mathematics at Lyon, pointed out that several of the professors in the Catholic faculties had belonged to the state faculties for a long time and in leaving the advantages they had acquired there they were not animated by feelings unworthy of the big family in which they had formerly lived. They hoped, by the example of a greater initiative, to continue helping their former colleagues to make the improvements and reforms in higher education that had formerly been their common aim.[8] A year or so later this conciliatory statement would be utterly meaningless because of the aggressive anticlerical policies of the government.

The Catholic position became untenable after Ferry, inspired by Condorcet and Quinet, attempted to replace the "old literary education" of the church with the "modern scientific education" of the republic. After some purges the University became an ideological rallying point for the new anticlericalism,[9] overtly scientistic in nature.[10] Republican politicians publicly scorned the idea that a believer could do serious science.[11] It was the fight against this ideology that Rector Baudrillart had in mind when in 1925 he referred to one of the achievements of the institute in Paris as having shown in the mathematical, physical, and natural sciences that an agreement between faith and science was possible in many controversial matters. "Les maîtres à qui sont dus de telles découvertes et de tels enseignements ne constituent-ils point par eux-mêmes comme une apologétique vivante de la divine vérité? Leur personne même démontre cette possibilité que nos adversaires contestaient de l'union complète d'une science étendue et d'une foi sincère et sans réserve."[12]

In the period of conflict during the nineteenth century the justification for establishing Catholic faculties of science was cogently stated in numerous speeches and pamphlets by the clergy. In a lecture given in December 1883, d'Hulst set forth the "Christian mission of science," which was essentially to combat the antireligious program of Jean Macé and the Ligue de l'enseignement, whose destructive work was being carried out in the name of science. Former heretics had based their arguments on theology, which produced only partial negations, for both Catholics and heretics were united on the common ground of the divinity of Christ. The new enemies of the faith, with the support of the science of man, declared God useless, miracles impossible,

Table 1. *Catholic institutes*

No. of diplomas earned (1876–86)

	Docteurs	Agrégés	Licenciés	Capacitaires
Paris	30	1	210	0
Lille	11	0	176	13
Lyon	20	0	296	12
Angers	20	0	191	0
Toulouse	10	0	90	10
Total	91	1	963	35

	Docteurs				Agrégés		Licenciés	
	Lett.	Scis.	Med.	Phar.	Lett.	Scis.	Lett.	Scis.
Paris	0	1	0	0	2	2	155	71
Lille	0	0	50	20	0	0	36	24
Lyon	0	1	0	0	0	0	59	16
Angers	5	4	0	0	0	0	106	14
Toulouse	0	0	0	0	0	0	30	3
Total	5	6	50	20	2	2	386	128

Summary

	Docteurs	Agrégés	Licenciés	Capacitaires
Law	91	1	963	35
Letters	5	2	386	
Sciences	6	2	128	
Med.	70			
Phar.	20			
Total	1,709			

No. of professors (1886)

	Theology	Law	Medicine	Letters	Sciences
Lille	8	13	22	7	8
Paris	9	17	0	11	8
Lyon	6	17	0	6	6
Angers	5	16	0	5	8
Toulouse	7	14	0	8	0
Total	201 (filling about 250 chairs)				

Source: M. d'Hulst, *Les Dix premières années des facultés libres* (Paris, 1886).

mystery inadmissible, and relegated Christianity to the imperfections of a dead past. The plan put forward for several years in the works of the founder of the Ligue de l'enseignement should not be a subject of amusement. The Genevan programs of Jean Macé were becoming a system of government. Although the league had the popular aim of spreading elementary education, its real object was to snatch religion from the soul of the people. The league, aping science in its procedures, popularized science after having poisoned it with atheism. Although the criticism of scientistic arguments by apologists was still useful, even the few people who read it were seldom convinced. Nor was it any longer sufficient to point to the religious beliefs of great scientists, few of whom were contemporary and fewer of whom bothered to defend Christianity. The only remedy was the creation of a university to form generations of Catholic scientists, who would create a science under the inspiration, guarantee, and protection of the Christian faith.[13] The eternal battle between believers and unbelievers had become more violent than ever, and the battleground had shifted to the arena of science. "Notre siècle est pressé, paraît-il, d'en finir avec le christianisme, et pour cela il a résolu de le chasser de la science."[14] The science faculties of the Catholic universities therefore had a utility entirely independent of the small number of students for whom such great sacrifices were made. (See Tables 1, 2, and 3.) Only in these "foyers scientifiques chrétiens" could a defense be secured against a contemporary science that "rules the world and wants to rule it against God and his Christ."[15] The danger lay in the "treaty concluded between higher science and impiety to deprive people of their faith." Salvation would therefore be in an alliance of higher knowledge and belief for setting minds right again. No longer could the Church concentrate solely on elementary and secondary education; the efficacy of those programs themselves depended on a solid scientific education deriving from the highest levels of science. To let the products of Catholic schools go to the state University and quaff the heady draughts of atheistic science was to risk throwing away the fruits of years of labor in Catholic elementary and secondary education.

> Et quand ces enfants, formés avec tant de soins, au prix de tant de sacrifices sortiront de ces écoles, ils tomberont dans un milieu social que nous n'aurons pas influencé; ils y respireront la science athée. On leur dira qu'il faut choisir entre savoir et croire et que les chrétiens ne savent pas. On leur montera les académies, les laboratoires, les bibliothèques, tous les lieux réservés d'où jallit la science, d'où elle part pour gouverner le monde; et on leur demandera quelle place les croyants occupent sur ces sommets. Combien en est-il qui résisteront à cette épreuve?[16]

Table 2. *Number of students in state and Catholic facilities, 1903*

	State	Catholic
Law	10,930	881
Medicine	8,104	121
Sciences	4,401	173
Letters	4,401	195
Pharmacy	3,590	15
Prot. theo.	110	—
Total	31,277	1,385

Source: Institut catholique de Paris: Cinquante ans d'enseignement libre: Mémorial, 1875–1925 (Paris, 1925).

These arguments became part of the common rhetoric of the defenders of the Catholic universities and especially of the hard-pressed bishops who had to find the funds to run the new institutions.[17]

In 1879 the *Bulletin de l'Université catholique de Lyon* justified the establishment of the Catholic universities on the grounds of contemporary irreligion and religious indifference.

> L'une des plus grandes plaies de la France, à l'heure actuelle, est incontestablement l'ignorance religieuse ou l'absence de croyances chez la plupart de ceux qui ont reçu un enseignement superiéur. Les uns ont oublié Dieu, les autres ont même perdu la foi dans le cours de leurs études. La science, qui naturellement devrait conduire à la religion, en éloigne, dans notre pays, un grand nombre d'âmes.

The influence of men of science, with a few exceptions, had been turned against religion. The church had no worse enemies than the alarming number of doctors, pharmacists, teachers, lawyers, and notaries with no religious faith who got the most influential positions in each town and village. Their example led others to conclude that religion, the enemy of science, is based on ignorance. There was also a great threat to Christian families in having a son pursue his studies under a nonbeliever. What was the cause of this fatal rift in higher education? The *Bulletin* concluded that it was not in science itself but in the manner in which it was taught and in the moral abandon of teachers. In higher education a small number were hostile to religion, but most were simply indifferent. The small number of Christian teachers in the state system could not fight successfully against these overwhelming odds.

Table 3. *Catholic institutions, 1964*

	No. of faculty	No. of students[a]	No. of vols. in library
Lille	350	4,588	500,000
Lyon	150	3,000	250,000
Paris	560	10,800	500,000
Toulouse	50	1,195	150,000

[a] In 1963–4 the state universities had 326,300 students.
Source: New Catholic Encyclopedia (Washington, D.C., 1967), vol. 5.

Besides, the state had to accommodate Jews, Protestants, and unbelievers in its educational system. After fifty years of worrying, the bishops had succeeded in getting the state to permit the establishment of a Catholic system of higher education.[18]

The dangers present in the godless state University were presumably avoided in the Catholic institutes by the imposition of certain moral requirements. An applicant needed a certificate of good conduct issued by the director of his *collège* or by a member of the clergy of his parish. Since the isolation of a young man in the city was dangerous, bad for studies and morals, a substitute was devised for the domestic hearth: a *maison de famille* with a gentle but firm director. During the first three years of study, residence in one of the *maisons* was compulsory. A curfew existed. Prayers were compulsory in the morning and in the evening. The provisional statutes of the institute at Lyon required all professors to profess the Roman, apostolic, and Catholic religion and to set an example in the fulfillment of religious duties. The requirement was strictly implemented, at least in the beginning.[19] Although there was room in the state University for the Catholic, there was no place in a Catholic university for those who were not practicing Catholics.

The antireligious threat was viewed as most serious in medicine. The countryside and towns had to be saved from the corrupting influence of doctors and pharmacists, whose prototype, Homais, had been given a brilliant caricature by Flaubert. Homais's reply to the accusation that he had no religion could have been subscribed to by many of the Third Republican secularists: "Mon Dieu, à moi, c'est le Dieu de Socrate, de Franklin, de Voltaire et de Béranger! Je suis pour la *Profession de foi du vicaire savoyard* et les immortels principes de 89! . . . les prêtres ont toujours croupi dans une ignorance turpide, où ils s'efforcent d'engloutir avec eux les populations."[20]

In his fund-raising campaign for the faculty of medicine and pharmacy at

Lille, the only one Catholics succeeded in establishing in France, Msgr. Baunard, the rector, always emphasized the religious and social influence of the physician in towns and especially in the countryside. One of the chief needs of the church was for a Christian order of doctors. Contemporary medical science was materialistic and atheistic; the church wanted to make it religious and Christian.[21] The medical school in Lille was fortunate enough to have at its disposal two wings of the Hôpital Sainte-Eugénie with at least 200 beds, an amphitheater, a dissecting room, and a consulting room for professors. In 1875 a perpetual contract was signed stipulating an annual rent of 140,000 francs. But in 1898 the municipal council voted to annul the contract and, if ordered by the courts, to assume the administrative costs of the hospital itself. After months of discussion and what the Catholics denounced as anticlerical mischief making, the Conseil d'état decided in favor of the Catholic institute, thus leaving the hospital at the disposal of the Catholic faculty of medicine. In 1902 the budget commission of the Chamber of Deputies expressed the wish that the Catholic medical faculty of Lille be no longer admitted to the hospitals of the city. Appeals to the tribunal of Lille, the Court of Douai, and the *Cour de cassation* upheld the contract of 1875 against the attempt of the hospitals of Lille to break it. Two learned societies held their meetings in the faculty of medicine: the Société des sciences médicales (1877–), which published a report of its proceedings and a bulletin, and the Société anatomo-clinique (1886–), which, as a result of the generosity of Féron-Vrau, awarded an annual prize. In 1892 there were twenty-eight on the faculty, of whom twenty-two were doctors of medicine.[22]

Arguments for the continuation of the work of building a Catholic medical faculty in Paris emphasized the same need to prevent young Catholic doctors from being corrupted by the official antireligious teaching of the Sorbonne. As the example of Louvain showed, there were considerable political advantages in having a Catholic university to supply towns and villages with professional people who had the right ideas about duty. A more ambitious aim of Catholics was the hope of reestablishing the lost prestige of French medicine among foreign students, thus recapturing those who went to Germany for study. In the last quarter of the nineteenth century this could be done only by modernizing medical science.[23] But this ambition remained not much more than a noble dream in Paris; it was left to Lille to try to correct Flaubert's *Dictionnaire des idées reçues*, where for "Physician" we find the damning entry "Tous matérialistes."

In recruiting scientists for their new universities, the Catholic organizers were able to turn to their advantage a number of dysfunctions in the French educational system.[24] The state faculties of science not having begun their period of expansion, it was easier to recruit faculty in 1875 than it would have been a decade or so later. The leading stars of the institute in Paris were the geologist Albert de Lapparent and the physicist Edouard Branly. Lapparent had been with the Service de la carte géologique de France for more than ten years and eagerly seized the opportunity to get a teaching position as a refuge from the tedious tasks of a government geologist. Not having a doctorate, this brilliant graduate of the Polytechnique and of the Ecole des mines could not satisfy his pedagogic urges by seeking employment in the state University. With enough doctors to satisfy the state requirements for the creation of a faculty, the institute was able to offer Lapparent a position that enabled him to serve science brilliantly as one of Europe's best-known geologists and to get elected to the Academy of Sciences.[25] The acquisition of Branly was most fortuitous. He had just resigned as *directeur-adjoint* of the physics teaching laboratory at the Sorbonne because he did not want to marry the eldest daughter of the *directeur*, Paul Desains. Branly was given 80,000 francs to organize his teaching and was promised a new laboratory. Branly's invention of the coherer made him a key figure in the development of the radio and turned his miserable little laboratory into a high-class tourist attraction. Mrs. Roosevelt visited it in 1934. It is probable that the fame of Branly was an important factor in getting the property lease for the institute renewed by the state in 1909. But Branly always bitterly reproached the institute for not having done more for him. The chief culprit was the rector, Cardinal Baudrillart – "Alfred" to Branly. "J'ai dit à Alfred: le jour où je publierai mes mémoires, vous n'aurez plus qu'à vous cacher dans les cabinets."[26] The institute was also able to use the economical services of a few abbés who had braved the "atheist" and "materialist" dangers of the Sorbonne to get their doctorates. Abbé Hamonet in chemistry was perhaps the best known in scientific circles outside the institute. Between 1875 and 1881 the Catholic university in Paris had the services of Georges Lemoine as professor of chemistry. Lemoine was an Ingénieur des ponts et chaussées whose excellent work in chemistry made him a superb addition to the faculty. Unfortunately, in 1881 the government forced him, like Lapparent, to choose between state service and the institute, which was not in a position to offer the security he needed.[27] Lapparent's financial and social position was sufficient to permit him to risk staying at the institute even though there was the danger it might be closed. Branly took out the insurance of a medical degree and practiced medicine for many years. Branly's medical thesis was an important and influential study: *Dosage de l'hémoglobine dans le sang par les procédés optiques* (1882). In addition to certain specializations (stomach ailments, fractures, rheumatism, and gynecology), Branly was interested in electrical treatment.

He had close relations with d'Arsonval at the Collège de France and some contacts with Gustave Le Bon. After the secularization campaign of the state began in earnest, a job at the institute did not seem the most secure situation in Paris.

The scientists at the institute in Paris brought it many triumphs. In 1894 Supan, the director of the important German periodical, *Petermanns Mitteilungen*, published a highly favorable review of the third edition of Lapparent's *Traité de géologie*. Supan heaped praise on the author for his deep knowledge of foreign literature, emphasizing the treatise's value for geographers. "Un ouvrage purement scientifique . . . Quel est le savant allemand qui pourrait se prévaloir d'un succès simplement analogue? Des chiffres comme ceux que nous venons de citer [8,300 copies sold since the first edition in 1883] sont la mesure des efforts réalisés pour l'éducation d'un peuple et, en vérité, pour nous allemands, la comparaison n'a rien de flatteur." In 1892 abbé Pierre-Jean Rousselot, professor of the history of the French language at the institute, defended his famous theses at the Sorbonne. It was an unusual defense involving the faculty of sciences as well as the faculty of letters. One of the jury members, Petit de Julleville, indicated the pioneering nature of Rousselot's work in experimental phonetics: "Vous présentez vos thèses devant la Faculté des Lettres. Nous nous sommes adjoint un professeur de sciences. Nous aurions besoin aussi d'un médecin pour discuter vos conclusions et d'un professeur de droit pour décider au jugement de quelle Faculté votre travail doit être soumis."[28] In 1895 Rousselot was president of the Société des parlers de France. A certain number of professors of the institute participated in the exposition of 1900. The results were quite happy for the institute. Both Branly and Rousselot won a grand prix; Lapparent won a gold medal; Abbé Hamonet won one of the three silver medals that went to institute professors; and Gendron, Branly's physics assistant, won a collaborator's bronze medal. Lapparent also won a grand prix for his exhibition in geographical science. Branly became a chevalier of the Legion of Honor. Rector Péchenard rightly paid homage to the spirit of impartiality and justice of the republic on this occasion. In 1892 Lapparent had pioneered a course in physical geography for history students at the institute. This move was soon followed at the Sorbonne. Lapparent's courses in geology had been dropped because of poor attendance. The advantages of taking the competing courses at the Sorbonne and the Ecole des mines, especially in securing state jobs, could not be overcome by any amount of pedagogical charisma at the institute.

Many of the scientists who went to the institutes in the 1870s were delighted with their support and status, especially as compared with their modest roles in a nonexpanding state system. But this situation changed rapidly after the 1880s because of the growth of the state system and the increasing difficulties of the Catholic faculties of science in financing the mushrooming demands of ambitious researchers. Branly's case shows this to some extent, but it is clearer in the case of the zoologist A.-L. Donnadieu, who left his unique chair of natural science in a lycée to join the Lyon institute. Obviously a lycée teacher would be overjoyed at the resources of the institute, but as late as 1879 Donnadieu could point out that his facilities were far superior to most of the state faculties in terms of offices, research tools and laboratories, library access, and scientific collections. Donnadieu thought that the state was neglecting the natural sciences; it had quickly abolished the only French lycée chair (Lyon) in the natural sciences when he resigned. In nearly all religious establishments, however, the natural sciences were held in high honor. The payoff in terms of scientific results and good pupils led Donnadieu to place his hopes on the viability and future of the Catholic scientific enterprise. But by 1891 Donnadieu had to resign his job at the institute because of the inability of the faculty to support his program and research. In 1892 Amagat also left Lyon for the Polytechnique because the institute could not afford to pay his laboratory expenses. An era had ended, at least in Lyon.[29]

In 1879–80 there was only one cleric at Lille, the well-known abbé Boulay (*sciences naturelles*), out of the ten faculty listed for the sciences, although by 1892 there were three abbés out of a total of eleven. This increase probably reflected the serious competition of opportunities for lay personnel then available in the state system.[30] Lille also proudly displayed the state pedigree of the physicist Jules Chautard, former dean of the faculty of sciences at Nancy, and of the chemist Antoine Béchamp. Béchamp and his son were lured to Lille from Montpellier as a result of a set of exaggerated recruiting promises that could not be kept and which resulted in Béchamp's suing the bishops of Cambrai (Conseil supérieur des évêques).[31]

At Lyon the new science faculty was fortunate in being able to take advantage of the dissatisfaction of the physicist Emile Amagat with his position in the normal school at Cluny.[32] The initial expenses for physics was 50,000 francs, which gave Amagat a real laboratory for the first time. Very important for Amagat's work was the hiring of Bénévolo as *chef de travaux* in physics. Bénévolo was a constructor of scientific instruments in Lyon. The chemistry laboratory was given between 8,000 and 10,000 francs. The collections in zoology, mineralogy, and geology started with between 20,000 and 25,000 francs. The sciences library began with a budget between 10,000 and 12,000 francs. The total budget for 1890–91 at the institute rose to over 246,000 francs, of which law got 69,300, letters 31,100, and the sciences 61,200, plus 17,061 francs for the library, the

231

FROM KNOWLEDGE TO POWER

collections, the observatory, and other things. Valson, who became dean, was paid 10,000 francs, the same salary as the state dean at Grenoble; the state dean at Lyon was paid 12,000 francs a year. An attempt was made to give professors who left the state system the same salaries and fringe benefits as they had in the state system.[33]

The new universities, including Angers, did very well in recruiting competent faculties in the sciences, which, in spite of their relatively small sizes, could compare quite favorably with the state system, especially some of the provincial universities. A substantial amount of the financing of the Université catholique de l'Ouest was done through a central committee in which the local gentry played a large role. By 1897 the committee had assured the university's existence by providing it with a total of 162,795 francs. Each year it collected between 20,000 and 40,000 francs; in 1907–8 it collected over 38,000 francs, one-third of the university's receipts. In 1923 the amount was up to nearly 50,000 inflated francs. The founders of chairs at the UCO were Mme la Comtesse de Quatrebarbes-Bourreau, Mme la Vicomtesse des Cars, M. le Marquis de Nicolaï, Mme la Marquise de Champagne, MM. les Comtes de la Bouillerie, Mme la Vicomtesse Say de Tredern, M. et Mme Yves Jallot, M. Armand de Maille de la Tour Landry, the duc de Plaisance, M. Emmanuel Eriau, S. Em. le cardinal Brossais Saint Marc, Archevêque de Rennes, and S.G. Mgr Freppel, Evêque d'Angers. The landed nobility also played a large role in the foundation of the Ecole supérieure d'agriculture – "de viticulture" was added later – d'Angers and got the Conseil général of Maine-et-Loire to vote a subsidy of 500 francs. Leading the support by the *grands propriétaires angevins* was le comte Henri de la Bouillerie; as president of the Congrès des syndicats agricoles he was in the best possible position to get a motion of support passed in favor of the Ecole at the meeting of the congress in Angers in 1895. The school began with six students in 1898 and moved to splendid new quarters in 1920 with 198 students and twenty-five teachers.[34]

One of the best scientists lured into the fledgling Catholic system was Antoine Béchamp, who in 1876 became dean of the Catholic faculty of medicine at Lille. Béchamp began his career under the patronage of the chemist, senator, and grand vizier of the politics of science in the Second Empire, Jean-Baptiste Dumas. An *agrégé de l'école de pharmacie de Strasbourg* (1851), Béchamp became professor of medical chemistry and pharmacy at the faculty of medicine in Montpellier. An ecstatic letter by Béchamp to Dumas on September 9, 1856, expressed the forlorn hope that Dumas would be able to get him into the faculty of sciences, although he confessed that he would not get all the votes of the faculty because of the lowly origins of his *agrégation*. In 1855–6, enlisting the political support of Joachim

Murat, deputy and secretary of the Corps législatif, and with the support of the rector, Béchamp tried for the chair in chemistry at Strasbourg. He also asked for the chair in chemistry at Marseille, when vacant. Although Dumas had a good opinion of Béchamp's work and presented some of it to the Académie des sciences for inclusion in its *Comptes rendus*, the best appointment that he could get was at Montpellier, where he stayed for nearly twenty years.[35]

Early in Béchamp's career (1856) the rector who wrote the report on him at Strasbourg mentioned the ardor of his religious convictions, as well as his application to work and his solid and sophisticated teaching.[36] The depth of these religious beliefs may be judged from a letter he wrote on September 25, 1870, to the Government of National Defense. "À MM. Trochu, Arago, Crémieux, Fauvre, Ferry," etc. "It is a law of history that no . . . people have raised themselves to any degree of intellectual culture whatsoever without knowing and adoring God. This belief, characteristic of man, is as necessary for humanity as air and food are for the individual." Béchamp noted the universality of the worship of God, a salutary belief that creates prosperity. Another law, the scientist continued, is that the decadence of a nation (in ethics, moral virility, and patriotism) always coincides with loss of respect for God and inexact observance of His law. It was in this area that the Second Empire had failed and in which the new government would, Béchamp hoped, give a different example. After long passages on and quotations from Lincoln, who conquered by the grace of God, Béchamp hoped for a purified and regenerated France under the Government of National Defense.[37]

Béchamp's concern over the course being followed by France had not changed by the second decade of the Third Republic. In a letter thanking Dumas for sending him a copy of his *Réponse à M. Taine*, he complained that Taine was still the same as ever – "le même esprit systématique." Béchamp recalled for Dumas his friendship with Liebert, Taine's friend, also an "esprit d'élite" and a mind gone wrong. As he used to tell Liebert, Béchamp thought that basically these two were intelligent, elegant, and diabolical corruptors of the French mind. No doubt, Béchamp reflected with some satisfaction, God would require of these wretches a strict account of talent so badly employed. But Taine was essentially only a juggler of paradoxical ideas, camouflaged by well-turned phrases, which were able to affect only those who had not drunk deeply at the Pierian spring or those who had the same outlook as the great corruptor before being exposed to him.[38] In addition to getting favorable academic and financial terms, Béchamp probably believed that he would be better able to counter the baleful effect of the godless group and better able to provide sustenance for the naturally righteous French mind by accepting

the call to the Catholic faculty at Lille.[39] Béchamp began his first course with
Veni, sancte Spiritus and *Ave Maria*. He then proceeded to declare that he
and his students would work for three great causes: the freedom of the
persecuted church, the salvation of France, and the dignity of science.

> En commençant de cours, je sentais le besoin de prier avec vous,
> Messieurs, pour implorer l'assistance divine et pour dominer mon
> émotion.
>   C'est qu'en effet, nous fondons, en ce moment, une grande oeuvre.
> Nous allons travailler ensemble pour trois grandes causes: pour la liberté
> de l'Eglise, pour le salut de la France, pour la dignité de la Science; –
> pour la dignité de la Science abaissée par quelques hommes; – pour le
> salut de la France malheureuse et tourmentée, – pour la liberté de l'Eglise
> persécutée et haïe ... Nous servirons trois grandes méconnues. Nous
> leur consacrons toutes nos forces et toutes nos intelligences: vous,
> Messieurs, au début de votre carrière; moi, hélas! bien près de terminer
> la mienne; vous, avec les ardeurs de la jeunesse; moi, avec l'enthousiasme
> qui me reste encore
>   Et maintenant, à l'oeuvre.
>   *Sursum corda.*[40]

But Béchamp's mission and afflatus at Lille were afflicted with several
incidents that made work difficult and unfruitful, ultimately ending in a
lawsuit against some of the Catholic hierarchy. In 1876 Féron-Vrau,[41] a
solid bourgeois social Catholic in the spinning business at Lille, who was the
agent of the organization set up by Catholics to recruit professors from the
state faculties for the new Catholic institute, supposedly painted a very rosy
picture of the academic future at Lille to both Antoine Béchamp and his son
Joseph Béchamp. A guarantee was made for eternity. It seems that overtures
were first made to the son, who was offered an annual salary of 7,000
francs; he earned 1,200 at Montpellier as assistant to his father. Later
Béchamp *père* was offered the deanship of the Catholic medical faculty,
which he accepted on the condition that his son also be given a position. The
son was made a professor of analytical chemistry and toxicology on the basis
of a ten-year agreement, which was approved by the Conseil supérieur des
évêques. By 1886 the attempts of the father to reorganize the faculty had
brought him into a conflict with the rector that led to his dismissal. On July
30, 1886, the professorships of both father and son were abolished and
"consolidated" to form a professorship of medical chemistry, which was
offered to the son, but Joseph refused to accept it on the grounds that it was
really the professorship that had been given to his father on a permanent
basis. The offer was then withdrawn. Negotiations with the father
concluded with an outright gift of 95,000 francs and an annual income of

6,000 francs until his death.[42] When Joseph's salary was stopped in November 1886, he brought suit before the civil tribunal of Lille (February 1888) against the bishops of the Catholic province of Cambrai for elimination of his professorship and against the Société civile of the Catholic institute for stopping his salary. One of his requests was for an award of 200,000 francs for damages. The institute pleaded only the need for reorganization and that the interests of the professors were adequately taken care of. The Béchamps were giving 132 chemistry lectures, the institute claimed, whereas the need was for only one general course of thirty-six lectures, the same as in the state system. Although the reorganization required Joseph to change from analytical chemistry and toxicology to organic and biological chemistry, and, it was claimed, the institute was violating the law by understaffing, with only eleven or twelve medical professors, the tribunal decided on March 1 to dismiss the suit against the bishops and the institute.

The end of his career at Lille probably confirmed Joseph Béchamp's earlier impressions about the Nord versus the Midi. A young man who had flunked out of medical school had once threatened Béchamp with a pistol and then had shot himself after a struggle in which Béchamp *fils* nearly got shot as well. In a letter to Dumas, Béchamp said that he had seen brutal passions much more excited in the Nord than in the Midi, where people were passionate but noble, in contrast to the cold and calculating northerners. "Although the Revolution and materialism have lowered moral standards everywhere, I prefer the Midi."[43] Boussingault would probably not have agreed: He would not have exchanged one good Alsatian for all the inhabitants of the Midi.

The chief function of the faculties of science in the institutes was to prepare candidates to pass the licence examinations in mathematics, the physical sciences, and the natural sciences. But this was not very different from the situation in most provincial faculties. Only a small number of doctorates was done; even the luminaries of the institutes spent most of their time, except that devoted to their personal research, preparing candidates to take the state examinations for the licence qualifying them to teach in the Catholic secondary educational system, which in 1891 enrolled 52 percent of the students in the *collèges*. Most of the graduates became teachers in the *collèges libres* and in the *petits séminaires*. One of the most significant exceptions to this overall assessment was the laboratory of the abbé Henri Colin (1880–1943) at the Paris institute, where after 1921, as "professeur titulaire de physiologie végétale," he directed nineteen doctoral theses.[44]

The Catholic universities were of necessity genuinely regional, in close touch with local industry and agriculture. In 1886 Lavisse paid tribute to the accomplishments of the institute at Lille, noting especially its services to agriculture and industry in the Nord by the establishment of special programs making science valuable in the eyes of practical people and showing its utility in the region. The two schools of higher agricultural and higher industrial studies clearly show Catholic interests in the intellectual and material needs of the Nord. The state faculties could well imitate Catholics in this matter, for the former, living in their own cocoons, had too long been abstractions that resembled one another no matter where they were located. The industrial school at Lille, which had been largely founded by the municipal government, should be connected with the state faculty and supported by an agricultural school. "Le jour où les Universités régionales se préoccuperont des besoins intellectuels et matériels de leurs régions, elles seront vraiment fondées. Il faut que chacune d'elles ait dans l'unité de la science et dans l'unité de la patrie, une sorte de spécialité." Lavisse was already out of date, especially at Lille. A great deal of nonsense has been written on the disdain of the European professor for industry and its problems. Nothing could be further from the truth in the case of the French universities in the late nineteenth and early twentieth centuries. In a period when close contact between the professors of chemistry and local industry and municipal government led to the establishment of several institutes of chemistry, many chairs of industrial physics and agricultural botany, and other projects directly related to industrial, commercial, and agricultural interests, the Catholic institutes played a similar role wherever feasible. On a local level the institutes were remarkably responsive to the development of industrial and agricultural programs. Among the first scientists associated with the Lille institute were A. Witz in physics, N. Boulay in botany, F. Bourgeat in geology, and E. Lenoble in chemistry. The practical emphasis is evident in Witz's treatises on gas and steam engines, his work on thermodynamics for engineers, and in Lenoble's attempts to solve the problems of local industry. Witz's treatise on the study and application of gas motors went through five editions between 1886 and 1924; his work on steam engines went through three editions between 1891 and 1913; and his book on thermodynamics for engineers saw four editions between 1872 and 1924. Lenoble's laboratory concentrated on determining the calorific power of combustibles, the use of the saccharimeter, and problems associated with the manufacture of products like ether, white lead, and zinc white (oxide). The Ecole d'études industrielles connected with the Lille institute was the brainchild of Féron, who wanted a sort of normal school for the full education of young managers from all over France.

The Ecole des hautes études agricoles at Lille was founded in 1886 for the purpose of producing Christian proprietors. It was assumed that the Catholic faculties of Lille could offer valuable resources to the school in their personnel, laboratories, and collections because the progress made in agriculture over the half century before the founding of the school was due chiefly to discoveries in chemistry and plant physiology. Following the study plan of the Institut national agronomique and the Ecole d'agriculture attached to Louvain, the teaching at Lille was theoretically oriented, although there was a great deal of complementary practical activity. Frequent visits were made to the "exploitations agricoles" of the region. These visits were facilitated by the Comice agricole de Lille, whose director was a member of the school's visiting agricultural commission. In addition to the certificate awarded to ordinary students, there was a diploma of *ingénieur-agronome* for distinguished students. A system was also set up whereby well-known Christian agronomists in various parts of France would take recommended students for certain periods to extend their training. The Société des agriculteurs de France extended its friendship to the school and even gave it 2,000 francs in 1891.[45]

The faculty of the Catholic university at Lille contained a large proportion of polytechnicians, engineers of the Corps des mines, and engineers of the Ecole centrale, although most of them also had doctorates. Two of the key figures in founding the Catholic university at Lyon were associated with Ponts et chaussées: Théodore Aynard, Inspecteur général honoraire des ponts et chaussées, and Louis Jacquet, Ingénieur en chef des ponts et chaussées. The work of the leading scientist of the Lyon faculty, Emile Amagat, related directly to the nineteenth-century version of the military–industrial complex. Ten years after the founding of the Catholic university of Lille in 1875 the Ecole des hautes études industrielles was opened with great fanfare. The director, Colonel Arnould, was a graduate of the Ecole polytechnique; on the Conseil de perfectionnement was Descottes, Inspecteur général des mines, and four industrialists from Armentières, Lille, Val-des-Bois, Roubaix, and Tourcoing.[46] The teaching personnel included artillery colonel Arnould as professor of industrial technology; Villié, Ingénieur au corps des mines, polytechnicien, dean of the Catholic faculty of sciences; Frémaux, Ingénieur des arts et manufactures; Witz, Ingénieur des arts et manufactures and professor in the faculty; Lenoble and Bernard from the faculty of medicine (industrial hygiene); Ch. Maurice (natural sciences applied to industry); and the well-known abbé Vassart, professor of dyeing. Vassart was one of the school's star attractions. "L'éminent professeur qui a créé naguère l'enseignement de la teinture dans le Nord de la France" was the way the university's catalog put it.

When the *Soleil illustré (du dimanche)* published on October 29, 1893, a long, illustrated, flattering report on the Catholic university, it paid special

Table 4. *Finances of the Catholic institutions in Lyon and Lille*

Lyon: subscriptions on April 30, 1881

Dioceses	Amount in francs
Lyon	1,691,130.82
Aix	11,500.00
Ajaccio	3,304.00
Annecy	6,000.00
Autun	170,557.00
Avignon	4,000.00
Belley	202,216.55
Chambéry	15,925.00
Digne	6,000.00
Dijon	50,150.00
Fréjus	17,000.00
Gap	4,000.00
Grenoble	65,123.20
Langres	25,500.00
Marseille	117,785.00
Montpellier	14,500.00
Moulins	25,200.00
Nice	1,535.00
Nîmes	41,000.00
St. Claude	35,200.00
St. Jean de Maurienne	4,286.00
Tarentaise	6,870.00
Valence	47,048.45
Viviers	31,800.00
Total	2,597,631.02

Lille: summary of contributions up to 1877 (francs)

	Clergy	Lay	Total
Diocese of Cambrai	866,358	4,645,604	5,511,962
Diocese of Arras	356,590	530,326	886,916
Outside the archdiocese	17,225	57,160	75,385
Total	1,240,173	5,233,090	6,473,263

Notes to Table 4

Notes: The only detailed subscription list I found for Lyon indicated that bankers, manufacturers and sellers of silk, and *rentiers* were prominent among those who contributed, chiefly in sums of 5,000 or sometimes 10,000 francs.

One Lille family gave 500,000 francs. Two anonymous gifts of 106,125 francs each went for two chairs in letters and in science; two more anonymous gifts of 100,000 each went for chairs in the faculty of medicine. A gift of 100,000 francs by the surviving family of Hippolyte-Anné-Julien, Comte de Docquer T'Serroelofs, established a chair of canon law in the faculty of theology.

Sources: Archives, Institut catholique, Lyon. The initial funding of the Lille institute is given in a published work of 583 pages, *Souscription pour la Fondation de l'université catholique de Lille* (1877).

attention to medicine, pharmacy, physics, and the two industrial and agricultural schools. The renewal of the decennial subscription for the Catholic faculties in Lille brought in over 2.75 million francs by 1889. The Midi matched the Nord in an outpouring of rhetoric, but it fell far behind in francs. Although subscriptions in the south to the collection of 1881 for the Catholic university in Lyon reached over 2.5 million francs, the Lyon diocese provided nearly 1.7 million of the amount. Lyonnais bankers, manufacturers, silk merchants, businessmen and *rentiers* were prominent among those contributing, chiefly in sums of 5,000 or sometimes 10,000 francs. (See Table 4.) In Lille the two brothers Philibert Vrau and Camille Féron-Vrau, pillars of the spinning industry, played a great role in getting the Catholic university under way. Camille was also a leader in the paternalistic Association catholique des patrons du Nord and took over *La Croix* in 1901 from the anti-Dreyfusard Assumptionists. Also prominent in the Catholic educational movement was the social Catholic leader, the industrialist patriarch of Val-des-Bois, Léon Harmel. Like other Catholic bourgeois of the Nord, he took a special interest in the Ecole des hautes études industrielles. Msg. Baunard, the rector of the university, would sometimes call the school "la Faculté industrielle," although the teaching personnel came from other faculties. The Ecole was housed in the spacious three-story building of the faculty of sciences, erected between 1883 and 1885.

What is striking about the Catholic industrial and agricultural schools at Lille is that they were both founded with a publicly articulated set of aims based on a clear ideology. One can consider on several levels the purpose of founding these institutions. The Catholic faculties of science did have a general religious purpose. Speaking in 1883 to the Assemblée annuelle des catholiques du Nord et du Pas-de-Calais, Msgr. d'Hulst defined the scientific

Table 5. *Origins of students entering the Ecole des hautes études industrielles (Lille), 1885–97*

Father's occupation	Number
Banquiers et assureurs	6
Brasseurs et malteurs	10
Céramistes	2
Chimistes	14
Directeur de magasins généraux	1
Etudiants	2
Distillateurs	5
Entrepreneur	1
Fabricants divers	5
Fabricants de papier et carton	10
Fabricants de sucre	4
Fabricants de tissus divers	33
Filateurs et filtiers	24
Imprimeur-éditeur	1
Inconnus	2
Ingénieurs divers	38
Métallurgistes (maîtres de forges, fondeurs, etc.)	7
Mineur	1
Minotiers	4
Négociants	10
Professeurs	3
Propriétaires	18
Tanneurs	2
Teinturiers	23
Total	226

Source: Bulletin de l'Association des anciens élèves (1897), p. 31.

role of the Catholic faculties in terms of trends of the Republican secular theodicy. "Il nous faut des foyers scientifiques, parce que la science aujourd'hui gouverne le monde et veut le gouverner contre Dieu et contre son Christ." Moving from the level of fundraiser to the level of fund giver, we discover a more precise type of aim. Next year, Léon Harmel, speaking to the same audience, declared, "L'usine est une oeuvre de la Providence; si elle est devenue un moyen de perdition, si elle est au pouvoir de l'enfer, c'est que la matière est plus avancée que l'intelligence et le coeur du patron." Delphic precision, perhaps.[47]

The prospectus announcing the opening of the school in November 1885 specified the uniqueness of the school's aims.

> Depuis longtemps on en sollicitait la création. Il existe, en effet, en France et à l'étranger, de nombreuses écoles d'ingénieurs, mais on n'en trouve aucune qui ait pour but de former spécialement les hommes que leur naissance met à la tête d'importantes exploitations, et dont le rôle social est tout autre que celui de leurs agents. – Généralement, les institutions de ce genre localisent leurs élèves dans les cadres particuliers, établis en vue des services de l'État, ou de quelques carrières indépendantes, mais qui exigent des connaissances spéciales. – Chacune comprend aujourd'hui qu'en dehors de ces cadres, une instruction développée est surtout indispensable à ceux qui, nés dans une situation aisée et dégagés d'inquiétudes personnelles, ont le devoir de consacrer en partie leur vie, soit aux intérêts publics, soit à des exploitations industrielles privées.

Although the organizers quoted the Duke of Devonshire and Sir Lyon Playfair, the Belgian example was more important, for Belgium was seen as profiting from English industrial losses. *Le Mouvement industriel belge*, a technical journal, showed how the really practical Belgian mind possessing *solid* theoretical knowledge enabled Belgium to profit from the ground lost by England. Yes, here was a school for bosses, or, more politely, the sons of industrialists. (See Table 5.) Harmel explained why.

> Faute d'une école appropriée au patronat, la plupart des industriels font passeur leurs fils par toutes les phases du travail, soit dans leur propre usine, soit dans une usine étrangère; mais le moyen est loin d'être parfait et offre bien peu de sécurité; il expose au danger de voir entretenir une routine fatale dans les industries se succédant de cette façon, et un tel enseignement ne peut rien pour élever le niveau intellectuel du jeune homme.

The ideology of the founders contained that peculiar insistence on the importance of science and technology leavened by the strong strain of paternalistic social Catholicism found in the Nord. The *Livret des facultés catholiques* (1892) was quite blunt: "C'est aux futurs patrons et directeurs d'établissements industriels que s'adresse l'enseignement de notre Ecole des hautes études. Elle pourrait s'intituler: *Ecole normale des patrons*, mais des patrons *chrétiens*." The program of *instruction* and *éducation* was subordinated to this aim. The original hope was that the school would have a national impact in becoming the center in France for the creation of a new education for industrialists. After a few years the school started to attract students from outside the Nord, thus partially fulfilling this aim.[48]

The basic aim of the school was to produce a new type of industrialist. The new man would possess the necessary scientific and technical know-

241

ledge – "l'instruction complète" – but he would also have the type of knowledge – legal, historical, geographical, literary – that elevates the distinguished and cultivated man above the trained practician. This aim was frankly associated with class responsibility. "Les connaissances . . . assurent à ces futurs maîtres la supériorité qu'on est en droit d'attendre des classes dirigeantes." The commission for the founding of the school (1884) included Witz, Féron-Vrau, Louis Delcourt, Gustave Théry, Trolley de Prévaux, Béchaux, and Villié, with Henri Bernard as president. It was Aimé Witz, the leading industrial scientist of the faculty, later dean, who, as secretary of the organizing commission of the school, stated its pedagogical aims.

> Pour former un bon industriel, pour lui fournir les moyens de diriger des usines avec compétence, avec autorité et succès, il importe de lui donner une solide instruction *scientifique* et *technique*: il a besoin d'un petit nombre de notions exactes et très nettes de droit pour gérer ses affaires; il lui faut enfin la connaissance des éléments de droit public et administratif et des principes de l'économie sociale pour remplir son rôle dans la société et faire bénéficier l'Eglise de l'influence que lui donneront son éducation, sa science et sa position de fortune.

The manifesto announcing the opening of the school made it clear that the possession of practical skill by the head of an industry would only put him at the level of his workers, a situation damaging to "l'autorité patronale." Only a general culture would let him see the whole of his enterprise while acting as its dynamic center as well as enable him to play a role in public affairs. "Sa responsabilité morale aussi bien que son rôle social, l'obligent à avoir des notions exactes de droit, d'économie politique, et, par-dessus tout, une solide instruction religieuse qui le protège contre l'envahissement des faux systèmes et des erreurs de l'époque, et qui lui fasse connaître ses devoirs envers les ouvriers."[49]

A curriculum was specifically designed to produce the new Catholic economic man. There were the traditional courses in mathematics, descriptive geometry, mechanics, physics, industrial and organic chemistry, commerce, drawing, and technology. The program contained lectures in civil law, commercial geography, the history of work, foreign languages, and, of course, French composition. This education would best qualify young men for whatever position they took in the world of industry and business. In order to furnish the Christian *éducation* needed for those who would take their places among the "autorités sociales" of the country, courses in religious demonstration or proofs and apologetics, political economy, and industrial hygiene were included in the curriculum. Thus would come about the reign of truth, justice, and peace, under an industrial umbrella.

To develop the professional part of the student's *éducation*, the school worked out a system of visits to industries, including weekly visits to big factories and manufacturing plants of the Lille region and an annual visit to the most remarkable centers of industry in France and abroad. About 400 different industrial establishments and coal mines could be studied by the students. Delegations of students under Arnould visited industrial exhibitions wherever they were held. They also visited industries in Bohemia and in the Rhineland, where they were welcomed by deputies of the Center party. Reports on these educational trips were published in the *Bulletin des facultés* and the *Bulletin de l'association des anciens élèves de l'école*. The first *Bulletin annuel* was published in 1892. In 1906 the association had a cash reserve of over 13,000 francs, which rose to nearly 19,000 in 1924, when its income was 27,580 francs. Foreign students also came to the school. The director was made a Commander of the Order of Charles III by the Spanish government as a reward for the services given to the Spanish students who came to the school. We are not dealing with large numbers here; by 1902–3 about 300 students had graduated from the school and gone, for the most part, into family enterprises; seventy-eight students were following courses in the school in 1902–3. A career structure was built into the recruitment program, but there was also a good demand for graduates, even from the railroad companies. The graduates do not seem to have had the same problems with devalued degrees as some of the public establishments were experiencing by the end of the nineteenth century. After the First World War the school's diploma increased in value. Although forty-eight of those who had been associated with the school were war casualties, the increase in value was not seen by the school in terms of a shortage but in the fact that during the war the graduates had shown their excellent scientific formation and fine moral behavior. A very dynamic and solicitous Association des anciens élèves was of great importance in making sure graduates got jobs – a typical pattern of many schools. Given the commercial and industrial connections of the families of the graduates of the school, it is not surprising that a Catholic writer could boast of the superior situation of the school's graduates as compared with similar public schools. In 1892 the director was pleased with the achievements of the first graduates. " . . . Nos premières troupes opèrent avec habileté dans l'usine, à Tourcoing, à Lille, à Cambrai, en Picardie, en Maine-et-Loire; notre avant-garde est en très bonne place sur les marches, je veux dire sur les listes des conseils municipaux . . . "[50] The Conseil de perfectionnement (1892) included Léon Harmel, Féron-Vrau, Amédée Prouvost (Roubaix), and Motte-Bernard (Tourcoing). Whatever the value of any such comparison of state and Catholic schools, it certainly was a good propaganda line for recruitment.

No doubt there are a good many peculiarly Catholic characteristics of this attempt by nineteenth-century *patrons* of the Nord to solve the problems of a society in the throes of industrial growth. One obvious intention of the capitalists was to press the fight against socialism among the workers; this was linked with the desire to maintain a traditional social structure while wholeheartedly supporting the latest innovations in technology. Nor can the fight against the program of secularization in the Third Republic be dismissed. As Nathan Rosenberg and others point out, technology cannot be simply defined in terms of simple "hardware" or "capital," but must be broadly construed as any improvement in the relations between inputs and outputs. Ideas are thus an important part of technology. Technology is not an isolated thing but part of the complex of natural and human resources. While Weber was interpreting modern industrial society within the framework of a more general theoretical analysis of the structure and functioning of social systems, Catholics of the Nord were westling with the problems of accommodating one with the other. "L'homme du Nord est un traditionnaliste qui ne pense qu'au futur." Certainly Catholics saw the problems of industrial society through a glass darkly, but they made a sincere if very limited and only partially successful attempt to solve them, more along the lines of *Rerum novarum* (1891) than of the Communist Manifesto, of course, although in Tawneyesque or puckish mood we do attribute Scholastic flavor to both panaceas. The Catholic school struggles bravely on, still under the patronage of Saint Michael, the leader of the angelic aristocracy.[51]

The pure sciences, with only seventeen students in 1889, remained somewhat subsidiary to medicine in the Catholic Institute at Lille. By 1890 the medical faculty had produced 100 physicians to look after livers and souls. By 1895 the institute's student population had grown to 700. Its students had the excellent success rate of 75–80 percent in state examinations, as compared with about 63 percent for state students. The Paris institute was also active: By 1900 it had turned out 119 doctorates, 97 of which were in civil law and 22 in letters and sciences, 26 *agrégés*, 1,619 licentiates (892 in law, 534 in letters, and 193 in science).[52]

There were difficulties in attracting the minimum number of students. Nine years after the founding of the institute in Lyon the rector sent a letter to those interested in Catholic higher education to point out that the continued attacks on the Catholic faculties were reducing the number of students because of a loss of confidence on the part of parents. Young people also feared that they might not get into the "liberal careers" if they were graduates of a Catholic university. Many Catholics still attached a great prestige to an education in a state institution. Not insignificant was the hope of students that one would be more likely to succeed in examinations where one's own professors were the examiners. The rector's chief justification for a parent's sending his offspring to the Catholic faculties was one calculated to inspire the confidence of the Catholic bourgeois – the success rate of about 75 percent in the state examinations for diplomas. A public vice produced a private good.[53] The Catholic institutions of higher learning showed more enthusiasm than the state institutions in accepting the traditional structure of nonreligious values in French higher education.[54]

Whether these results justified the enormous financial burden placed on Catholics is a question that split the Catholic world itself after the great financial losses of the separation of church and state in 1905. The finances of the institutes took a sharp turn for the worse after the 1870s. In 1879 d'Hulst reported that the financial situation of the Catholic university in Paris was not bad: About a million francs had been placed in trust for professorships and, along with fees, would produce about 100,000 francs income. It was expected that the thirty dioceses associated with the university would average at least a contribution of 11,000 francs annually. Already nine Paris parishes had undertaken to found chairs with a capital of 100,000 francs each. Yet funds were so limited for the purchase of scientific journals in 1898 that Lapparent was giving his own copies to keep up the collection.[55] But by the time of the separation, when the entire Paris institute had a budget of only 400,000 francs, such optimism was no longer justified. Articles in the Lyonnais review *Demain* and in the *Journal des débats* argued that there was no need to have faculties of letters, of sciences, of law, and of medicine in the Catholic institutes, which should restrict their programs to theology and the religious sciences. Péchenard, rector of the institute in Paris, and Battifol, rector of the institute in Toulouse, entered the debate to defend the institutes. The rectors of the five institutes were against any reform to reduce the scope of their institutions. A professor at Lyon was fired for arguing in favor of a reduction in programs. Professor J. Calvet (Toulouse) argued that the assumption of the founders that there are two sciences, one antireligious and one Catholic, had been an unfortunate tactic putting Catholics both outside and in opposition to the French nation. There are not two truths and there are not two sciences. The institutes did not have enough resources and money to compete with the state University. "L'enseignement supérieur libre, parce qu'il a été imaginé comme un enseignement de combat, se trouve dans une impasse." The paucity of results obtained did not justify the resources used. In the state University, however, the faculties commanded respect because of their scientific spirit and their competence. There was no reason for Catholics to deprive themselves of the benefits partially supplied with their own taxes, for these state facilities were supported by the money of all citizens. An added advantage of a single system would be that the clergy, now separated from and ignorant of their own times, would become enlightened by being educated in the state University. Calvet ended by suggesting that French Catholics follow the German example of using state universities while adding special programs of teaching and tutoring to supply the religious deficiencies of the state institutions.

The bishops rejected this idea, and Msgr. Péchenard, the rector of the institute in Paris, stated with especial clarity the reasons for maintaining the institutes in their totality, in spite of the eternal threat of financial catastrophe. He denied any idea of rivalry with the state, claiming rather that the aim of the institutes was to assure the clergy a more solid intellectual formation, thus enabling them to resist scientific objections to religion. Equally important was initiating youth into a life founded on faith, science, and good morals. It was all very well to say that there are not two ways of teaching science and to ask what was the harm in learning mathematics from Poincaré or medicine from Brouardel. The fact was that there were professors in the state universities who were atheists, rationalists, materialists, and enemies of Catholicism. Péchenard agreed that there are not two sciences, but he held to the argument that there are two ways of presenting science. The same scientific facts could be interpreted very differently. Could a physician who is religious and another who is materialistic teach the same doctrines even though they accepted the same facts? In the face of the growing atheism that had become official, it was only in the Catholic institutes that refuge could be found for the science that finds its support, its inspiration, and its ideal in the idea of God.[56] In 1907 the papal secretary of state, Merry del Val, wrote the archbishop of Paris to support the rectors' view because the Holy See had received complaints that some clergy were following courses at the Sorbonne rather than the institute. In the modernist period, this practice was viewed as presenting grave dangers to the faith and, if too many went to the Sorbonne, a threat to the future of the Catholic institutes themselves. The Sorbonne should be used only in cases of necessity, and the bishops should be difficult, especially in dangerous subjects like philosophy and history. A contribution of 100,000 francs from the Holy See showed how seriously Pope Pius X viewed the issues raised in the letter. After World War I the financial situation was desperate: Lille had a deficit of 268,766 francs (budget of 868,228 francs) for 1919–20. In 1925, with a grant of 64,000 francs from the Pasteur fund for its chemistry, zoology, and geology laboratories, the Institut catholique at Lille received public money for the first time.[57]

The Catholic faculties of science had not been able to supply the Catholic educational system with the required number of science teachers. Péchenard wrote a letter in 1906 to the superiors of educational establishments saying that soon the Falloux law would be replaced by the Chaumié law, which would require the degree of *licencié* or an equivalent diploma for modern languages in the case of all the masters who headed *collèges* and all teachers of higher classes. Péchenard thought that this was a result of the separation of church and state. The Catholic educational system would probably not be hurt in letters, philosophy, and history. Sciences and languages were in serious difficulty, for the legal requirements could not be met at that time. If the law passed, the *collèges* and seminaries would be threatened. The cause of the Catholic deficiency was well known; in most of the private establishments sciences and languages lagged seriously behind.

> Ces études n'ont été ni assez prisées ni assez soignées – on a manqué de professeurs spéciaux et compétents; on ne s'est point assez préoccupé d'en faire former dans nos universités catholiques, créées dans ce but. De là, qu'est-il arrivé? Tandis que les établissements publics réalisaient de réels progrès sur ce double terrain, un trop grand nombre de nos établissements libres se confinaient dans une regrettable infériorité.

The fact of Catholic inferiority in these two areas was so well known, Péchenard warned, that careful judges were saying that it was one of the principal reasons that Catholic families were not sending their children to Catholic estabishments or were, in some cases, taking them out. This was one factor behind the general drop in the number of students in Catholic establishments. The inconveniences attendant on having been educated in a Catholic establishment probably included serious if subtle discrimination on social, political, and economic levels, a consequence not ignored by the Catholic middle classes. Péchenard concluded that the number of science and language students had to be doubled or tripled if Catholic secondary education were to be saved at all.[58] Although disaster was averted, no great improvement in the situation was evident in the next quarter of a century.

Considering the attention that was given to the Catholic system of higher education by its opponents, the involvement of its faculty in national affairs, its competition with the state, especially in some local situations, it is surprising that historians have paid so little attention to the topic.[59] It is curious even if we consider only the importance of the system in the history of Catholicism. From the wider perspective of modern French history, one can argue that the existence of the Catholic system was a factor in policy making within the state system of higher education, not only on a local level, but in the inner sanctum of the Section permanente of the Conseil supérieur de l'instruction publique. In 1907 the Section permanente discussed a transformation of professorships at the state University in Lille that would leave the medical faculty with only one chair of clinical surgery. Bayet, one of the lions of republican secularism, declared that he was afraid that if one of the chairs were suppressed, the Catholic faculty of medicine would make it known through friendly newspapers that the state faculty was declining, thus scoring a propaganda victory against its opponent. Louis Liard, then president of the Section, pointed out that teaching would be effectively carried on by a *chargé de cours* and this objection was groundless; "il ne faut pas avoir le fétichisme des chaires." The faculty request for the transformation of chairs was approved, but the question asked by Bayet was one that was always in the minds of the members of the Section when dealing with Lille, where the state medical faculty had been established only in 1875. The National Assembly, following the advice of Paul Bert's report of 1874, refused to establish a medical faculty at Lille, but in 1875 it voted several millions to establish a Faculté de médecine et de pharmacie and 244,000 francs annually for twelve years. The foundation of the state faculty of law at Lyon took place, in similar circumstances, as a reaction to the new Catholic faculty.[60] Another type of impact of the establishment of the Catholic universities on the state system may be seen in a decision taken by the astronomer Le Verrier in August, 1875. He wrote to the dean of the faculty of sciences at the University of Paris to inform him that he thought that as a result of the organization of the Catholic universities the state would soon be putting great pressure on professors who did not teach their own courses.[61] The existence of the Catholic system played some role in the argument for reform that was characteristic of French higher education from the inception of the Third Republic.

The importance that the Catholic system of higher education had in the thought of the republican educational establishment was clearly shown in Ernest Lavisse's famous article on "The Question of the French Universities" in 1886. The institute at Lille, whose strong support by the Catholic bourgeoisie and proprietors of the Nord made it the cynosure of the other Catholic institutes, was the most successful of the new universities and therefore alarmed republicans more than the competition of the other four institutes. This success seemed all the more striking in a period of difficulties, expansion, and reorganization for the state faculties at Lille. Lavisse was not an enemy of freedom in higher education, but he would have preferred that the intellectual opposition between Catholics and republicans manifest itself in the state University rather than in the clash of two sets of institutions. Since the war between the state and Catholics made this impossible, the existing situation had to be accepted. Lavisse was struck by the important role given to the theological college in Lille, most of whose professors had been trained by the Roman College in Italy. A mentality quite different from that of the republicans revealed itself in the declaration by theologians that they were always ready to enlighten their colleagues in the other faculties on the difficult and delicate questions that could arise out of the study of law, medicine, science, and letters. Lavisse interpreted this as a declaration against the freedom of science but respected it because even declarations against the freedom of thought must be tolerated if one accepts this principle. The Catholic universities should be left free to act, for they were directed by good Frenchmen who believed they were serving their country in the best possible manner. But republicans should serve France in their own way, which Lavisse believed to be better. Defend the principles of modern thought against the Catholic attack. Build school to compete with school. It would be incomprehensible for the state to build, at great expense, its lycées to compete with the ecclesiastical *collèges* if it did not have its University to compete with the Catholic university of Lille. An additional reason for creating a great state University at Lille was its big population: 188,000 in Lille, 91,000 in Roubaix, and 52,000 in Tourcoing. Rich library, archival, commercial, and industrial resources existed. Learned societies abounded. In order to strengthen the state University, the government decided in 1887 to shift the faculties of letters and law, existing separately at Douai, to Lille itself. In the face of the powerful Catholic university that had developed, a weak state institution could offer little competition. Lavisse noted that in the founding period, from 1875 to 1882, Catholics had spent 8,765,000 francs for their university. A new subscription begun in November 1883 had reached 2,254,421 francs by September 1886. Catholics justified these sacrifices by noting they had no trouble employing their young doctors of medicine and that their students were gaining entrance into the church, the army, law, industry, and commerce. At the time Lavisse wrote, 2,000 students had already come out of the Catholic university. An organization of alumni existed. The Collège Saint-Joseph, built on university grounds, had quadrupled its student body in a few years, ensuring a steady flow into the university. In 1896 only five of the seventeen state universities were bigger than the Catholic Institute in Paris, which then had 651 students; the five were Paris, Toulouse, Bordeaux, Lyon, and Poitiers. A formidable challenge to the state University required a response from the state.[62]

Republican forces had to rally to the defense of their interests. Lavisse's great hope was that there would be a true state regional University in Lille, intimately tied to the region. Although it was unlikely that republicans could organize the type of support that Catholics had in the Nord, there was no reason why in this rich area several big manufacturers and big proprietors belonging to the liberal group should not follow the example of supporters of the Catholic university. If the republicans of the Nord did not see the necessity of fighting school with school, and therefore unity with unity, if they did not fight their adversaries for control over the education of youth, they would be condemned to see the defeat of their ideas in future struggles. Lavisse argued that the victor in the battle between the two great parties for the Nord would be science and France. He ended by appealing to the example of Germany, where the national importance of universities had been convincingly demonstrated. German universities attracted the youth of all countries and spread among people a very favorable opinion of France's enemies. France's system of defense against Germany would be complete only when she had built "great intellectual fortresses." The task was not impossible, for a solid foundation already existed.[63]

In the beginning the Catholic system gave opportunities to some scientists that would have been unavailable for years in the state system, and, in the case of some of the graduates of the *grandes écoles*, opportunities that were not possible in the state University. But in the end, due to the prerogatives jealously guarded by the state, the faculties of science, like the institutes themselves, were permitted only a limited development. It may not be too much of an exaggeration to say that, once the threat of republican scientism had passed, or, from a practical viewpoint, had peaked with the separation of church and state, the administrations of the institutes did not show the same enthusiasm for making great sacrifices that had been exuded by Msgr. d'Hulst. This explains the sign on Branly's door for fifty years: "PHYSICS. TEMPORARY LABORATORY – SINCE 1875." But this should not obscure the creditable if modest achievement of the Catholic faculties of science in the nineteenth and twentieth centuries. The work of Lapparent, Branly, Rousselot, Hamonet, Amagat, Senderens, Boulay, Valson, le comte de Sparre, Lepercq, and Colin was recognized as solid by the scientific establishment and even beyond.

7

SCIENTIFIC PUBLICATION
THE FLOOD OF MONOGRAPHS, BOOKS, AND JOURNALS UNLEASHED BY THE NEW RESEARCH IMPERATIVE IN SCHOOLS AND SOCIETIES

More than half of modern culture depends on what one shouldn't read.
– Wilde, *The Importance of Being Earnest*

Monographs and books

BUFFETED by conflicting ideologies, numerous intellectual movements, prolific authors, and cheap printing, the nineteenth century unleashed an unprecedented flood of the printed word. With the growth of education, professors took on a greater importance in publishing materials for the educational system and in writing works for the educated public. Among the most colossal scholarly works, although not the most perfect, were Migne's editions of the *Patrologia latina* in 221 volumes (1844–55) and the *Patrologia graeca* in 166 volumes (1857–66). The growth and popularization of science led to series like Reinwald's *Bibliothèque des sciences contemporaines*. The social sciences and philosophy had a famous outlet in the publisher Alcan's *Bibliothèque contemporaine*, many of whose volumes were written by his fellow normaliens. By 1904 Alcan's *Bibliothèque scientifique internationale*, under the editorship of Emile Alglave, included 103 volumes, many of which were adopted by the Ministry of Education for inclusion in the libraries of lycées and collèges, although only about fifty of the volumes (including translations) were in physics, chemistry, biology, and physiology. Much of Alcan's list was made up of excellent works in the *haute vulgarisation* of science; Alcan did not achieve the same distinction in scientific publication as it did in philosophy, history, and psychology. Alcan also had a *Bibliothèque utile* of 125 volumes (1904), which included the physical and natural sciences; in this series, Ferrière's book on Darwinism was in its seventh edition by 1904. The series covered a miscellany of subjects ranging from elementary chemistry for farmers to tuberculosis, and it included an abridgement of Debidour's history of church–state relations between 1789 and 1871 (1904). Quarrels over relations between science and religion gave birth to two Catholic series: Palmé's *Nouvelle librairie scientifique* and Bloud's

251

famous *Science et religion*. Bloud's series reached about 400 modestly sized volumes by 1906. Science and religion were important ingredients in the intellectual ferment that kept printers busy, but a much greater number of duller scientific works was produced by teachers and researchers in faculties and schools.

By the end of the century new encyclopedic works appeared to synthesize the result of the century's "knowledge explosion"; *La Grande encyclopédie* was a fitting monument of erudition and utility to the age. (*La Grande encyclopédie* – "inventaire raisonné des sciences, des lettres et des arts, par une société de savants et de gens de lettres sous la direction de MM. Berthelot, Levasseur, Marion," etc.) Perhaps the culmination of this vast movement was the founding in 1921 of the Presses universitaires de France, a "maison d'édition française" – legally a cooperative limited liability company – with its typically nineteenth-century aim of spreading French culture at home and abroad. Its many collections of scientific, historical, and philosophical works started with the "ancien fonds Alcan"; its most famous contribution to popularization was the "Que sais-je?" series.

The publication of scientific and mathematical works posed special problems of editing and, above all, of financing. In spite of its near monopoly of government-financed scientific publication, the firm of Gauthier-Villars – "Imprimerie et librairie pour les mathématiques, les sciences et les arts" – had to make its profits in other areas of publishing. Jean-Albert Gauthier-Villars (1828–98) is one of the great names in the history of publishing. Son and grandson of printers, a graduate of the Ecole polytechnique, after a short period as a telegraph engineer, he bought the printing establishment and bookstore founded by Courcier in 1791 and continued by Bachelier (1821–53) and Mallet (1853–64). Gauthier-Villars developed his business into the leading French scientific publisher with a near monopoly on the publication of mathematical works. He published government-subsidized works for the Bureau des longitudes, the Polytechnique, the Observatoire de Paris, the Ecole normale supérieure, and the Académie des sciences. Under the auspices of the Ministry of Public Education and the Académie des sciences, Gauthier-Villars also published the collected works of some of the giants of French science and mathematics – Cauchy, Fermat, Fourier, Lagrange, and Laplace. This costly collection began in 1865 with a plan to publish the *Oeuvres de Lagrange* in seven volumes, of which the fourteenth and last volume came out in 1892. Even the original plan called for the ministry to subscribe to 300 copies of each volume for an average of 10,500 francs per volume. The history of the process of publication reveals the mixture of science and politics inspiring one of the greatest ventures in the history of scientific publishing.

In a note (February 19, 1877) to the Commission administrative de l'Académie des sciences, Gauthier-Villars pointed out that it was very costly to publish such works with the care that they deserved. True, these great scientific works gave the firm its reputation, but in order not to be bankrupted by publication costs, he had to create other sources of income, to publish other works, and, especially, to add to his "Imprimerie mathématique" a large printing establishment to produce profits from administrative manuals and catalogs. Science fiction made a considerable contribution to the diffusion of science by Gauthier-Villars, for he also published the works of Jules Verne. Lagrange's work brought honor to the firm, but it had to be paid for with profits from Verne. Yet scientific publication could not exist only as a financial parasite on science fiction.

By the 1880s scientific publishing could profit from the role it played in the political game of international science, in which each seminal work could represent a battle gained in the struggle to establish national works as major inspirations for a discipline's paradigm. One of the Third Republic's weapons in this game had to be the works of the unregenerate Catholic royalist Cauchy. In a letter to the Ministry of Public Education (January 6, 1882), the permanent secretaries of the Académie des sciences urged the ministry to subsidize Gauthier-Villars's publication of Cauchy's works. Published under the auspices of the Académie des sciences, Cauchy's works would be "si honorable pour la science française." As science progressed, the secretaries argued, Cauchy became more significant, and European geometers, who had all become his students, "today place our illustrious compatriot in the ranks of the greatest discoverers in the history of mathematics." The Académie des sciences would be emulating other European learned bodies in patronizing Cauchy's work, some of which could not easily be found. Gauss's works had been published by the academy of Göttingen, Abel's by the Norwegian government, and Jacobi's were being published by the academy of Berlin. The secretaries made a list of the various academies and eminent geometers to whom the ministry should send copies, and also noted that it would be a good idea to send copies to the libraries of university faculties and of large towns. An alert anticlerical in the ministry placed a question mark opposite the name of the Catholic faculty of sciences in Lyon; perhaps he did not know that a mathematician of that faculty, Valson, was a collaborator on the project. The determination of the republic to promote science and to keep France in the forefront of scientific work was an important factor in financing expensive scientific publications.[1]

In no area did scientific advance and the expansion of faculties of science combine to produce more striking growth in publication than in physics, especially by the 1890s. Reviewing the situation of physics textbooks in

1892, Aimé Witz took stock of general physics books.[2] There were two categories of treatises. First, there was a group of advanced texts, extensive in coverage and excessively erudite in their treatment of the material. Their general defect was they were too big and too expensive: even the most enthusiastic students were discouraged by 3,000 pages in octavo, and shocked by a price of seventy francs. To this class of elite texts belonged the course at the Ecole polytechnique written by Jamin, revised and completed by Bouty. Comparable to this work were Wüllner's *Lehrbuch der experimental Physik* (with *Compendium*), Verdet's *Cours de physique professé à l'école polytechnique* (1868–69), and Violle's *Cours de physique* (4 vols., 1883–92). The second class of texts, which students would be tempted to substitute for the unsatisfactory first group, was unfortunately at too low a level for licence students. Drion and Fernet, *Traité de physique* (1861; 12th ed., 1893), Boutain and d'Almeida, *Cours élémentaire de physique* (2 vols., 1862; 5th ed., 1884), Angot, *Eléments de physique* (4 vols., 1881), Pellat, *Cours de physique* (2 vols., 1890) were the better-known texts written for students in secondary education. Because they did not use calculus they were sadly deficient for students taking a licence in the physical sciences. In 1891 the gap between the two groups of texts was filled by the three-volume *Leçons de physique générale*, running to nearly 1,400 pages, published by Gauthier-Villars. This work satisfied Witz's criteria: "un livre renfermant un exposé clair, court et précis de toutes les belles et hautes théories de la physique moderne, faisant une place modérée aux travaux des maîtres, tout en développant leurs conclusions: c'est un ensemble de leçons ... plutôt qu'un Traité complet."[3]

There was a considerable choice of textbooks in physics at the secondary-school level. Some authors became famous because their texts went through from fifteen to twenty editions. Most elementary texts in French developed the programs for the baccalaureates and for admission to the Ecole Saint-Cyr and the Ecole centrale, although a few covered the material needed for courses at the Ecole polytechnique and the Ecole normale supérieure. But nearly all were more manuals than treatises. Obsession with examinations had corrupted the great pedagogical traditions of Biot, Lamé, and Pouillet. Instead of emphasizing the general ideas and doctrines making up the foundations of science, these books gave numerous statements of laws and facts with minute descriptions of instruments and apparatuses. "Les procédés étroits des *chauffeurs* d'examen tendent à prévaloir." For a distinguished engineer-physicist like Witz, professor at the Catholic institute in Lille, it was more important to make sure that a student who took a year and a half of science got an understanding of the basic principles of natural science than be forced to learn the laws of physics. Professors should train students' minds and strengthen their judgment rather than prepare them for examinations. "Il import de . . . former son esprit, de fortifier son jugement et de le dresser aux méthodes inflexibles du raisonnement . . . j'estime qu'une éducation reste incomplète, si elle ne termine pas par cette gymnastique puissante de cours de sciences, qui accoutume l'esprit à la netteté, à l'ordre, à la précision, à la rigueur, à la logique en un mot." Witz believed than an elementary textbook by Father van Tricht was a model text, embodying all the virtues he found lacking in so many similar works.[4] Jesuit pedagogical virtues were not without their advantages in teaching modern science.

There was a great deal of concern in Catholic education over the quality of textbooks. In physics, unlike biology, this was not a matter of excluding the secular republican ideology: There was no Catholic position on electromagnetism, whereas there was one on evolution, at least concerning man. So when Edouard Branly, professor of physics at the Institut catholique in Paris, published a text of the most elementary sort, aimed only at the baccalaureate students, the Jesuit physicist J. Thirion took the occasion to set forth the qualifications of a model textbook writer. He should be a true scientist and an experienced teacher with a gift for precise, sober, and methodical exposition. In order to keep his text exact and rigorous while being simple and easy, he must know much more than he puts on paper. Without these qualities the writer cannot make scientific principles clear, develop ideas one from another, foresee and forestall difficulties, separate the important from the incidental, or prevent essential truths from being lost in a sea of obscure details. The superior knowledge of the author must be held in check, used only to enlighten, not to dazzle, to throw light on his subject matter; the best use of his talent for exposition is in hiding the difficulties he conquers in achieving simplicity and rigor. If to the knowledge of the scientist and the talent of the teacher the writer is able to add literary skill, and if he can find a publisher concerned with typography and eager to provide clear and expressive diagrams, he will win widespread acceptance of his work. Such was the textbook by Branly, worthy of comparison with the higher-level textbooks by Brisse and André, by Pellat, and by Moutier.[5] Branly's book (*Traité élémentaire de physique*) was part of the collection of classics published by the Alliance des maisons d'éducation chrétienne.

There is considerable evidence that the French-speaking scientific community believed its model of a scientific monograph or textbook to be superior to English and German models. The classic statement of the intellectual grounds for this intellectual and linguistic imperialism is found in Pierre Duhem's intriguing analysis of national styles of thinking, which, applied to Oliver Lodge's use of mechanical models in physics, resulted in a famous, if wrongheaded, judgment. "We thought we were entering the tranquil and neatly ordered abode of reason, but we find ourselves in a factory."[6] A more typical, and less striking, opinion was that of Victor van Tricht, S.J., who in his review of translations of Jenkin and Maxwell complained of the excessive comprehensiveness and lack of order in English and German scientific books. "Tout y est, rien n'est passé sous silence, ni un fait, ni un instrument, ni une loi, ni une théorie, c'est une compilation de choses vraiment prodigieuse; mais que cela est peu rangé! et comme on a peine à s'y retrouver! Passez de là à un chapitre de Lamé, à un traité ou à un mémoire de Verdet, c'est une vraie jouissance!" When the Belgian mathematician Paul Mansion reviewed the French translation of *An Essay on the Foundations of Geometry* by Bertrand Russell, a paragon of clear thinking in the Anglo-Saxon world, he found the second chapter so obscure that he was reminded of the poem "Pleine mer" in Victor Hugo's first *Légende des siècles*: "Abîme, obscurité, ténèbres, cécité, immense nuit, c'est bien cela."[7]

By the 1870s and especially by the 1880s, continuing publication on electricity and magnetism had totally revised this area of science. New treatises on electromagnetism appeared with some regularity, in spite of the existence of remarkable works such as Gordon's *Traité expérimental d'électricité et de magnétisme* (translated by Reynaud; 2 vols., 1881), and Mascart's and Joubert's *Leçons sur l'électricité et le magnétisme* (2 vols., 1882–86). In 1884 Maxwell's *Traité élémentaire d'électricité* and in 1885 Fleming Jenkin's *Electricité et magnétisme* were published by Gauthier-Villars. Van Tricht did not regard them as the ultimate statement of the new paradigm in magnetism. So the stream of words would keep flowing from the presses until electromagnetism received its classic moulding by a great master. When a new Verdet, or a Lamé, or a Jamin, gave the renewed science its definitive form, the opportunity would only remain for successors to write copies of the classic treatise.[8]

Publication of a translation of Maxwell's great mathematical treatise on electricity elicited similar strictures from Philippe Gilbert in his eleven-page review of the work. The first English edition (1873) did little to spread Maxwell's ideas in France, although they were known through a few works (e.g., Mascart and Joubert). The translation of the second edition (1881), which was revised by Niven after Maxwell's death in 1879, appeared in French, bristling with critical notes and complementary studies by Cornu, Potier, and Sarrau (*Traité d'électricité et de magnétisme*, by J. Clerk Maxwell, 2 vols., 1885–9). G. Seligmann Lui, the translator, was a graduate of the Ecole polytechnique. The greatness of Maxwell's work was frankly recognized, but French physicists thought it prudent to present it to the French public in a quasi-Gallic guise. Some of the problems resulted from obscure language, artificial and illogical exposition, and proposed impossible, or inconclusive, experiments – problems present not only in the part on electrostatics but also in the section on electromagnetism, which contained most of his original ideas. Sarrau even showed how quaternions, a stumbling block for French readers, could be integrated into the general theory of complex quantities, more familiar to the French. Much other experimental work by Mathieu, Mascart, and Delsaulx was necessary to bring Maxwell up to date. A general attempt was made to relate Maxwell's work to continental research and mathematical work; so the translation turned out to be a "more modern and better-ordered" work than the original. "En général, lorsqu'on passe du texte de Maxwell aux notes de MM. Cornu et Poitier, on éprouve la même sensation qu'en sortant d'un fourré pour entrer dans un parc aux avenues bien tracées."[9] The attempt to clarify the work of the "prince of nineteenth-century physicists" was firmly grounded in the French experimental tradition, although the science produced was not comparable with that of Hertz and Boltzmann.[10]

While works on the scientific and industrial application of electricity streamed from the presses, the theoreticians – likely to be "physicists-geometers" in France – continued to try to understand the phenomenon completely through imposing a mathematical order on it. In reviewing J. Bertrand's *Leçons sur la théorie mathématique de l'électricité*, Gilbert looked forward to the day when this area of physics would be as certain and clear as celestial mechanics. Meanwhile the work of Riemann, Kotteritsch, Betti, the Neumanns, Clausius, Maxwell, Gauss, Green, Thomson, Helmholtz, Kirchoff, Mathieu (*Théorie du potentiel avec ses applications à l'électricité et au magnétisme*, 2 vols., 1885–86) and Duhem (*Leçons sur l'électricité et le magnétisme*, 3 vols., 1891–93) was laying the foundations for the future research of the happy geniuses who would flood the dark maze of electricity with the light of mathematics. The book by Bertrand, professor at the Collège de France and a permanent secretary of the Académie des sciences, culminating over thirty years of work, did not make any theoretical advance on what was already known, but it did state sophisticated theories with great elegance and made a serious attempt to connect them.[11]

More often than not, the type of method and clarity demanded by the French could only be found in books which, however useful and even necessary, were distinctly peripheral to that elusive entity "the advance of science." The first French work to deal extensively with the theory of units, *Introduction à l'étude des systèmes de mesure usités en physique* (1891) by Pionchon, professor at the faculty of sciences in Bordeaux, was of great use to anyone who wanted to study modern physical theory. L. L. Godard could render the flattering Gallic judgment, "Il est fait avec beaucoup de méthode et de clarté."[12]

Some modern historians of science have commented on the strange "decline" of physics, particularly mathematical physics, in nineteenth-century France.[13] Perhaps it would be more accurate to limit the "decline" to the generation between the 1840s and the 1880s. This curiosity was noted in the nineteenth century itself. In 1884 H. Résal, well known for his *Traité de mécanique générale* (7 vols., 2nd ed., 1895), brought out in book form many articles that had been published in his *Journal de mathématiques*; these appeared under the title of *Physique mathématique: Electrodynamique, capillarité, électricité, élasticité*. Gilbert regarded Résal's work as filling a most unfortunate gap in French scientific literature. One of Résal's aims was to revive the tradition of Laplace, Fourier, Poisson, Ampère, Fresnel, Cauchy, Saint-Venant, and Lamé – great names in the creation of mathematical physics – and to generate some encouragement for the few people, like Boussinesq and Mathieu, who continued to labor in the field without much recognition.

> Raviver le goût de ces belles études, en mettant à la disposition des jeunes géomètres un exposé rapide des principaux résultats obtenus par les grands hommes . . . avec les simplifications et les améliorations que peuvent inspirer une rare habileté dans l'application de l'analyse aux questions mécaniques et une profonde expérience de l'enseignement, tel est à la fois le but excellent et le caractère du traité par M. Résal.[14]

Besides consecrating the paradigms of the creators of the various areas in mathematical physics, Résal's book served as a pioneering and model text in higher education, where there was provision for an examination in the subject without a program or syllabus. If the student was not examined by the professor whose course he took, he could end up being examined in areas not covered in his course because of the vast number of topics possible. The two-volume second edition of 1887 nearly doubled in length, adding essential topics left out of the first edition – the theory of light waves, the mechanical theory of heat, and thermodynamics. As complete as such a treatise could be, it played a very useful role in University courses; with Résal as their *cicerone*, professors could now avoid the dilemma of covering all topics superficially or treating a few in depth.[15]

After the 1870s a steady increase of interest was evident in the area of thermodynamics. By the 1880s thermodynamics was a significant subject for specialists in mechanics as well as physicists, and theoreticians as well as engineers. A diverse literature catered to these varied interests. In 1885 Jules Moutier, who was known for his work on capillarity and for his *Eléments de thermodynamique* (1872), published *La Thermodynamique et ses principales applications*, which was palatable to chemists, geologists, and even many physicists usually frightened by the formidable mathematical baggage of the subject. By contrast, *Thermodynamique* (1887), by J. Bertrand, was a work at a very high level: "une série d'etudes critiques, pleines de vie, de variété, de pénétration et d'originalité, sur les points principaux de cette branche de la science et sur ses applications aux problèmes élevés de la physique."[16] By 1887 Clausius's *Théorie mécanique de la chaleur* was out in a second edition, translated from the third German edition; this was the same year that Gustav Zeuner's *Technische Thermodynamik* came out in its third edition. In 1892 J. Blondin edited Henri Poincaré's *Cours de physique mathématique: thermodynamique*, which was viewed by physicists as a fascinating, if exotic, treatment of thermodynamics by a mathematician. Pierre Duhem made an amusing comparison between Poincaré's treatment of physics and a possible commentary on European civilization by a visiting Brahmin. The scepticism the physicist imagined he found between Poincaré's lines could be most disconcerting. Physicists, after all, eternally sought objective reality and useful applications of their work, in contrast to the mathematician's usual concern with "unreal truths." One of Duhem's comments would give little comfort to those who thought thermodynamics difficult because of its mathematical difficulty: In no branch of theoretical physics was the role of mathematical analysis more limited and the distinguishing qualities of the physicist – precision of definition and detailed criticism of the sense and extent of experimental data – more in demand. Yet the judgments of one of the most powerful and original minds of the age could be of profit and interest to the physicist even if they were surprising and sometimes outrageous.[17]

The tremendous boost given to scientific publications by the growth of institutions of higher learning was evident in nearly all areas of science. But there were exceptions. Curiously, meteorology, in whose development France had played a leading role, was of low priority in research and publication well into the twentieth century. In his *Traité élémentaire de météorologie* (1899), Alfred Angot attributed the decadence of the discipline to the lack of any regular French teaching of the subject or of its more general context, geophysics. Belgium was possibly worse off than France. Except for the Institut agronomique, meterology was not in the program of French institutions of higher learning; in other countries the subject was part of the curriculum in at least some universities and other institutions. Although Angot exaggerated the position meteorology held abroad, French publication in the area compared poorly with its neighbors because it lacked the stimulus to research and publication inherent in higher education.[18]

In the last quarter of the nineteenth century no field changed more rapidly and proliferated more subspecialties than did chemistry. New chemical theories flourished in the lush environment provided by a fast-growing discipline with a strong empirical bent and a heavy applied emphasis. In the late 1870s the purely theoretical part of chemistry could be summed up in a few general ideas, which, insufficient to justify a separate publication, were usually outlined at the beginning of chemical treatises. By the late 1890s chemical theory had become a recognizable branch of the general discipline, with its own separate literature. This was strikingly evident in physical chemistry. The classics in the field were Ostwald, *Lehrbuch der allgemeinen Chemie* (2 vols., 1885–6), Lothar Meyer, *Die modernen Theorien der Chemie* ... (1876), Berthelot, *Essai de mécanique chimique fondée sur la thermochimie* (2 vols., 1879), and, above all, Ostwald's and van 't Hoff's journal, the *Zeitschrift für physikalische Chemie* (Leipzig, 1887–). Because Duhem's *Traité élémentaire de mécanique chimique* (4 vols., 1897–9) was incomprehensible to many chemists, Hermann brought out his *Thermodynamique et chimie: Leçons élémentaires à l'usage des chimistes* in 1902. Those who wanted a quick and reliable summary of physicochemical theories had their wish fulfilled in 1896 when A. Etard, a tutor at the Ecole polytechnique, published a volume of 196 pages in the *Encyclopédie scientifique des Aide-Mémoire* on *Les Nouvelles théories chimiques*. A manual by Monod on stereochemistry performed a similar function. French-language chemists – the phrase was current in the nineteenth century – needed a short work, based on periodical literature, dealing with the theories of Le Bel and van 't Hoff, as well as with the work of Fischer, Baeyer, Guye, and Friedel. Regarded as the youngest of chemical theories at the end of the nineteenth century, stereochemistry, in spite of its brilliant advances, was burdened by the accusation that atomic theory lacked an empirical foundation. Even a propagandist for stereochemistry like Monod could admit the possible unreality of it all.

> Avant tout, insistons sur ce point, à savoir que la stéréochimie n'est rien autre chose qu'un système de notation commode. Jamais on n'a prétendu avoir trouvé la forme réelle, exacte, d'une molécule; on a seulement pris, pour la représenter, un symbole nouveau; et tout ce que l'on peut affirmer, c'est que ce symbole est plus rapproché de la réalité que l'ancien, puisqu'il est plus d'accord avec les phénomènes connus.[19]

Nowhere was the impact of theory on publication more pronounced than in chemistry with the eventual triumph of atomic theory in France. Resistance to atomic theory on the part of the scientific mandarins like Berthelot, whose power over chemistry teaching was evident at all educational levels, had meant that the theory was absent from the texts. But with the substitution of the atomic for the dualistic theory and the system of equivalences in scientific work, including that done in faculties of science, the Ministry of Public Education finally authorized the teaching of atomic theory in secondary education. Ministerial authorization meant a general recommendation, leading to wholesale adoption of atomic notation in place of notation in equivalents. In a period of pedagogical transition, books were published for the University students who had not learned the atomic theory in the lycée. Some authors took the easy way out in revising their texts by simply interpolating the atomic theory on a secondary level; some put equivalent notation on a secondary level; the ambitious and modern-minded rallied to the atomic theory alone. What was needed temporarily was the type of book written by E. Lenoble, chemist at the Catholic faculty of sciences in Lille (*La Théorie atomique et la théorie dualistique: Transformation des formules: Différences essentielles entre les deux théories,* 1896). His work showed students how to translate easily an equation given in atomic notation into the language of equivalents; it also gave partisans of atomic notation the system of deciphering books using only equivalents. As Goossens observed, French chemists had been under pressure for a long time to abandon the dualistic theory and, above all, to drop notation in equivalents – at a theoretical level the two were not identical. Berthelot and his disciples used a system of individual notation. The only contemporary merit of the system being sacrificed was that it was French. The adoption of atomic notation had the added Gallic virtue of simplification, not much evident in this debate.

One of the distinguishing features of the French educational system was its emphasis on pedagogy – cynics would say examinations – with considerable attention given to the perfecting of courses expected to embody elements of originality, as well as to conform to national norms. Many scientific books published in France were basically the courses taught by professors. Mathematics was given a solid and perhaps dominant part of the market. The system of competitive examinations to determine entrance into the *grandes écoles* generated its own series of works, particularly in mathematics. One work could serve to prepare for a number of examinations. B. Niewenglowski's *Cours d'algèbre à l'usage de la classe de mathématiques spéciales et des candidats à l'Ecole normale supérieure et à l'Ecole polytechnique* (1889) is an example of this sort of text. In two volumes of nearly 900 pages, it was published in a second edition by Armand Colin in 1891. Most courses were published by Gauthier-Villars. Among the famous mathematical courses were those of Mannheim, Serret, and Hoüel. As might be expected from a "Chef d'escadron d'artillerie" who was a professor at the Polytechnique, Mannheim's *Cours de géométrie descriptive de l'Ecole polytechnique, comprenant les éléments de géométrie cinématique* (1880) was a solid and beautiful text. "Enfin, par la beauté de l'exécution, par la correction du texte et l'élégante clarté des figures, l'ouvrage est tout à fait digne de la maison Gauthier-Villars et de la haute réputation dont elle jouit dans la librairie mathématique." J.-A. Serret's *Cours d'algèbre supérieure* (1849), in two volumes of nearly 1,400 pages, was one of the best-known classic texts, so successful that Gauthier-Villars brought out a fifth edition in 1885; the first four editions had sold out. Charles Sturm's *Cours d'analyse de l'Ecole polytechnique* (1888), which came out in a ninth edition in 1888, was revised and updated to reach the clientele of the new licence program. The first edition had come out in 1857. A good text, even at high levels, had a steady, if unspectacular, market.

Rarely did a textbook have much to recommend it except the usual qualities demanded by the French in their manuals. But the first of the four volumes on calculus by J. Hoüel, professor of pure mathematics at Bordeaux, did stand out from other excellent mathematical texts because of its capability of inspiring the reader to think of pursuing his own original research. Paul Mansion lavished his praise on it because of the qualities distinguishing it from the common herd of manuals: "la clarté, la rigueur dans l'exposé des principes fondamentaux, le choix habile des applications, la bonne ordonnance de l'ensemble, enfin l'esprit scientifique qui y règne presque toujours, et qui est si éminemment propre à exciter chez le lecteur le goût des recherches originales." The first volume of Hoüel's *Cours de calcul infinitésimal* (1878) also sold for the reasonable price of fifteen francs.

Gauthier-Villars published a series of works of high-level popularization – if the term *haute vulgarisation* is not too absurd in this case – for an entity called the mathematical public. A group of mathematicians, well-known for their original work, prepared in didactic form a complete survey of contemporary knowledge in the different branches of higher mathematics. The works of Jordan and Picard on analysis, Halphen on elliptical functions, Darboux on the geometry of surfaces, Tisserand and Poincaré on celestial mechanics, and Appell on rational mechanics were among the "bel ensemble de traités magistraux."

The second half of the nineteenth century was one of the most fertile periods in the history of French mathematics. Following the work of Weierstrass and his disciples, G.-H. Halphen's *Traité des fonctions elliptiques et de leurs applications* (1886) was a classic in the series on the "theory of elliptic functions and their developments." H. Laurent added to the ninth edition of Sturm's course an elementary theory of elliptic functions when it came out in 1888. It was to prepare students to read the original works of mathematicians like Abel, Jacobi, Weierstrass, and Hermite, and the treatise of Halphen, that Jules Tannery and J. Molk began publication in 1893 of a four-volume work on the *Eléments de la théorie des fonctions elliptiques*, of which the first volume dealt with differential calculus. Sometimes a book could provide a definitive introduction to a new topic while incorporating much of the author's original work on the same topic; this was done by the polytechnician and Deputy C.-A. Laisant in his work *Théorie et application des équipollences* (1887). In 1870 Laisant had translated into French the pioneering work of Giusto Bellavitis. A more important example of the type of publication exemplified in Laisant's book was Gaston Darboux's four-volume *Leçons sur la théorie générale des surfaces et les applications géométriques du calcul infinitésimal*, the first volume of which Gauthier-Villars published in 1887. Together with mathematicians of the caliber of Poincaré, Picard, Hadamard, Appell, and Painlevé, Darboux made up the brilliant group that elevated the Sorbonne to one of the world's great mathematical centers. The careful publication of their research and courses by Gauthier-Villars made that firm a household word in scientific circles and the prima donna of the publishers of mathematical works. It is curious that the work hailed by Maurice d'Ocagne as the ultimate in typography and a real model of its kind was Bertrand's *Calcul des probabilités* (1889), which succeeded in the improbable task of allying wit and mathematics. Few French scientists could resist admiring a mind that produced sentences as beautiful as equations in a book hailed for its delicate and brilliant wit.

During the last two decades of the nineteenth century, scientific publications gave an increasing amount of attention to technical and practical matters. Looking at science as a whole, and especially physics and chemistry, one can detect a sort of intussusception between technology and science, in spite of the obvious barriers separating the two enterprises.

A substantial amount of publication dealt with the important and increasing role of photography in science. Literature on the application of photography to various branches of science, especially astronomy, was evident in the 1880s and more so in the 1890s. Mouchez wrote a long analysis of astronomical photography in the *Annuaire du Bureau des longitudes* (1887). In 1887 an international conference was held at the Paris observatory on photographic methods of establishing a general map of the heavens. Also in 1887 G. Rayet, director of the Bordeaux observatory, published his *Notes sur l'histoire de la photographie astronomique*. This brochure of sixty-four pages was the fourth number of the *Bulletin astronomique de l'Observatoire de Paris*, published by Gauthier-Villars. Nearly ten years later publications were numerous on both the theory and application of photography. In 1895 M. le Comte de la Baume Pluvinel published *La Théorie des procédés photographiques*, which combined exact scientific data with useful practical details to give "the quintessence of photographic science." The same year saw the publication of *Applications scientifiques de la photographie* by G. H. Niewenglowski, already known for many works on the subject. His book of 1895 dealt with the technique of scientific photography as well as with applications. The camera had become one of the most useful aids to science. Long an indispensable instrument in astronomy, it was coming to have key uses in other sciences, especially in experimentation. In 1889 Ottomar Anschütz took daylight photographs of projectiles with an initial speed of 418 meters per second; this work was continued by Mach and Salcher, and then by Boys. Another area of physics of obvious importance for photography was optics, especially spectroscopy: Cornu constructed a chart extending the ultraviolet solar spectrum of Angström. Little wonder that Gauthier-Villars had a special series entitled *Bibliothèque photographique* to honor the new technique.

The books by the comte de la Baume Pluvinel and Niewenglowski, along with an earlier one by E. Wallon on the *Choix et usage des objectifs photographiques*, were published jointly by Gauthier-Villars and G. Masson as part of an *Encyclopédie scientifique des Aide-Mémoire*, which spewed from the presses at the rate of thirty to forty volumes a year. Typical of works published were H. Lecomte, *Les Textiles végétaux, leur examen microchimique* (1892); L. Lindet, *La Bière* (1892); M. Berthelot, *Traité pratique de calorimétrie* (1893); Dr. A. Broca, *Traitement des tumeurs blanches chez l'enfant* (1892); and R. R. Koehler, *Application de la photographie aux sciences naturelles* (1893). A real scientific library in itself, the series was edited by Leaute. It was probably the best of such series and of most use to biologists and engineers. Written for scientists, it was beyond the comprehension of the ordinary reader of the *Bibliothèque des merveilles* and other series written to popularize science for the general public.

Another high-level series, the *Encyclopédie des travaux publics*, founded by Georges Lechalas, won a gold medal at the universal exhibition of 1899. Intended for engineers, many works in the series were probably at the level of the Centrale student, higher than that of the Arts et métiers student and below that of the Polytechnique student. Such works were of use for the graduate of the Arts et métiers if they were in his specialty, and even for the more theoretically trained polytechnicien if they were sufficiently specialized. In 1889 the series included the course of D. Monnier at the Ecole centrale in *Electricité industrielle, production et applications*, a work satisfying the need of electrical engineers for a book somewhere between a theoretical treatment and a popularization of electricity. A second edition of *Hydraulique* by Falmant, *inspecteur général des ponts et chaussées*, came out in 1900. The series was published by Béranger. In 1899 Bernard published the third volume of Aimé Witz's *Traité théorique et pratique des moteurs à gaz et à pétrole et des voitures automobiles*, which incorporated the advances of the preceding four years. The bicycle was not neglected: In the same year Gauthier-Villars brought out *La Bicyclette, sa construction et sa forme,* by C. Bourlet. In the technical area translations were important. H. Lorenz's work on refrigeration was translated from the German by Petit and Jacquet: *Machines frigorifiques, production et application du froid artificiel,* whose practicality enhanced its value in an area where most of the works were theoretical. Another series, the *Encyclopédie industrielle,* also founded by Georges Lechalas, was made up of similar works. E. Rouché and L. Lévy, *Analyse infinitésimale à l'usage des ingénieurs* (2 vols., 1900–2), and E. Deharme and A. Pulin, *Étude de la locomotive: la chaudière* (1900) were among the series' monuments to scientific utility published by Gauthier-Villars.

The increase in scientific publication in the nineteenth century is chiefly the result of the great increase in the number of scientists, most of whom worked in areas whose embryonic forms were evident in the eighteenth century. But some of the scientists worked in new areas and even in "new sciences." The insatiable demand by engineers for works on the latest scientific research with practical relevance produced a vast literature, constantly revised to incorporate appropriate scientific advances. (E.g.: A. Madamet [director of the Ecole d'application du génie maritime], *La Thermodynamique et ses applications aux machines à vapeur,* 1889; J. Boussinesq [professor of mathematics at the Sorbonne and formerly at the University of Lille and the Institut industriel du Nord], *Cours d'analyse infinitésimale, à l'usage des personnes qui étudient cette science en vue de ses applications mécaniques et physiques,* 2 vols., 1888; Maurice Lévy's four volumes on *La Statique graphique et ses applications aux constructions,* 2 vols., 1888; and Paul Appell's *Eléments d'analyse mathématique, à l'usage des ingénieurs et des physiciens,* "Cours professé à l'Ecole centrale des arts et manufactures," 1925.)

A new species of scientific literature was produced by the growing interaction of various scientific disciplines. By the end of the nineteenth century, physics had come to play an important role in medicine. Sometimes, as in the case of X-rays, optics (including general spectroscopy of the blood), and acoustics, this meant better medicine; sometimes, as in the case of electrical shock, it meant atrocity. In 1894 Dr. Moeller waxed enthusiastic about the wide-scale therapeutic use of physics and the important lessons to be derived from physics for physiology and pathology.[20] Treatises applying physics to medicine could be immensely useful to both scientists and doctors. From the physician's point of view, however, most of the standard works – e.g., Wundt's *Handbuch der medicinischen Physik* (1867; translated into French, 1871) and Gavarret's works – covered too much at too high a level. The need for concise, clear, and methodical statements of the principles of physics that were usable by medical men was met by works like that by J. Bergonié, *Physique du physiologiste et de l'étudiant en médecine,* of which volume one dealt with *Actions moléculaires; acoustique; électricité (Encyclopédie scientifique des Aide-Mémoire*, 1894). Often the interaction of disciplines showed itself in the synthetic nature of works: Stanislas Meunier's *La Géologie comparée* (Bibliothèque scientifique internationale of Alcan, 1895) was a type of synthesis of physical astronomy, geology, meteorology, and "planetary physics." The later nineteenth century also saw the birth of new sciences as well as the interaction and division of old ones into specialized subdisciplines. French scientific literature had no specialized book on the new science of oceanography until 1890, when J. Thoulet, a professor in the faculty of sciences of Nancy, published his *Océanographie (statique)*, based on French and foreign published research as well as on his own scientific voyages.

One of the obvious results of the publication of scientific results in book form was that it nearly always produced a detailed public scrutiny and criticism from fellow professionals. The French scientific community had divisions based on genuine professional issues. In biology the hostilities between Giard and Delage were just as serious as, and possibly more virulent than, the battle between Virchow and Haeckel in Germany. Generally, as one would expect of the products of a "normal science" that abhors surprises, the books of fellow professionals were favorably received by the scientific communities, although even an enthusiastic welcome did not preclude ritual minor criticisms of obscure details and minor omissions, and there were a few notorious exceptions to this happy communion of fellows. The exception can also be significant when the review raises the issue of the basic direction of research in an area of science, although reviews rarely do this because of their very origin and nature as part of "normal science." Of course, one could always be criticized for rushing into print. Duhem regretted that Poincaré let two of his disciples publish his lectures (*Théorie mathématique de la lumière*, II, 1892) as they were given rather than rethink and rework them into a definitive study. "Si, au lieu de laisser ses auditeurs livrer à la publicité des recherches à peine écloses et toutes chaudes encore de l'improvisation, M. Poincaré consacrait sa puissante intelligence à poser sur des bases solides des principes de la théorie de la lumière et à en dérouler méthodiquement les conséquences, quelle belle optique il écrirait!"[21] Duhem's reviews represented, of course, the best of that usually dull art form produced by the mutual admiration societies making up the nineteenth-century scientific community.

In France, as in Belgium, the creation of a Catholic scientific structure introduced a new tension among scientists. After the creation of the active Société scientifique de Bruxelles, armed with two scientific journals, one of which was for high-level popularization, secular science at last found a worthy opponent in the century's battle for minds. Within the context of this organization the scientific work of an active secularist like Berthelot could be subjected to a merciless criticism from both scientific and philosophical viewpoints by a Catholic scientist like Duhem.

Duhem was in agreement with his arch-enemy Berthelot, the "dread Commander" of the scientistic gang, on the idea of separate spheres for science and religion. Duhem separated the two to save religion, Berthelot to destroy it. The general religious implications of Duhem's scientific work, insofar as there were any, were favorable to religion, or at least opposed to antireligious implications drawn from other scientific paradigms; the reverse was true of Berthelot's scientific work. Explicit antireligious obiter dicta in scientific works were more likely to show up in geology and paleontology than in physics and chemistry. Biology was often exploited as a basis of scientism, especially after Darwinism came to be utilized as a foundation for materialism (cf. Emile Cartailhac, *La France préhistorique*, 1899, and Stanislas Meunier, *La Géologie expérimentale*, 1899, both published by Alcan). Of course, a great deal of the quarrel between Duhem and Berthelot was about real scientific issues, which can be considered on their own merits, separate from their philosophical and religious implications, although science, philosophy, and religion were often notoriously inseparable in nineteenth-century intellectual history.[22]

Jacques Ellul tells us that "in our society the two great fundamental myths on which all other myths rest are Science and History." Science and history are the basis of propaganda, the source of its success. "The progress of technology is continuous and propaganda must voice this reality."[23] It is difficult to imagine the possibility of stating such dogmas without the fossil remains of nineteenth-century science as the immediate source of these *Ur-myths*.

Societies and journals

The growth of research communities and scientific societies led to an astounding increase in the number of scientific journals. "The proliferation of the *sociétés savantes* is one of the most startling and neglected cultural phenomena of nineteenth-century France."[24] And behind nearly every journal there was a society. Most important learned societies had their headquarters in Paris, but their scope was national. The network of societies had a common link in the Comité des travaux historiques et scientifiques of the Ministry of Education. It organized an annual congress of learned societies. The government granted an annual subsidy of 95,000 francs for the societies to support research and publication. Most of the work of the departmental societies was historical and archeological, some of which promoted antiquarian interests and proclaimed local glories. The *Annuaire des sociétés savantes de Paris* listed 360 societies for the capital; in the 1890s there were probably more than a thousand outside Paris. The Comité des travaux historiques et scientifiques was linked to about 600 of these provincial societies, few of which had anything more than local or regional significance.[25]

Some of the material sent to the sciences section turned out to be worthy of further encouragement and, on rare occasions, of publication. The work of Désiré André, of the faculty of sciences at Dijon, on the functions of Weierstrass was worthy of support, and so was the work of L. Pierre, director of the botanical garden of Saigon. Many works were sent in by lycée teachers. Most of the works were returned to the authors, sometimes because the problem had already been treated in another publication unknown to the author. There was also a fair amount of crackpot material sent to the committee. The setup was a serious one, however, with the section including scientists like Alphonse Milne Edwards, Wurtz, Blanchard, Darboux, Bert, Puiseux, and de Quatrefages. It is probable, of course, that most competent scientists were able to use other existing means for the support of their work and its publication in professional journals.[26]

The establishment of new societies and the growth of old ones stimulated research and scholarship through the establishment of prizes, the imposition of high standards, the provision of means of publication, as well as the intangible benefits of personal contacts in meetings. Often the society also established a specialized library, as did the Société géologique de France. This society also gave the annual Viquesnel prize of 300 francs. Not at all unusual was the society's dual aim of increasing geological knowledge and utilizing it for the benefit of industry and agriculture. The Société de biologie showed that it was aware of the need for public relations by reserving a table for the press at its weekly meetings. This society, which also had its own library and archives, awarded the Godard prize every two years for the best paper in biology.

Nor were the social sciences and humanities moribund. Ernest Godard established a prize awarded by the Société d'anthropologie, which also awarded the Broca and Bertillon prizes. The Thiers foundation was a special institution with an original capital of nearly four million francs, which spent well over a 100,000 francs a year to feed, lodge, and maintain fifteen young scholars for three years while they pursued their own work. The entire system of higher education and learning may have been bourgeois, but opportunities for a poor, bright young man to break into the system did exist and were substantially expanded by the Third Republic. And the emphasis of the new support system for scholarship was on the production of articles and books.

What was the role of these scientific associations in the development of professional science? A new type of society emerged in the early nineteenth century because of the development of disciplines, the growth of educational institutions, and the desire of professional scientists to employ the group's resources in promoting their own work and the growth of their discipline through publication and funding, including as much government support as possible. This desire led scientific groups as well as individual scientists to the necessary paradox of adopting neutrality in the rather frequent political changes in France while supporting the regime in power.

One of the first of the new associations – in Fox's typology, a national society – was the Société géologique de France, founded in 1830 with a nucleus of forty people meeting in the same place as the Société philomatique de Paris. Early speeches by members emphasized the many advantages for the progress of geology and its applications to industrial techniques and to agriculture that could be derived from the establishment of such a society.[27] As in other societies that "managed to achieve disciplinary purity" – the Société entomologique (1832), the Société météorologique (1852), and the Société botanique (1854) – most members were not professional researchers in the discipline represented by the society. Although books in geology did not outsell novels, as they supposedly did in England, the subject was popular, and academic and nonacademic geologists, especially mining engineers, met in the society; the Ecole des mines was a common bond between them. The first president was Cordier; the four vice presidents were Alexandre Brongniart, de Blainville, Constant Prévost, and Brochant de Villiers; among council members were Delafosse and Deshayes. One of the two secretaries was Elie de Beaumont. But Fox points out that nonacademics usually did "the donkey-work as secretary and treasurer." The top geologists did publish in the *Bulletin* and in the *Mémoires*, but of course the *Annale des mines* and the speedy *Comptes rendus* of the Academy of Sciences were also used. Many types of people – academics, engineers, amateur geologists – were interested in extending knowledge of the soil of France. Mining companies were also keenly interested, especially in the later part of the century, and this capitalistic curiosity would be important in establishing geology as an academic subject in provincial universities. By 1831 the geological society had about 150 members, including, among foreigners, Buckland, Murchison, Sedgwick, and d'Omalius d'Halloy.[28] An illustrious beginning was assured on August 25, 1830, when the society was presented to the new bourgeois monarch, Louis Philippe. Constant Prévost (Sorbonne) thought that after the Revolution of 1830 it was appropriate to speak on liberty and science. The king was more interested in artesian wells, or had been briefed on this topic for this meeting, for his message was that they should go forth and dig.

Societies in Paris could reach an ostensibly high level of specialization early in the century. The Société entomologique de France (1832), whose micro-motto was "Natura maxime miranda in minimis," began with the ambitious program of covering the natural history of *Crustacea, Arachnida*, and insects. Among the founding members were Ardouin, Milne Edwards, Latreille, and Duméril. Even specialization was ecumenical in this period, and it was prudently encouraged to be. "L'histoire naturelle ne fait guère d'infidèles, et elle a sur les femmes l'avantage de pourvoir être courtisée à la fois par bien des amans sans faire de jaloux: tâchons donc de n'en faire une prostituée."[29]

Although the *Journal de mathématiques*, edited by Jospeh Liouville, and the *Bulletin de la Société géologique* were founded in the 1830s, comparable journals in chemistry and physics were not established until the late fifties and the early seventies. Important scientific journals, frequently the organs of disciplinary societies, were usually devoted to one science, which further bifurcated into subspecialties with their own journals as time brought increasing scientific specialization. One striking exception was the journal founded by Pasteur when he was the head of the sciences section of the Ecole normale. By 1864 the Normale had become such a good scientific school that its graduates were doing excellent research work. Pasteur thought that it would be useful and "glorieux" to have a periodical publication for the best work of professors and graduates of the school, then among the young scientific elite in France. The editorial committee was made up of brainy and illustrious *maîtres de conférences*: for mathematics, Briot, Hermite, and Puiseux; for physics, Verdet; for chemistry, Sainte-Claire Deville; and for natural history, Delesse, Des Cloizaux, Lacaze-Duthiers, and Valenciennes. Pasteur, Mascart, Darboux, and Terquem published in the early volumes. Most of the articles were by young men in the faculties and lycées.[30] In 1872 Sainte-Claire Deville became editor of the second series. Still going strong today, the journal is a place where even an "Anglo-Saxon" would willingly publish – and now the article may be in English!

The Société chimique de Paris was officially founded on June 4, 1857, after discussions in May between three young chemists: Arnaudon, an assistant of Chevreul in the Manufacture impériale des Gobelins, who later moved to Turin; Collinet, assistant of Dumas in the research laboratory at the Sorbonne; and Ubaldini, who was then working at the laboratory of the Collège de France but later went back to Florence. Their plan was to meet as a group and to establish a series of lectures making the study of chemistry easier for themselves as well as to keep abreast of the rapid progress being made. At the first meeting there were only ten members, but soon there were sixty, as scientists from the faculty of science, the Collège de France, the Ecole de médecine, and the Conservatoire des arts et métiers joined. Foreigners doing chemistry in France gave their enthusiastic support, for they saw in the society a means of establishing a connection between chemistry in France and chemistry in their own countries. The simple educational function of the society soon changed: Meetings were devoted to the communication of original papers, the examination and discussion of papers already published, and lectures on specific topics in chemistry. Progress was its most important aim. "Elle ne connaît qu'une école, celle du progrès, de quelque côté qu'il se fasse jour." Substantial financial support came from business and industry, especially in the 1880s.[31]

The Société chimique de Paris – "de France" after 1906 – was "the first society of national standing to benefit from the new availability of informed audiences for physical science at the research level."[32] Assistants started the *Bulletin* as a poverty-tainted sheet. Then in 1858 Dumas, probably France's most politically powerful scientist, became president of the society, and Wurtz, one of Europe's leading chemists, became secretary. Wurtz, the joint editor of the *Annales de chimie et de physique*, incorporated his *Répertoire de chimie pure* into the *Bulletin*. Wurtz promoted vigorously the obvious connection of chemistry with industry. In 1864 an intellectual alliance was cemented when the *Bulletin* absorbed Barreswil's *Répertoire de chimie appliquée*. The alliance with industry and business was evident in the rise in membership from 400 in 1880 to 1,000 by 1900 and especially in the financial resources of the society. In an age of chemical ebullience and profit, the Société française de physique had more members but far less funds than the Société chimique.

In 1872 the physics community achieved a new level of professionalization with the founding of the *Journal de physique théorique et appliquée* by Charles d'Almeida, professor of physics at the Lycée Corneille and then the Lycée Henri IV. In 1878 he was joined by Bouty, Cornu, Mascart, and Potier. Their aim was to give a new impulsion to the study of physics. The project involved presenting to readers the newest or least-known theories, the experiments on which they were based, and an indication of the easiest way of repeating them, as well as to show the daily progress made by physics in France and abroad. Thus teaching would be infused with new life, the research ethos would be stimulated, and new discoveries encouraged. The journal initially tried to reach professors of physics whose poor resources prevented the development of their work. But all those who had a "scientific profession" – industrialists, engineers, military men, doctors, and others –

were included in the public of the journal, for, after all, famous physicists had come out of their ranks. A special appeal was made to the younger generation of scientists. Love of science was not the sole motive inspiring the founding of the journal. Patriotism was of considerable importance: French intellectual forces would be developed by work and moral forces would result from the disinterested union of common efforts. When d'Almeida died in 1880 his four associates emphasized that they would continue to be guided by his dual principle of love of science and country. By 1890 the journal enjoyed the collaboration of the leading French physicists, including H. Becquerel, Blondlot, Branly, Brillouin, Duhem, Lippmann, and Raoult in the group of sixty-five scientists listed on the title page as collaborators.[33]

D'Almeida was also one of the founders and first general secretary of the Société française de physique (1873), which had a total of 946 members by 1902, of whom 433 were in Paris, 333 in the provinces, and 180 in other countries. The society came out of semi-private meetings Bertin organized at the Ecole normale after 1867. D'Almeida, Cornu, Gernez, Lissajous, and Mascart prepared the statutes of the society. Fizeau was the first president. Close relations inevitably existed between the society and the *Journal de physique*. The council of the society was made up of four members resident in Paris, headquarters, and four nonresident members, including some non-French members. Foreign honorary members included Kelvin, Bell, van der Waals, Hirn, Rowland, and Stokes. Many of the members were industrialists, engineers of Mines or Ponts et chaussées, and constructors of instruments. For a long time, budget deficits could only be balanced by generous donations from rich supporters. The deficit was over 1,100 francs in 1894. The beginning of subsidies from the government changed this situation. For 1894–95 the total income was nearly 17,000 francs, including a 1,240-franc subsidy by the Ministry of Public Education. Expenses of over 14,000 francs, nearly 6,000 to the printer Gauthier-Villars, left a surplus of nearly 2,500 francs. The society had assets worth 66,533 francs, mostly in railroad securities. Substantial donations kept the society in a healthy financial state.[34]

Early nineteenth-century *sociétés savantes* were not usually concerned with the advance of scientific knowledge, especially in biology, as it came to be understood by scientists in the second half of the century.[35] It is not true, however, that the provinces remained exclusively anchored in the old view of biological science as essentially natural history based on bird watching and plant gathering, while the new view, based on laboratory experiments and theory, emanated solely from Paris. The Société d'histoire naturelle de Toulouse, which celebrated its centenary in 1966, began in a mood of revisionism. There were only limited possibilities in the ancient Académie des sciences, inscriptions et belles-lettres for the publications and activity of natural scientists, who needed "a body of obvious utility." Initiative in founding the society came from the Toulousain chemist Edouard Filhol, who was interested in the geology of caves. Professor at the faculty of sciences, director of the school of medicine and pharmacy, he created a Museum of Natural History in 1861. The enemies he made as imperial mayor ensured that he would lose his monopoly over the museum after the fall of the Second Empire. Operating behind the scenes, he got naturalists to meet and found a society in 1866. The twenty-three founding members were not the "scientific authorities of the city," although many of them already had a certain "fame."[36] But this had the advantage of giving a new direction to the society, ensuring that it would not duplicate in membership and certainly not in aim the venerable academy of Toulouse.

The bulletin of the society was organized in sections of zoology, botany, and geology, with a miscellaneous section (including notices on scientists); each section had specialized subsections. Among the elect of this society were the lepidopterologist Auguste d'Aubuisson, Emile Bonnal, editor of the *Revue de Toulouse*, Paul-Emile Cartailhac, the famous prehistorian, Charles Fouque, a mineralogist from the ceramics dynasty, the head of the Ecole hydrologique des Pyrénées, some physicians, and representatives from mycology, orinthology, entomology, and botany. After a vagabond existence in various buildings the society settled in the old building of the faculty of letters, where it stayed between 1897 and 1920. Its library was integrated into the sciences-medicine section of the University library but was still owned by the society. Through a series of exchanges the society received 350 periodicals. So the move added an important research tool to the University. The *Bulletin* also offered the possibility of diffusing the work done in faculty laboratories, which meant that the journal became a sort of "Annales des sciences naturelles," alongside the existing annals limited to mathematics and the physical sciences. After 1928 the link with the faculty of sciences was cemented when its office was established in the faculty with the help of the dean, Paul Sabatier.

The development of the Société d'histoire naturelle took place at Toulouse during an extremely favorable intensification of general interest in natural science. On the basis of a strong eighteenth-century tradition a true school of Toulousain natural history had developed: The culture hero of natural history was Picot de Lapeyrouse (1744–1818). As the first dean of the faculty of sciences, he established the structure for an interaction between the older tradition of science and University science, which was characterized by increasing reliance on experiment and theory. Among the other heroes of Toulousain science were the best naturalists of the region of the Pyrenees, the zoologist Nicolas Joly, the botanist Alfred-Dominique Clos, Moquin-Tandon, and the prehistorian Edouard Lartet, all of whom had their bands of disciples. There were enough patrons for young scientists to follow or rebel against. The numerous intellectuals and scientists in Toulouse spawned a highly respectable number of societies, especially in the second half of the century. In 1872 the energetic Filhol founded the Société des sciences physiques et naturelles (de Toulouse), with its inevitable *Bulletin* to publish work in botany and in vertebrate paleontology, but the society ended soon after the death of Filhol in 1883. Then a Société française de botanique was founded; it published a *Revue de botanique* beginning in 1882. The Société de géographie (de Toulouse), which began in 1874 (officially in 1882) as a section of the Société d'histoire naturelle, the Société photographique de Toulouse (1875), and the Société entomologique de Toulouse et de la région pyrénéenne (1896) were spawned from the prolific mother Société d'histoire naturelle.

Associations provided an effective means by which scientists could lobby governments. In its session of March 11, 1870, the Société d'histoire naturelle de Toulouse, goaded by the near total absence of the teaching of natural science in lycées and collèges, named a commission to send a petition to the minister of education. On March 25 the Society sent an approved text to various scientific societies for their support in getting natural science put into secondary education. This movement would eventually have the result of producing students for the natural sciences in the faculties, which would increase job opportunities for natural scientists and create jobs in secondary education for students trained in the faculties. In short, this was the attempt of natural science to establish itself in the productive circularity of a discipline. Other societies were favorable to the initiative taken by Toulouse: the natural history society of Colmar, the industrial society of Mulhouse – both shortly to become part of the Second Reich – and the imperial society of sciences, agriculture, natural sciences, and useful arts of Lyon. Nothing much happened for two years because of the long sacred vacations in French education and the Franco-Prussian war. Then in 1872 support came from Olympia: E. Blanchard, professor of entomology at the Muséum and member of the Academy of Sciences, and the Association française pour l'avancement des sciences gave their support. In 1873 the Conseil supérieur de l'instruction publique named a commission for the revision of baccalaureate programs. Some results would be evident within a decade, but not all natural scientists would agree on their desirability.

Among the strongest of French professional societies of the nineteenth century was the Société des sciences physiques et naturelles de Bordeaux, whose quality depended largely upon the faculty of sciences. In 1850 professors, students, and "friends of scientific research" got together in order to "develop the taste for serious studies," especially natural science, by founding the Société d'histoire naturelle de Bordeaux. Physics soon wanted to get in. In 1853 d'Abria, professor of physics in the faculty of sciences, got the Society to change its name to reflect the inclusion of mathematics and the physical sciences. By 1898 there were about 107 regular members, seven honorary members, and twenty-four corresponding members, most of whom were Bordelais, although some members were in Paris and Lille. Those who published under the auspices of the Society made up an impressive galaxy of stars in nineteenth-century science, many of whom were birds of passage at Bordeaux: d'Abria, A. Baudrimont, Bazin, Paul Bert, Brunel, Darboux, Delbos (the geologist), Duhem, Gayon (who also collaborated with Dubourg), Dupetit, and Laborde, Hadamard, Hoüel, Joannis, Lespiault, Micé, Millardet, J. Pérez, Pionchon, V. Raulin, Rayet, Royer, Paul Tannery, Turpain, and Vèzes.[37] Many of these laborers in the vineyard have now fallen into the inevitable limbo to which an ungrateful science relegates its less than brilliant practitioners.

Colmar also provided a model of success during the 1860s. After several years of activity that could be best carried on within the confines of an association, a group of physicians, some scientists of sorts, and a few politicians founded the Société d'histoire naturelle de Colmar with the support of the prefecture of Haut-Rhin and the mayor of Colmar. Science and politics seem to have had a happy marriage in Colmar. It is definitely comforting to read in the minutes of a meeting (May 27, 1867) that "M. le baron Ponsard, préfet du Haut-Rhin, occupe le fauteil de la présidence." Industry and business were not far behind the prefecture. Henri Schlumberger, "premier adjoint au maire de Colmar," was the president of the society in this period. In 1859 the Society had 195 members, whose numbers swelled to 398 in 1867; of these, 316 were regular members, 217 of whom lived in Colmar. Income was made up of grants from the Ministry of Education, the department, and the city. In an eight-year period the Society spent nearly 30,000 francs, most of which went for the establishment of a Musée d'histoire naturelle, with a library of more than 800 volumes, and a laboratory with an assistant after 1864. The bulletin of the society published some serious scientific work by well-known people. G. A. Hirn, Charles Grad, J. Nicklès, Scheurer-Kestner, and Godron were among the scientists who gave their support to this local effort.[38] Like Mulhouse and Strasbourg, Colmar was a serious scientific loss to France in 1871.[39]

Certainly, it would be quite misleading to jump from the fact that a first-rate scientist like Pasteur published in the Lillois Society's memoirs to the conclusion that they were important scientific publications. Pasteur and others made sure that their work also appeared in the right Parisian publications. But the Society itself was perhaps more important than the publication. Pasteur did go to meetings and he did make *several* communications.[40] The memoirs published literature as well as science – just the sort of interaction eighteenth-century Jesuits and *philosophes* did well and we spend millions trying to imitate.

Some societies probably should not be judged from the perspective of how much they promoted laboratory research. The membership of the Société des sciences historiques et naturelles de l'Yonne attracted many of the intellectual elite of Auxerre: physicians, lawyers, clergy, and a heavy dose of graduates of Ponts et chaussées. Included were Paul Bert, the local glory, and Charles de Kirwan, a prolific writer on scientific topics who was a "sous-inspecteur des eaux et fôrets" at Auxerre. Corresponding members included Albert Gaudry, François-Jules Pictet, Sylvestre de Sacy, and the comte de Saporta. The *Bulletin* did not include much "hard" science, although in the section of natural sciences there was lots of archeology and anthropology.[41]

Even the most naïve of natural history societies could have its classificatory scientific value along with a real function in the community. The Société d'études scientifiques d'Angers began when six founding members were inspired by Alexandre Boreau, well-known expert on the flora of the *centre* of France, during one of the walks he conducted in the botanical garden. He took care of the material organization of the society. The city paid 6,000 francs for his herbarium, a collection of 20,000 species, including specimens from all over the world. Boreau was made honorary president of the Society. Members were very young, but the Society of students and seekers emphasized that it was interested in the old-fashioned aim of studying nature, not in becoming a so-called *société savante*. Each would pursue his specialty in the context of the common activity that would become a synthesis of these specialties. Many of the thirty-four original members did not live in Angers but were students in Paris. Activity in 1872 included a herborization at Brissac undertaken by Boreau that resulted in a detailed note on the chief plants gathered. A charming botanical and zoological note resulted from a scientific stroll along the banks of the Maine with M. Chauveau. Other activity was a continuation of a catalog of the mosses of Maine-et-Loire and some work on entomology. The *Bulletin* also printed a letter supporting transformation of species. All of this was solid activity, good for the mental well-being of the young members. It was not even dangerous for science. The mayor and the prefect could approve the formation of the society without any qualms.[42]

Departmental and regional societies varied greatly in significance. Some of the societies in the provincial cities with important centers of learning had genuine importance, reaching far beyond the region. Whether called a society or an academy, the chief departmental organization usually included some combination of sciences, letters, fine arts, and, in some cases, history, archeology, and agriculture. In the case of societies centering on faculties like those in Lille and Bordeaux, the meetings and publications were important largely because of the contributions of the scientists in the faculties. The nature of the activity of many of these societies can hardly be narrowly defined, especially if they were inspired by polymath physicians. Consider the Société linnéenne de la Charente-Inférieure, founded in 1874 by Guillaume Alexandre, who later joined the medical faculty at Bordeaux. The early activities of the society were helped enormously by Alexandre's avocation of printer. It had ambitious aims: to popularize knowledge of natural history, to furnish free collections to public museums and to departmental schools, and to establish relations between individuals carrying out special studies. It sent some members to the meetings of learned societies in Paris and organized scientific lectures and exursions from its home base of St. Jean d'Angély. It created a museum for local collections and established a library. With a membership of 200 in 1879, it published a trimestral bulletin and a special organ, *La Chronique charentaise*, dealing with science, local history, and literature. It was from the local society that one learned of the capture of a new bird in a department, as, for example, in the *Revue des oiseaux* of the department of Saône-et-Loire, which had its Société des sciences naturelles. The Société botanique de Lyon (1871) published *Annales* worthy of note by top botanists.[43] The network of locally based societies was an important part of the fabric of the natural sciences and also of regional studies. There was a direct relation between the significance of a society and the proximity of the faculties of science.

One of the biggest publishing enterprises in the natural sciences was the *Annales des sciences naturelles*, in its fourth series by the 1850s. The fourth series (1854–63) of twenty volumes carried some of the best-known names of the day: van Beneden, Bernard, Bert, Blanchard, Claparède, Dareste, Henri and Alphonse Milne Edwards, Flourens, Gervais, Gratiolet, Lacaze-Duthiers, Pasteur, Pictet, Pouchet, de Quatrefages, and Marcel de Serres. With each new series a new generation of natural scientists – although some of the authors, like Bernard, would haughtily reject such a vague designation – would begin its insertion in the mainstream of scientific publication, with some of their material showing up in the *Annales*. In the fifth series, ending in 1874, new names appeared: the Agassizs, Chatin, Marion, Perrier, and A. Sabatier. By the 1880s new currents in zoology were evident in the work by Charles and J. Barrois, but the old type of classificatory work was still honored: Hesse's "Description des Crustacés rares ou nouveaux des côtes de France" was a mammoth series of thirty-six articles. By this time the *Annales* was a bit of a relic because its aims had not changed much since the days when, edited by H. Milne Edwards for zoology and Ad. Brongniart and J. Decaisne for botany, it advertised itself as including the comparative anatomy and physiology of the two orders as well as the history of fossil organic life. True, by the 1870s the bifurcation into specialized publications, with zoology and paleontology in one series, recognized this trend, but the fidelity to comprehensiveness remained: "zoologie et paléontologie comprenant l'anatomie, la physiologie, la classification et l'histoire naturelle des animaux." This tradition, extraordinarily strong in French biology, would be quite ferociously defended by Giard as late as the first decade of the twentieth century. It was not entirely without virtue to claim that man could understand nature; it was no longer the program of scientific research in the late nineteenth century.

There certainly was no shortage of respectable, well-known journals in the natural sciences, most of which carried the scholarly seal of approval. The 1870s saw a flurry of expansion in the field of scholarly journals. Some were destined to stand as landmarks in the history of experimental science, while others were fated to fade into the limbo of the history of classificatory natural science of a pre-Bernardian variety. Some publications did not last long, although perhaps a decade is not a short time in the general history of journals. The *Revue des sciences naturelles* (1872–84) of Montpellier started under the editorship of E. Heckel, doctor-pharmacist, head of Montpellier's hospitals, and Ernest Dubrueil, a founder of the journal and son of the well-known anatomist. Dubrueil edited the journal for nine of its thirteen years of existence. It published original work in zoology, botany, and geology. The first article was by N. Joly in embryology. Advertised as collaborators over the years were some of France's leading scientists. Gervais, Godron, Joly, Ch. Martins, Périer, Robin, Sabatier, de Saporta, Flahault, Perrier, and Giard were in the rotating crew of the first few years. Several factors doomed the journal: the death of Dubrueil, the absence of a strong biological tradition outside medicine at Montpellier, and the competition of better journals in which to publish new experimental research. No wealthy donor appeared with a big gift, no support came from the financially pinched faculty. And perhaps, as Conry argues, a Protestant-fueled, anti-Darwinian ideology prevented biology in Montpellier from being fertilized by the century's greatest research paradigm.[44] A fatal combination.

Some journals with distinguished collaborators failed to make it into the small list of great scientific publications because of a practical emphasis that was frequently of only local interest. The *Revue des sciences naturelles de l'Ouest* came forth proudly to its readers in 1895 with a list of strong collaborators:

E. Bettremieux, president of the Société des sciences naturelles de la Charente-Inférieure, Directeur-conservateur du Muséum Fleuriau at La Rochelle, Raphaël Blanchard, professor at the Paris faculty of medicine and general secretary of the Société zoologique de France, Giard and Bonnier from Lille, Meunier and Georges Pouchet of the Muséum, and a few others, including faculty and lycée professors and teachers from Poitiers, Bordeaux, and Brest. Pouchet reported on sardines. If one did not care to learn about edible mushrooms, there was enlightenment on the love life of the lobster and the disappearance of salmon from the rivers of Brittany. (The salmon were the victims of the works of the Génie maritime in straightening out the rivers and of Ponts et chaussées in cleaning out weeds and other plants vital to the fish.) The title of the journal, whose office was in Paris, eventually put an added emphasis on agriculture and fishing.[45] There was little to distinguish this journal from many others in France, journals that did a competent job, carried interesting articles, and were useful to several groups of producers or growers. But it was not generally the product of experimental science, it was not the organ of a research group with a mission to spread new views, and it had no institutional ties. It could be only a good regional journal.

An extremely successful but practical journal in the natural sciences was the *Bulletin de la Société zoologique d'acclimatation*, later called *Revue des sciences naturelles appliquées*, begun in 1854. In the 1840s scientists and rich farmers showed great interest in the formation of an association for the improvement of domestic animals. In 1849 Isidore Geoffroy-Saint-Hilaire published his work *Sur la naturalisation des animaux*, and early in the Second Empire he became president of the new Société zoologique d'acclimation (1854), which included farmers, naturalists, men of property, and other intelligent people. By the end of the founding year the Society had grown from 80 to 550 members. The leaders (or at least the *grosses légumes*) of the scientific community appeared in profusion: de Quatrefages, Pouchet, Rayer, E. Becquerel, Blanchard, Coste, Dareste, Duméril, Duvernoy, Milne Edwards, Fremy, Moquin-Tandon, who were joined by Saint-Marc Girardin, a fine classical stylist in literature, and a herd of counts and barons. By the 1880s membership was up to nearly 2,000. The *Revue* carried scientific articles of a useful nature, divided into sections covering such divisions of the natural world as mammals, birds, plants, and insects. Articles on the beaver, Dutch cows, ponies, and cheese were complemented by news on mobile zoological stations in Scotland and Denmark, a chronicle of the activities of scientific societies, and bibliographical information.[46] A remarkable example of the successful collaboration between scientists and enlightened men of the upper classes.

The flourishing state of the natural sciences encouraged the development of specialist journals, especially in botany and zoology. Botany took about a generation to shake off its antiquarianism. A small group met at Antoine Passy's in 1854 to discuss whether a society should be formed with the aim of contributing to the progress of botany and to solidify its useful relations with science. Passy, an ex-prefect and former deputy, was an original member of the Société géologique de France, founded twenty-four years earlier and still growing. It seemed wise to emulate success by adopting the regulations of the Société géologique as the principal basis of the statutes of the Société botanique. Adolphe Brongniart was the first president; vice presidents included J. Decaisne, François Delessert, and Moquin-Tandon, all members of the Academy of Sciences. Other scientists of the caliber of P. Duchartre, A. Chatin (pharmacy, Paris), and Victor Raulin (faculty of sciences, Bordeaux) mingled with a few physicians and former politicians. The Society, with 420 members in 1860, did not change much in the next twenty years, and then declined to 371 by 1910.[47]

In 1889 the firm of Librairie Paul Klincksieck began publication of a *Revue générale de botanique*, edited by Gaston Bonnier, professor of botany at the Sorbonne. This was evidence of the new role of the faculty of sciences in Paris and the ambition of Bonnier to transform botany. It also shows that learned journals could be commercially successful, or at least not big money losers, although most commercial science publishers were always begging the Ministry of Education for a subsidy, especially if the subject was an expensive one to print. In two pages at the end of the first volume, Paul Klincksieck explained the aim of the journal. A strong movement in favor of the natural sciences, botany in particular, justified the appearance of a new journal. Memoirs and notes would appear on all parts of this science: anatomy, physiology, classification, geographical botany, paleontology, and plant chemistry. Scientific applications of botany would not be ignored. Both French and foreign botanical work would be reported in *comptes rendus*. Reviews of works published in the different branches of botany would be edited by well-known scientists: Bonnier on plant physiology, Léon Boutroux on bacteria and fermentations, J. Costantin on mushrooms, L. Dufour on technical procedures and methods, Charles Flahault on algae, Leclerc du Sablon on anatomy, and the marquis de Saporta on paleobotany. Edited by Bonnier, a leading Lamarckian, the journal was also a manifestation of the scientific nationalism of late nineteenth-century France, which would certainly adopt its own culture hero rather than one from across the channel.[48]

The influence of regional factors on the growth of science was nowhere more evident than in the natural sciences in Marseille, where a high-quality

Annales du Musée d'histoire naturelle de Marseille appeared nearly a decade before the annals of the faculty of science. Both publications were funded by the city's municipal government. This municipal generosity was a clear testimony to the fact that the "sciences of observation" were regarded as having a distinctive characteristic in the galaxy of nineteenth-century sciences: They seemed to be the very expression of the region they studied, borrowing from it an original stamp and a special flavor. Unlike physicists and chemists, except those working on industrial applications, naturalists were bound to the soil. Their research and even their *chef d'oeuvre* could be no more than a reflection of the habitat in which they worked. In this innocent epistemology science was the mirror of nature. The successful proposal of the director of the museum, A.-F. Marion, for a municipal publishing subsidy in 1881 specified that the *Recueil* would promote the publication of studies relating to the zoology and the paleontology of the French Mediterranean basin. Division of these areas into subspecialties, especially in the case of zoology, led to an increasing dependence on governmental funding because of the need for substantial matériel and numerous personnel with individual functions. Important growth in provincial scientific research and publication depended more on local than national money.

In the case of the museum in Marseille the generosity of the city was partly inspired by the prevalent idea – not entirely illusory – that biology was not only making rapid progress in expanding the frontiers of knowledge but was also influencing philosophy, psychology, social economy, and ethics. In the optimistic opinion of the scientific elite of Marseille, the scientific *devotio moderna* responded to the new concern of an enlightened public for the promotion of "all the sciences trying to solve the most serious problems facing the human mind." Filled with enthusiasm by their first flush of public funding, biologists found it curious that "the more transcendental sciences" had been so successful in getting expensive astronomical observatories built before biologists managed to join them as junior partners in the enjoyment of the rations available at the public fiscal trough. Marseille participated in the general scientific movement, but its privileged position on a gulf teeming with marine life was vital in attracting the attention of specialists. Its rich museum was attached to the faculty laboratory of marine biology. The zoological station at Endoume, founded with credits of 65,000 francs from the city and 20,000 francs from the Ministry of Education, increased the importance of Marseille as a center of biological research. The *Recueil zoologique du Musée de Marseille* was chiefly devoted to marine zoology, a rubric hiding general embryogeny, with its interest in the evolution of organisms, and a subject with the odd title of zoological geography, which studied the modifications that animal forms underwent through time and space. Marine research dealt with the great biological issues of the day, and for the professors who established the laboratories this was as important as the more practical interests of politicians.[49]

The faculty of sciences was founded in Marseille in December 1854 with the aim of establishing a complete system of education in the Academy of Aix. For over a decade a few teachers pandered to the pleasure taken by the public in hearing the faculty's mandarins expound popular science. Only with the beginning of preparation of serious candidates for the licence and the doctorate did there appear students with a taste for serious research. It was not until 1869, as a result of the support of the newly founded Ecole pratique des hautes études, that true technical work was begun in the natural and physical sciences. And the faculty did not begin its own publication until 1891, when the *Annales de la Faculté des sciences de Marseille* began its irregular appearance. The high-quality scientific work published in the *Annales* included the research of J. Macé de Lépinay, Appell, and Fabry in physics and Edouard Heckel in natural history.[50]

Some journals began in an optimistically ecumenical spirit by including nonscientific areas. Often science and medicine split off as a separate series, as in the case of the *Annales de l'Université de Lyon*, which began a new series of short scientific and medical monographs in 1899. But it was not until 1924 that the bifurcation took place at the University of Grenoble, whose annals included law, science, letters, medicine, and the technology of paper-making. The split was for purely economic reasons. Presumably the community of thought originally used as the justification for an all-inclusive annals still existed, even if separate publications were cheaper. "En même temps que la vie scientifique devient plus intense dans notre Enseignement supérieur, la communauté de l'effort, malgré la diversité des sujets d'études rapproche les esprits." Among scientists committed to the Grenoblois ethic were Paul Janet, F.-M. Raoult, L. Barbillion, and Raoul Blanchard. "On se pénètre de plus en plus du sentiment de l'intérêt moral dont la prosperité et la dignité commune sont l'objet." There was a particularly lively atmosphere in Grenoble with a widespread taste for intellectual matters and after 1870 a generosity showing public appreciation for the benefits of instruction.[51] A nineteenth-century version of Rohmer's *Ma nuit chez Maud* would have been as well-placed in the ancient capital of the Dauphiné as the actual film is in modern Clermont-Ferrand.

France's major provincial universities proudly produced similar varieties of annals in the late nineteenth century. These journals and yearbooks carried a good deal of significant scientific and mathematical research, an eloquent testimony to the growth of science in the chief provincial cities. The *Annales de la Faculté des sciences de Toulouse*, beginning in 1887, published by an editorial committee of professors of mathematics, physics, and chemistry, with the patronage of the minister of education and the city of Toulouse, was also aided by the general councils of the departments of the Haute-Garonne and the Hautes-Pyrénées. The inspiration for the journal was strongly nationalistic. "Nous comptons . . . sur le concours de tous ceux qu'intéresse le développement de la Science française." The first result of public support for science should be a strong increase in "our country's scientific production." The level of scientific publication in Toulouse's annals was unusually high: Picard, Appell, Goursat, Sabatier, Koenigs, Garbe, Andoyer, Brillouin, Painlevé, Hermite, Stieltjes, and Duhem were among the stars appearing in the first volumes. Like the *Annales de l'Ecole normale* and the *Bulletin des sciences mathématiques*, this was an important mathematical publication. A *Bulletin astronomique* was added to it by Tisserand. Few university annals were of Toulousain quality or significance.

The profusion of publications in natural history made it seem that an organization devoted exclusively to zoology would be clearly redundant. When Raphaël Blanchard tried to make a success of the Société zoologique de France, founded in 1876, he had a tough thirteen years before the Society, with its *Bulletin* and memoirs, amounted to much in the scientific community.[52] Along with the Bernardian prejudice still operative against admitting zoology to the rank of experimental science, there was the fact that ambitious zoologists wanted to start their own publications. And physiologists could publish in the *Archives de physiologie* or the *comptes rendus* of the Société de biologie, as well as in the well-known standard outlets like the *Comptes rendus de l'Académie des sciences*. It was the enthusiasm generated by the experimental virus that lay behind the founding of the *Archives de zoologie expérimentale* in 1872 by Lacaze-Duthiers and the *Bulletin scientifique de la France* . . . in 1874 by Giard. A great deal of the important zoological work done in France appeared in these two publications.

There was a certain conceptual muddle in biology well into the twentieth century as biologists tried to work out a general definition of their subject. Some biologists believed that monographs had not succeeded in giving a clear identity to the subject of general biology. That would be the function of a journal, not of more monographs. The word monograph meant "originally, in Natural History, a separate treatise on a single species, genus, or larger group of plants, animals, or minerals" (*Oxford English Dictionary*). Examples in the OED come from Hooker on Darwin – "It was monographing the Barnacles that brought it about" – and the *Encyclopedia Britannica* – "Few existing birds offer a better subject for a monographer [than the kakapo]." The obscure monographs of Cuvier, based on his dissections, remained models for the biologist late in the nineteenth century, while his general works were only of stylistic interest. Partly because of the advances in microscopic technique by the end of the century, most of the journal articles and monographs were strikingly similar: studies in the anatomy and embryogeny of plants and animals, histological studies of tissue or organ systems of plants and animals, experimental work on animals, and analyses of plant excretions. All of the scientific results could be summed up as perfecting, extending, or modifying facts already presented in the works of masters like Milne Edwards and Lacaze-Duthiers, who had also used the basic techniques still employed at the end of the century. In short, this was Kuhnian normal science.

Yves Delage found this approach particularly unsatisfactory in general biology because it did not produce "useful [scientific] progress" – the general conception of biology remained the same. Delage's hope for turning research in the right direction lay in promoting general biology through the study of the cell, basically the physical and chemical analysis of protoplasm, which was already well under way in Germany and England. Delage hoped for the publication of a work that would change the direction of basic biological research in France, with cell studies occupying the place of honor. Perhaps he dreamed of his own work, *La Structure du protoplasma* (1895), as providing the impetus for a paradigm switch.[53] Had such a shift in research direction taken place it would have been comparable to the shift in mathematics leading to the emphasis on the geometrical applications of analysis, which occurred partly as a result of the publication of Darboux's *Cours de géométrie: Leçons sur la théorie générale des surfaces et les applications géométriques du calcul infinitésimal* (4 vols., 1887–96). But it was more probable that such a shift would come from a piece of research written up as a journal article rather than in the classic magisterial treatise. Certainly most of the functions of the monograph had been in establishing and in continuing fields of research. But innovation was more likely to be the function of the journal paper, although its functions were not entirely separate from those of the monograph. As biology grew closer to the other sciences the role of the monograph diminished. This tendency was encouraged by more specialized research on limited subjects and the pressures for quick publication stimulated by the desire for priority in scientific discovery.

The monograph clearly fulfilled an important function in establishing an area of scientific investigation as a discipline, or at least as a differentiated subdiscipline. As was evident in biology, the journal could then absorb a large share of the role of the monograph. A certain usurpation of function was evident in the founding of *Biologia generalis*, the international archives of general biology, in 1925, for it was assumed that only with the creation of a journal did an area of investigation finish its mutation into a "new science" with a respectable pedigree. The justification for the founding of *Biologia generalis* is useful in helping us understand the different, but mingled, roles of journals and monographs. It may be that general biology did not have a special journal until the 1920s because it had been specially cultivated in medical faculties, both as a teaching subject and as a "science d'investigation exacte" – to use the prestige-laden French phrase. Scientists in medical faculties did not have the same espirt de corps or share the intellectual cohesiveness characteristic of older groups like chemists, physicists, and zoologists. Hence it took longer for the processes resulting in a clear sense of a discipline to come to fruition. And, certainly, a great deal of general biology existed in the various biological subspecialties. Whatever the explanation, by the 1920s the time had come for a "special journal for this new science."

The basic reason for creating a new journal was that publications relevant to general biology were spread out over the periodical literature of the other sciences, partly because of the special relations general biology had with other sciences, both organic and inorganic. Work in general biology was therefore made unnecessarily difficult, for a researcher could not easily ascertain the state of contemporary research. The journal would concentrate most original biological work and give "impartial reviews" of the rest. *Biologia generalis* hoped to publish all the original works in general biology, following the triparite division of morphology, physiology, and ecology. Metaphysical works were excluded. It was recognized that this key step in the development of the discipline rested in considerable part on the solid foundations that had been laid by collections of monographs related to general biology. Among the most famous were those edited by Wilhelm Roux (*Vorträge und Aufsätze über Entwickelungsmechanik der Organismen*, Leipzig, 1905–), by Julius Schaxel (*Arbeiten aus dem Gebiet der experimentellen Biologie*, 1921–), Jacques Loeb et al. (monographs on experimental biology), and Maurice Caulley's *Bibliothèque de biologie générale*. Perhaps the one great function of the monograph that could not be absorbed by the journal was its teaching function, which was directly related to discipline growth. The great influence of Darboux's magisterial treatise on the geometrical applications of differential calculus gave this subject an increasingly larger role in higher education toward the end of the nineteenth century. That type of function could not be usurped by the journal, and it continued to guarantee the monograph's raison d'être, although the cutting edge of research was honed in journals.

The mummery of number

In his discussion of Lotka's law of productivity (the number of authors publishing just N papers is proportional to $1/N^2$), Derek Price notes that "the distribution of productivity among scientists has not changed much over the whole three hundred years for which papers have been produced." And "over the years there have been about three papers for every author." Although it makes statistical sense "to define a scientist as a man who writes at least one scientific paper in his lifetime," some scientists are prolific producers. Lord Kelvin's lifetime average was one paper per month.[54] Students of the growth of scientific productivity in France have drawn some significant as well as trivial conclusions from their data. I shall summarize the significant. Craig Zwerling found that the scientific productivity of a scientist trained at the Ecole normale rose from an early nineteenth-century low of 11.3 papers over a career to 46.7 papers per career by the end of the century, with a peak of 61.2 papers produced by students who were stimulated into research when Pasteur was scientific director at the Normale (1857–67). The productivity of scientists at the Ecole polytechnique went down from an average of 40 papers per career for the period 1793–1821 to 29.9 papers in the period 1890–1903, with a sublime peak of 71.1 papers per career in the 1830s.[55] But it might be perversely reactionary to conclude that this development was the result of a lowering in the social origins of students, for it seems to be more a case of changing functions at the Polytechnique and shifting resources within the educational system. Besides, the students of lower social origin did quite well in research and publishing at the Normale and in the faculties of science. George Weisz has found that for professors in medical faculties scientific productivity also "increased significantly during the second half of the nineteenth century."[56] Even at the Muséum, where much of the research continued to be old-fashioned natural history instead of newfangled experimental biology, production of articles by researchers rose strikingly.[57] And the titular professors continued to produce per capita the same number of books as earlier in the century. Productivity had come to be recognized everywhere as "the key to institutional survival and to individual career success,"[58] a powerful bourgeois fetish in the academy as well as in the factory.

In this bright new universe of publication, there is perhaps one black hole. Terry Shinn's compilation of the number of books and articles written by 372 professors in the faculties of science between 1808 and 1914 reveals a pattern comparable to the findings of other scholars for various institutions. During the years 1896–1900, Shinn found that there appeared 84 books and 1,916 articles dealing with basic research topics and 41 books and 157 articles on topics in applied research. "Between 1901 and 1914 science faculty productivity in most areas of pure research fell sharply and applied research stagnated. From 1911 to 1914 only 12 books and 121 articles were produced in the domain of fundamental research and 12 books and 62 articles in that of applied inquiry."[59] Weisz, rightly astonished by this drop, demands a recount, with the single source of Poggendorf supplemented by other inventories and lists. Chemistry professors at the Sorbonne, at Montpellier, and Nancy kept up a respectable rate of publication from 1909 to 1913. So Shinn's threnody for Marianne may be premature, even if some decline in publication did take place.[60] No doubt some feeble readers, without abstracts and printouts, welcoming a drop in the volume of books and articles, felt like the Duke of Gloucester when he received the second volume of the *Decline and Fall*: "Another damned, thick, square book! Always scribble, scribble, scribble! Eh, Mr. Gibbon?" And few researchers were Gibbons of the scientific empire. In any case, the drought of the prewar period was of short historical duration: the exponential river of publications soon resumed its flow, with the dip of the sigmoid curve nowhere in sight.

8

SCIENCE FUNDING IN THE TWENTIETH CENTURY
LAYING THE FOUNDATIONS OF THE SCIENCE EMPIRE

Comme il est bien, dans la requête au Prince,
d'interposer l'ivoire ou bien le jade
Entre la face suzeraine et la louange courtisane.

– Saint-John Perse, *Amers*

Ruf mir Yoshke, ruf mir Moshke,
Aber gib mir die groschke.

– Saul Bellow, *The Victim*

The trouble with prizes

THE history of the funding of science in France is pervaded by the theme of poverty, eloquently stated by a line of lamenters from Pasteur to Maurice Barrès and on to Jacques Monod. So convincing has been the brief for poverty that historians, ever victims of their sources, have generally been mesmerized into repeating this litany. Not that Pasteur was wrong in 1868: Dumas, Foucault, Fizeau, and Boussingault had private laboratories because there were no funds specifically earmarked for research in the educational budget. In 1884 Fremy echoed Dumas's plea of 1881 that those with money support science, but few followed Fremy's example of making a gift of 5,000 francs. Twenty-seven years later Georges Lemoine deplored the insufficiency of funds for the support of the scientific community, a vastly larger entity than it was in 1868, but seemingly in the same penury. Not really, of course; and it is to the steady, solid, increasing financial support for scientific research that the historian must also direct his attention. Recent scrutiny of the science support system in France has already produced some surprises, especially in the studies by Shinn, Crawford, and Weart. My focus will be on two funding organisms that have generally been ignored or relegated to peripheral roles in the growth of the research empire. To some extent, this ignorance is part of the mythology spread by the jealous creators of the new funding mechanism that eclipsed the primitive system of their poorer ancestors.[1]

In the early 1850s prizes awarded by the Academy of Sciences amounted to only 34,000 francs a year; by the second decade of the twentieth century the amount was up to 160,000 francs; and by 1925 stood at the substantial figure of 204,256 francs – not so impressive as it would have been without inflation, of course. As Crawford has shown, "the prizes of the Academy evolved into a system of material awards for scientific achievement." The total amounts distributed were considerable, enough to make a great impact on the funding of research. From 1871 to 1880, 650,000 francs were bestowed on lucky scientists, and from 1901 to 1910, 1.5 million francs were given to scientists for their achievements. Perhaps less than half of the prize money trickled back into research. And after 1890, funds for grants took on considerable importance, rising from 392,400 francs in the period 1901–10 to 461,900 francs in the period 1911–15. A few of the prizes were large enough to make some difference in a scientist's work in the days of small science, with a range from 3,000 to 10,000 francs, the equivalent of anywhere from one-third to a whole of a researcher's annual salary. The Osiris prize, dating from 1899, was 100,000 francs every three years, not quite up to the five annual Nobel prizes of 200,000 Swedish crowns each, created in 1901, but a handsome amount anyway. Several of the grant funds of the academy – Bonaparte, Debrousse, and Loutreuil, e.g. – were large enough to weigh heavily in the research balance. The Loutreuil gift to the academy was 3.5 million francs. Crawford thinks that the French system of scientific prizes was unmatched anywhere in Europe; in 1900 the academy's payout for physics alone was three times as much as that of the Royal Society of London or of both the academies of Berlin and of Göttingen.[2] Even after due allowance is made for the Gallic *esprit critique*, how then do we explain the bitter complaints of the French scientific community about the inadequate funding of research?

The key is in the distinction between a prize and a grant, along with the increasing importance of the grant in funding research. In France at the turn of the century, prize money was nearly nine times greater than grant money; in Germany and England, the French proportion of prize to grant money was nearly reversed. The grant was a much better way of getting more research for the same amount of money. Obviously a prize could be used for the same purpose, as the Curies did with their Nobel prize, but the fact is that not everyone copied the Curies. Scientists realized this very early. Scientific journals were full of high praise for the type of funding of science that other countries had initiated, while France lagged behind, enmeshed in the system of the old régime. After a visit to Copenhagen, the physiologist Charles Richet pointed out the ideal method of proceeding. In 1876 J. C. Jacobsen, to perpetuate the memory of his son Carl, annexed to his brewery a laboratory for research in the chemistry and physiology of alcoholic fermentation. This was the origin of the Carlsberg foundation. Jacobsen, interested in the natural sciences, followed Pasteur's experiments on fermentation with more than the usual brewer's interest. One of the first-rate scientists he employed in his brewery, E. C. Hansen, encouraged him to use pure yeast in manufacturing beer; the practice came to be adopted in all fermentation industries, replacing the old empirical methods that had given so much trouble and bad beer. The foundation's laboratory had chemical and physiological sections whose work was published in the journal *Meddelelser fra Carlsberg laboratoriet*, which carried summaries of its articles in French for scientists suffering from rusty Danish. The foundation also made grants to scientists and supported expeditions. In 1887 Jacobsen bequeathed his brewery to the foundation, which eventually became a sort of big general scientific funding organization administered by the Royal Danish Academy of Sciences with funds equivalent to about eight million francs to promote research in science and mathematics. By 1933 the annual budget was about 300,000 francs, two-thirds of which was supplied by the Danish government. The foundation was much admired in France.[3]

As in most countries, the biological sciences had greater difficulty than physics and chemistry in getting big funding in France. In 1891 Emile Duclaux, a member of the institute, the head of the Institut Pasteur, was still complaining bitterly about lack of funding in a country afloat in prize money.[4] This was a complaint of a quite different nature from Giard's diatribe against the deadweight of the old-boy network's system of intellectual incest maintained through its prizes, although Giard certainly had the same opinion of the total inadequacy of science grants. The issue was how to put into research the maximum of money budgeted for science.

Duclaux's analysis began with the inevitable tragic overture dominated by the leitmotif of foreign generosity and French parsimony in funding bacteriological institutes. The Prussian Landtag had given the Koch Institute an annual budget of 165,000 marks (206,000 francs). The Institute of St. Petersburg got a subsidy of 132,000 rubles (about 350,000 francs). Scandalously, the Institut Pasteur did not have even 100,000 francs of revenue or endowment; so the budget could be balanced only because Pasteur had left it the right to profits from the sale of vaccines and the self-sacrificing heads of departments had enough other income to give up their institute salaries. But no expansion was possible. A new danger had appeared: Public donations, which paid for all expenses and accounted for about one half of annual revenue, were drying up at a fast rate. For Duclaux it was a colossal irony that, while the Academy of Sciences had so much prize money it did not know what to do with it, many laboratories had to struggle along with a lean budget. An illogical result of this situation was that the amount and quality of scientific work fell at the same time the number of rewards for doing it was increasing. Fortunately, the Institut Pasteur worked against this system: Nearly all its budget went for laboratory expenses. Because its request was impersonal, the institute could make its complaint public. Duclaux could not refrain from adding that the plea was solely in the interests of science.[5] No doubt, but this private institution always managed to appear as the victim of its own virtue within a scientific estate that was nearly exclusively dependent on mother *état*. Less than a century later the Institut Pasteur would have to be rescued from imminent bankruptcy by becoming heavily dependent on state funds. It was the end of a long flirtation.

Gifts of a million francs were rare – Legs Debrousse (1905), Fonds Commercy (University of Paris, 1908), Legs Prince de Monaco (Albert I, 1923) – and the Legs Loutreuil of 1912 was unique, 3.5 million to the academy with 2.5 million to the universities. Gustave-Adolphe Commercy was a Parisian *rentier* who left four million to the University of Paris for scholarships and to promote the advancement of science. It is impossible to guess what was the exact scientific value of all the prizes established by pious widows and relatives to perpetuate the memories of their dearly departed. The unsatisfactory nature of the system so impressed one of the leading benefactors of French scientific research that he took the deliberate step of counteracting its baneful effects by giving 100,000 francs to the academy in 1908 for a small number of immediate grants. Prince Roland Bonaparte told his fellow academicians that he did not think that new prizes would do much for scientific progress. What was needed was not money for work done, but money to facilitate the research of people who had already shown their capabilities by publishing original work and yet lacked the resources available to members of the academy. "Maecenas atavis edite regibus": Henri Becquerel lapsed into his lycée Horace when he read the prince's letter to the academy.[6]

A prize could enable a scientist to spend more time on research; a grant could equip a laboratory; a grant could buy some radium or an X-ray machine; but neither prize nor grant was the type of support needed for running a permanent laboratory, and that was what worried the scientist-administrators like Duclaux who had to keep the operations going whether or not they got prizes or grants. Everybody knew what needed to be done. The academy tried at the end of the nineteenth century to counteract the evil influences of the prize system with a set of reforms. It tried to keep the terms of gifts as general as possible, thus giving flexibility in the distribution of funds; it divided up prizes; it would award several prizes to the same scientists, who then had a shaky equivalent of a renewable grant. Awards were made to much younger men than was the case for medals and prizes in other countries, and this had the advantage of giving support at the age when it was possible to maximize chances for a creative return on the prize.[7] Such concessions, however significant for the changing influence of the academy itself, were inadequate to meet the needs of the scientific community of 1900. A mutation was required in the evolution of funding itself.

This mutation was the establishment of the Caisse des recherches scientifiques in 1901, although important changes had taken place in funding by then. Crawford argues reasonably that peer control of the prize system and the introduction of research grants from specific funds requiring formal applications, as in the case of the Loutreuil money, laid the groundwork for the rules of modern science funding, which emerged clearly with the Caisse. Here we have the "transition from informal to bureaucratic forms of control."[8] The main sources of funds were no longer the scientists and academics themselves, but rather the industrialists and businessmen who could profit from the research they supported, although this certainly did not preclude some altruistic support for pure science. The growth of the scientific establishment was much faster than the growth of available prizes and funds. It was not easy for scientists to understand why the state, their creator, did not properly fund them. After all, the Third Republic deemed modern science to be a basic component of its ideology and the foundation of national prosperity. But there were no previous models for this sort of thing. Perhaps the remarkable thing is that a funding mechanism developed at all, at least this early. The old system had grown in a striking manner and seemed to be adapting to new demands on it. Who could tell that it would be unequal to a new task?

By 1925 the French research budget, that is, money for exclusively research purposes (personnel), had swollen to over twenty-five million francs, which might be the equivalent of a quarter of that sum in francs of prewar vintage. Fourteen of the twenty-five million was supposed to come from the new *taxe d'apprentissage*, a tax on the industrial payroll intended to support technical training. Emile Borel, mathematician turned deputy, got the tax raised, with the increase – a penny for the laboratories – destined for scientific research. The two big spenders were the ministeries of agriculture and of education. Nearly all of the research funds of the Ministry of Agriculture were funneled into its fifty-eight laboratories and agricultural stations through the Institut des recherches agronomiques. The total amount spent by agriculture was 5,389,300 francs, of which 600,000 went to laboratories and stations not under the ministry and 150,000 to scientists not belonging to the Institut. A rough guess would be that about 2.5 million francs of the total amount went into the French personnel research budget, which was the same amount injected into it by the Ministry of Education out of its expenses of about 5.5 millions. Within the educational budget the big recipients were the Muséum (1,330,000), the Office des recherches scientifiques et industrielles et des inventions (1,400,000), the Caisse (462,000), the Collège de France (750,000), and sciences in Hautes études (336,000). Although it spent nearly 900,000 francs annually on science and health, the City of Paris put only about 150,000 of its subsidies into research. The Academy of Sciences, with its eighty-one prizes worth 204,256 francs annually, put about 426,871 francs into the annual research budget, compared to 111,460 francs for the Academy of Medicine. In the private sector, a handful of institutions and organizations had an input of some importance into research. The Institut Pasteur had a budget of 2,440,000 francs. The Fondation Edmond de Rothschild, founded with ten millions in 1921, had an annual revenue of 400,000 francs. That of the Fondation Curie was 320,000 francs. In 1923, to celebrate the centenary of Pasteur, the Chambre de Commerce de Lyon collected 680,000 francs from industrialists and businessmen for the laboratories of their city. Instead of handing out a lot of small grants from the interest on the capital, the chamber decided to make six or seven grants of about 100,000 francs each in order to have a substantial impact on scientific research.[9] If we add to this exclusive support for scientific speculation the support for "impure" research done in industrial laboratories and the time siphoned off in the universities, where there was no line for research per se in the budget, and a host of lesser research enterprises, it is evident that the research imperative had become firmly rooted in the French scientific community. (See Table 1.)

The institutionalization of funding: La Caisse des recherches scientifiques

In the history of science funding in the Third Republic, the Caisse des recherches scientifiques holds a place of honor as a significant failure.[10] It was not heavily funded by the state. Worse than its failure, some purists censoriously point out, was its constitutional inability to support "pure" science, whose lavish endowment has become in scholarly circles the touchstone of a virtuous science policy. But within the context of twentieth-century French politics it might be better to judge the founding of the Caisse as a definitive step in the direction of a permanent commitment to providing state funds for the type of science the state wanted done. It would be easy enough after World War I, and especially after World War II, to extend the range of scientific research deemed vital to state interests. If principle triumphs, can bureaucracy and funds be far behind?

293

Table 1. *Annual budgets and revenues for scientific research personnel (1924–5)*

Establishment	Amount in francs
Académie des sciences	426,871
Académie de médecine	111,460
Société d'encouragement pour l'industrie nationale	27,773
Société chimique de France	2,500
Société géologique de France	3,000
Société botanique de France	700
Société entomologique de France	1,050
Société des ingénieurs civils	14,000
Association française pour l'avancement des sciences	28,000
Société française de navigation aérienne	11,000
Radio-Club de France	10,000
Société française des électriciens	7,000
Société française de physique	3,000
Société de pharmacie de Paris	1,750
Société française de photographie	1,000
Société centrale de médecine vétérinaire	3,500
Société de médecine de Paris	366
Société zoologique de France	200
Académie d'agriculture	5,500
Touring-club de France	2,000
Société de biologie	8,000
Académie Stanislas, de Nancy	260
Académie des sciences, belles-lettres et arts de Rouen	270
Société de médecine et de chirurgie de Bordeaux	350
Université de Paris	145,000
Université de Bordeaux	2,500
Université de Montpellier	1,200
Université de Nancy	1,850
Université d'Aix-Marseille	300
Université de Rennes	7,555
Université de Lyon	23,000
Université de Lille	6,200
Université de Clermont-Ferrand	195
Ministère de l'Agriculture	2,500,000
Ministère de l'Instruction Publique	2,500,000
Institut Pasteur	800,000
Fondation Ed. de Rothschild	400,000
Fondation Barès	15,000
Fondation Rosenthal	60,000
Fondation Curie	300,000

Table 1 *(continued)*

Establishment	Amount in francs
Institut et musée océanographiques	102,000
Chambre de commerce de Lyon	120,000
Souscription du journal Le *Matin*	180,000
Institut Marey	18,000
Société de photographie transcendentale	6,000
Ligue Franco–Anglo–Américaine contre le cancer	30,000
Société hydrotechnique de France	40,000
Syndicats des fabricants d'extrait tannant et tinctoriaux	5,000
Les Amis de l'Université de Montpellier	4,500
Académie des sciences et lettres de Montpellier	4,000
La Ville de Bordeaux	1,000
Fondation Claire-Georges Dieulafoy	26,000
Ville de Paris	150,000
Fondation Marquet	5,000
Prélèvements de l'état sur les salaires industriels et commerciaux	14,000,000
Institut scientifique Chérifien	250,000
Total	25,393,850

Source: Albert Ranc, *Le Budget du personnel des recherches scientifiques en France* (Paris, 1926), pp. 91–3.

The politician chiefly responsible for the creation of the Caisse was a deputy, later senator, from the Loire, Jean-Honoré Audiffred, an apostle of national progress and a leading light of the Musée social. There were three key dates in the history of the Caisse: July 14, 1901, date of its creation as an autonomous organization under the political umbrella of the Ministry of Education; April 30, 1921, when the activities it could support were enlarged; and December 29, 1922, when it was made into a province of Breton's empire, the more powerful and richer Office national des recherches scientifiques et industrielles et des inventions. A parliamentary charter burdened the Caisse with a grandiose aim it was never adequately funded to fulfill: "endow all sciences with sufficient means of action; ensure that no serious researcher will have to stop work for lack of money." At least a safe source of research money was found in basing support on horses. The state used part of its lucrative cut of pari-mutuel funds to make payments to the Caisse. More money was diverted from sin to science when gambling profits also became a resource. In 1912, on the basis of a report by a special commission, the Ministry of the Interior gave for the first time a grant of

100,000 francs of the revenues derived from "jeux de hasard." An administrative council of politicians and bureaucrats managed the Caisse, but a technical commission, mostly scientists from the establishment – the academies of science and of medicine, the Muséum, the Collège de France, and the faculties of science and of medicine – met at least twice a year to decide on the distribution of funds. It was a scheme not much different from our own incestuous but indispensable system of peer review. This setup had the advantage of bringing politicians, administrators, and scientists together formally for the purpose of deciding on priorities in research.[11] The interaction of three groups seriously interested in promoting scientific research resulted in the creation – symbolic though it was at the time – of the CNRS on the eve of World War II. Fortunately for science, symbols have an irritating habit of metamorphosing into very expensive programs.

Believers in science support systems were delighted at the government's decision to base funding of research on pari-mutuel and gambling profits; their only regret was that the amount was so low. Here at last was a decent and practical use for the large profits earned annually by the state from activities it ought to be opposing instead of encouraging. A previous scheme for the use of these tainted funds gave 6.5 million francs to 525 communes for the adduction of their drinking water, but it seems that much of the money was frittered away on projects of dubious salubrity. It was thus easier to argue that the Caisse should receive more, on the grounds that scientists were less likely to waste money than local politicians – not an unreasonable assumption because the scientists had less money. Intellectuals and journalists – in France the distinction is somewhat oversubtle – were enthusiastic over the double philosophical aim of the Caisse: to prolong human life and to increase agricultural yields, two clearly articulated political aims in France since the Enlightenment. What more could be said of such an excellent program, except to link it with a civilization of increasing distinction: "c'est celui d'une civilisation qui s'élève et prend conscience des problèmes essentiels au point de vue pratique."[12] Civilization, problem solving, practicality: the hallmarks of French intelligence.

Up to 1921 the Commission technique, which awarded the grants to suppliant scientists, was divided into two sections, one for biological sciences and another for the rest of the sciences. The biological section funded research on disease in man, animal, and plant life. The Commission technique, presided over by the Directeur de l'enseignement supérieur, was convoked by the minister of education for the purpose of distributing annual grants. The budget was prepared by a manager employed for fiscal work. A report had to be made to the Conseil d'administration, which had a supervisory authority over decisions made by the commission. Each year the

Commission technique met as an assembly to divide the budget into two parts, one for research in biological (medical) sciences and Cinderella's part for the other sciences. The first section of the commission, responsible for the biological sciences, met separately to distribute the lion's share of the loot. In the beginning 125,000 francs were taken yearly from pari-mutuel funds for studies of disease; the 100,000 francs for the second section went into research on water pollution. Calmette, head of the Institut Pasteur (Lille), was given 60,000 francs to organize, with the help of big industrialists in the Nord, a series of pratical studies aimed at cleaning up water polluted by factories. Given the extent and serious nature of pollution in the Nord, the award was a modest one.

Calmette was not only the most successful practitioner of the embryonic art of grantsmanship, he may also have been the only scientist suspected of malfeasance in the use of funds. An unsigned note to the director of higher education, written about 1912, pointed out that since the Caisse had supported Calmette's research on waste waters for seven or eight years, it might be in order to find out if the funds were always used for scientific purposes. There were two reasons for raising this astounding question about one of France's leading scientific administrators. In 1904 or in 1905, according to the note, 4,000 francs were paid to Serpollet at the same time that Calmette bought one of its automobiles for his personal use. Later Calmette and two or three of his collaborators used the same grant to pay 3,000 francs for a trip to Austria and Germany. The note ended with the ominous prediction that a more complete check would reveal other abuses. Perhaps. It must have been assumed that the automobile and the trip were legitimate expenses, or that the suspicions were unfounded, or that nothing could be done – except perhaps a private, friendly warning – because nothing seems to have happened. Calmette went on to greater glory as head of the Institut Pasteur in Paris and continued to get his grants.

Most grants ranged between 500 and 2,000 francs. Members of the first section in 1904 were Marey, Ranvier, Schloesing, van Tieghem, Brouardel, and Lanceraux – obviously a strong representation of basic medical, especially physiological, research, which occupied an important place in French biology. The nonbiological section was reasonably representative of chemistry, physics, mathematics, geology, and zoology: Berthelot, Bouquet de la Grye, Darboux, Fouqué, and Perrier. Leading research scientists continued to work in the Caisse with administrators and politicians. By 1911 Chauveau, Bouchard, Lannelongue, Landouzy, Forgue, Troost, Lippmann, d'Arsonval, Maquenne, Appell, and Barrois were in harness. The academic establishment (dean, faculty of medicine, Paris; secretaries of the academies of sciences and of moral sciences; director of the Muséum) was joined with representatives of the Conseil d'état, the Senate and the Chamber of Deputies, ministries (finance and education), the Conseil supérieur du commerce et de l'industrie, the Conseil supérieur de l'instruction publique, and the director of higher education. It might be that never was so little handed out by so many, but it is also true that the presence of these people in the Caisse indicated that the idea of institutionalizing and increasing state funding of scientific research had strong support from several powerful groups other than the recipients of the research money.

Over seventy scientists had received subsidies of 1.4 million francs from the Caisse by 1911, but over 900,000 francs had gone for biological research and over 400,000 francs for research on control of water pollution. The big grants in 1910 went to Calmette (IPL) – 30,000 francs for work on polluted waste water, Trouard-Riolle (head of the agricultural Ecole de Grignon) – 15,000 francs also for research on water pollution, and Arloing (veterinary medicine, Lyon) – 10,500 francs for work on serums and vaccines against tuberculosis. There were three smaller grants of 3,000 francs each: Courmont in Lyon for the study of cancer in mice; Gastou of the Hôpital Saint-Louis in Paris for research on syphilis; and Gley of the Collège de France for research on immunity to toxic serums. In 1910 the annual income of the Caisse was up to over 450,000 francs with expenses of only 250,000 francs, including prehistorically small administrative costs of 4,540 francs.

By 1912 some scientists were convinced that the grant-awarding mechanism of the Caisse had developed a dangerous flaw. Its paltry treasury was being increasingly used for purposes other than the stimulation of original scientific research. Perhaps it was becoming clear that support of the typical humdrum activity of scientists is difficult to distinguish from the rare brilliant achievement, especially in the research stage. And the French, notoriously intolerant of their own failings, became restive in an effort to work out a method to separate "l'obscur" from "l'éclat." The whole debate centered on a variant of the eternal issue of *cumul*.

The problem of *cumul* of grants was pointed out by the director of higher education (Bayet) in a letter he wrote in 1911 to a permanent secretary of the Academy of Sciences (Darboux). Grants had been made by the Academy from the Fonds Bonaparte to André Broca, Houard, and Charles Moureu for the same project subsidized by the Caisse. In 1911 the Fonds Bonaparte distributed 30,000 francs to eleven winners out of thirty-four researchers asking for grants. But the director did not protest against a researcher's having grants from several sources; it was enough to call attention to the fact, presumably on the assumption that the establishment would prevent any abuse by a greedy grant getter. In 1910–11 Houard and Broca received 2,000 francs each, bringing the total to 14,000 francs for the period 1906–11.

Funds of the Caisse were given to scientists – not to organizations – who made a request based on a plan and a program. Requests were judged according to the scientific advantages that the Commission technique thought should result from research. According to the limits imposed by its resources, the Caisse made awards on the principle of one sum of money for each issue to be worked on. In the course of a year the recipient sent in a report on work done and results achieved together with any publications on the funded activity. If the work were not finished in one year, the scientist could ask for another grant, which was considered in the same way as the first request. Apparently simple.

It often happened that the renewal of a grant for a project would give the unfortunate appearance of an annuity. Even worse, recipients of renewed grants had fallen into the habit of expecting them annually. One member of the Conseil d'administration believed that renewed grants were considered most of the time as supplements to laboratory expenses and even to salaries instead of for original and temporary research. One way out of this difficulty was proposed by the president of the Caisse, A. Picard: Instead of numerous small grants, give some relatively big ones for significant research. Up to 1912 it seems that the Caisse was reluctant to turn down a request for a renewal, although the renewal would be given with a warning, or the amount requested would not be granted if it was determined that the research was not fruitful or the method defective, with the risk that the research would be sterile. If a second report showed no improvement, the grant was not supposed to be renewed, but weak-hearted colleagues often did renew their fellow scientists' dubious requests. In order not to make a premature condemnation of the research, the Caisse would sometimes make a substantial reduction in the amount of the grant. The significance of such reductions was rarely understood but probably seen in terms of the reasonable, ego-soothing explanation that there were limited funds for many applicants.

Spurred by the worries of some members of parliament, the Commission du pari-mutuel, and the Commission des jeux about the dispersion of the resources of the Caisse among a large number of researchers, most of whom got small renewed grants, and also worried about the nonresearch use of funds, Picard convinced the Conseil d'administration of the Caisse to adopt a new policy, set forth in several simple rules (1912). Scientists would submit a complete program, specifying the research that the grant would be used for, and, in so far as possible, the probable length of time and expenses required for the work. Grants would be made only for research of real importance. Renewal of a grant would depend on an end-of-the-year proof of serious efforts and achieved results or, at least, of work justifying hope for success. In no case should renewed grants become a supplemental contribution for laboratory expenses. Although much of this policy would require great subtlety and vigilance to implement, the probability of its success was considerably increased by a decision of the Conseil d'administration not to grant subsidies for more than four consecutive years. After four years, any grant coming up for renewal would be specially examined by the administrative council itself; this procedure arrogated to the council a prerogative reserved by law to the first section of the Caisse – making, diminishing, or ending grants – but done under the ultimate supervisory authority of the council, which was also responsible for the budget. At least it was a serious effort to prevent the funding trough from always being emptied by the big hogs of grantsmanship.

Many people believed that small grants were useful; some people even had a philosophy of the usefulness of small grants. All of this was quite sensible in an era between the nineteenth century, when grants, at least in our sense of the word, were nearly nonexistent, and the post–World War II period, when science through grants became the rule rather than the exception and a symbol of the Moloch worshiped by committees with an ardor formerly reserved for the Holy Ghost and scientific genius. The following justification for maintaining small grants was stoutly upheld within the inner sanctum of the Caisse, the Conseil d'administration. Big discoveries do not necessarily come from big grants, nor only small scientific achievements from small credits. Sometimes smallest produces most. Even when big grants produce big scientific results, it is only after a long time – the implication here is that small grants would eventually produce similar results. Naturally, no one with a big grant, or hoping for one, could accept this heretical dogma enjoining fiscal small thinking, with its sly implication that big money frequently produced small thoughts. Big grants could have undesirable consequences. Abundant resources lead to big organizations, whose researchers have no hesitation in buying expensive equipment, which remains unused in laboratories after the experiments are ended. And the startup time of such experiments is always long; in 1912 quick payoff was a virtue. Even hard workers with a big grant could fail to come up with results over some years of experimenting. It was conceivable to defenders of small grants that the work would never end. As in the market, money invested does not always bring a return. According to the means one wishes to use and the particular program, a big or a small grant could be required to continue study of a thorny problem or to pursue what promised to be a big or small discovery. The Council concluded that only the particular circumstances could determine which course of action was better, but small grants had a definite role to play in scientific research.

The historical example of the funding of physiology could be invoked to show that the particular time period of a discipline and a scientist's type of training were important related issues. Vastly different amounts of money had been required as research styles evolved and research tools changed. The beginning had been expensive. Vivisection and experimentation shifted from big to small animals – horse and ass to dog to rabbit to guinea pig to mouse and frog – and then, in a conceptual leap, to isolated cells living in liquid. No further need for big stables, large spaces, and numerous personnel, at least until experimentation became entrenched in schools of veterinary medicine. Certainly, types of research centering on the use of poisons and chemical and physical agents required smaller laboratories with reduced expenses. But experimental pathology, which consumed large numbers of rabbits, guinea pigs, and mice, could be expensive, especially in the area of human infectious diseases. Then the discovery that these diseases could be transmitted to other primates led to a hundred-fold increase in laboratory expenses. Monkeys have always been expensive. Some types of socially desirable research could also require heavy expenses over long periods. Research on bovine tuberculosis was neither cheap nor quick. Experiments on cancer in mice could go on for years at great cost to animals and purse. It hardly made any kind of sense to limit such experiments arbitrarily to four years. So the Caisse would essentially have to continue its policy of making both large and small grants with the possibility of renewal. Such a policy of support for scientific research was within the context of the law of July 14, 1901.

If for no other reason than inflation, grants had to increase in amount. In 1903–4, when the Caisse spent just over 303,000 francs for the liberally ecumenical category of biological research, the big winners were Calmette, 8,000 francs (in addition to his 60,000 for water studies); Marey, 6,000; d'Arsonval, 5,000; Roger, 3,000; Guignard, 1,500; and Jungfleisch, 1,500. In 1924 the second section gave 248,900 francs to fifty-six scientists, of whom eighteen were in physics, fifteen in geology and physical geography, and twelve in chemistry. The large grants went to Guinchant (physics, Bordeaux) for work on gas, 20,000; Villard (laboratory at the Normale), 15,000, which was the second part of a grant for 30,000; and Madame Curie and her students, 22,000. By 1927–28, when the second section alone spent 451,000 francs, five big grants stood out: Perrin and Curie, 40,000 each; Croz, 35,000; and Fabry and Urbain, 20,000 each. In 1929, with a purse of 434,200 francs, the section gave Curie and the

Table 2. *Operations of first section, La Caisse des recherches scientifiques, 1924*

Subjects	Number of reports	Amount allotted (francs)
Anatomy	7	21,500
Embryology	3	8,000
Histo-physiology	8	28,000
Physiology	13	60,000
Flight of birds	1	50,000
Experimental psychology	1	6,000
Experimental pathology	14	45,000
Pharmacodynamics	1	5,000
Physics	6	37,000
Chemistry	6	20,500
Flora and fauna (explorations) (includes 60,000 for voyages of Charcot's yacht)	20	149,500
Protozoa and invertebrates	13	33,500
Botany	13	41,500
Plant pathology	4	12,500
Paleontology	2	3,000
Human pathology	14	47,500
Animal pathology	1	5,000
Tuberculosis	4	23,000
Cancer	11	106,000
Subtotal	142	703,900
No results given	14	54,000
Total	156	757,900

Source: Archives nationales, F[17] 17437.

Perrins two grants of 40,000; Gutton, 3,500; and P. Weiss, 2,000.[13] Even the grants of the second section were large and regular enough to have some impact on the research programs of France's leading scientists and trainers of researchers.

Grants from the first section were of even more importance for the fertile effect they had on a wide spectrum of disciplines in the biological sciences with related research from other scientific areas. In 1924 the sum of 757,900 francs went to 156 scientists; only 136,000 francs went that year for work on TB and cancer. At the end of the nineteenth century, TB was responsible for about 25 percent of the deaths in Paris – at least this was true in July 1888. In 1899 about 150,000 French died from TB, as compared to

302

Table 3. *Grants by La Caisse des recherches scientifiques, 1924*

Organization or journal	Amount (francs)
Société chimique de France	18,000
Journal de chimie physique	8,000
Fédération française des sociétés de sciences naturelles	8,000
Bibliographie scientifique française	6,000
Journal de physique	4,000
Société de chimie biologique	4,000
Société de biologie	2,000
Archives de zoologie expérimentale	2,000

Source: Archives nationales, F[17] 17437.

only 90,000 Germans. Deaths from cancer increased dramatically in the second half of the nineteenth century. Richet estimated that deaths from cancer in England went up from 360 to 606 per million inhabitants in the ten to twenty years before 1888. By 1906–10, the French figure was relatively low for the annual cancer morality per 100,000 inhabitants: 75.5, compared to 125.9 for Switzerland, 94 for England, 84.1 for Germany; lower than France were the United States (65.7), Italy (63.3), and Spain (49.9).[14] There were two big subsidies in the first section: 20,800 francs to Richet to permit him to finish his studies in nutrition; and 50,000 francs for Magnan to continue his studies on the flight of birds. Table 2 summarizes operations of the first section in 1924.[15]

One of the most important funding activities of the Caisse was the program of subsidies to journals and to societies for various kinds of publications. Many local societies with publications of limited influence were funded, but the largest grants went to the top journals of international reputation. Some of the significant grants in 1924 are listed in Table 3. By 1929 amounts had increased considerably: The Société chimique de France got 22,000 francs; the *Annales d'anatomie pathologique*, 18,000; the Société de chimie physique and the *Journal de physique*, 15,000 each; the *Archives de zoologie expérimentale*, 12,000; the Société zoologique de France, 8,000; the Société de chimie biologique, 6,000; and the Société d'ethnographie, 50,000 francs. A publication that had an obvious propaganda value, like the *Bibliographie*, an organization that did national work, like the Fédération, or a publication that was prestigious enough to attract foreign authors, like the bulletin of the society of biological chemistry or the *Journal de chimie physique*, would not die for lack of funds.

303

A problem more serious than the number and size of grants arose from the fact that funds from the pari-mutuel source could legally be used only for biological research carried out in establishments that did works of charity. This requirement sometimes caused difficulties. Although the law establishing the Caisse specifically favored scientific work related to the discovery of new methods of treating diseases affecting man, domestic animals, and agricultural plants, the first section of the Commission technique interpreted this provision to favor biologically related research, especially pathology very widely defined. No mention was made of the charitable establishments so piously specified in the law on the use of pari-mutuel funds, where it was perhaps assumed that a virtuous end justified an illegal means. The issue came up in the Chamber of Deputies as a consequence of the connections established between the two laws. It was clear that the law of June 2, 1891, required that money from pari-mutuel funds had to be used in establishments with some claim to charitable activity, such as an institution that prepared and distributed serum against diphtheria. The Senate recognized that it was not the intention of the authors of the law on pari-mutuel funds to exclude research done in noncharitable establishments from access to these funds. Fortified by political recognition of the need to ignore this provision, the first section of the Commission technique continued to subsidize institutional research in Paris, Lyon, and Lille on tuberculosis. The spirit killed the letter of the law.

In 1921 the government divided the Commission technique into four sections. The first section examined requests for funds for all biological research and for research on the purification of polluted waste water. The second section still handled all requests for research funds in all nonbiological sciences. The two new sections dealt with requests for subsidies for the publication of scientific works and of scholarly legal and literary works. In 1922 the third section allocated 92,500 francs for scientific works; 32,000 francs were left for legal and literary works. Although this subsidy for publication was a continuation of an old policy of governmental aid for scholarly publication, its institutionalization in an organization with the exclusive purpose of financing scientific research was a happy innovation. In 1930 publications in science cost the Caisse 325,000 francs and publications in the human sciences 158,400 francs. Total expenses for 1930 were nearly two million francs. The Caisse was a modest, limited, but not insignificant operation.

During most of its existence, from 1901 to 1934, the Caisse was heavily dependent on pari-mutuel funds. In 1920, out of an income of over one million francs, 720,000 francs came from this source. This proportion declined with an increase in subsidies from the Ministry of Education. In

1930, out of an income of 2,830,527 francs, only 850,000 francs came from pari-mutuel funds and game taxes, with 445,000 francs from the Ministry of Education.[16] The frugality as well as the prudent fiscal management of the Caisse can be seen in the basic figures for the period 1904–30:

Year	Income	Expenses	Surplus put into reserve fund or carried over to next year (francs)
1904	145,312.24	94,772.78	82,473.44
1913	1,270,872.29	1,237,909.36	316,429.41
1928	1,855,270.24	1,785,898.97	507,401.93
1929	2,023,173.40	1,810,871.98	719,703.35
1930	2,110,824.51	1,949,261.21	881,266.65

The budget for 1933 was still heavily dependent on the horses:

State subsidies	385,000 francs
Subsidies from the apprenticeship tax	850,000
Departmental subsidies	100
Grants taken from pari-mutuel profits	600,000
Grants from gambling profits	300,000
Back payments on investments	139,380

There were six sources from which the Caisse could derive its credits: subsidies from the state, the departments or communes, the colonies, and other public establishments; gifts and legacies; individual or group contributions; allocations from pari-mutuel funds; allocations from taxes on games of chance; and the interest on unspent funds, which French organizations of the period seemed adept at accumulating if they did not have to spend their funds within the rigid limits of a fiscal year with no possibility of carry-over into the next year. Contributions from departments, communes, businessmen, industrialists, and all individual sources made up less than 10 percent of the resources of the Caisse. When the Caisse began, d'Arsonval set a good example by asking a newspaper that offered him money to do experiments on radium – a research rage of the age – to give the money to the Caisse. Few followed d'Arsonval in this gesture, and fewer still were as generous as Zaharoff, who, like Solvay, was one of the few big supporters of scientific research in France. In 1921 Zaharoff gave the Caisse the sum of 100,000 francs, of which 75,000 francs went for scientific publications (third section) and the rest for legal and literary publications (fourth section). Three other big gifts were railway bonds given by Audiffred in 1903 with great fanfare to encourage other friends of science to do likewise, but the modest return was less than 2,000 francs annually; the Matté legacy (1911) of over 5,000 francs for biological research; and money from the Loutreuil legacy (1913) of a million francs for active scientific research. Funds were invested in many types of businesses and properties with a return of 3 or 5 percent. So rare were these gifts that small contributions elicited profuse thanks and excessive praise.

Why this remarkable parsimony? Laments of pessimistic politicians and administrators who tried to entice the bourgeoisie into Yankee liberality could lead one to believe that France was a nation of Harpagons and Grandets. But the explanation may be more cultural than genetic, and therefore intrinsically less interesting. Since the ancien régime, governments had developed a system of public subsidies for culture and scholarship. It avoided the formation of powerful private foundations, whose independent thinking the centralizing state found disagreeable, and ensured that scientific research took place outside the conservative and unmalleable University.[17] The state's policy was so successful that when the Third Republic tried to encourage donations by departmental and local governments and by individuals, it did not stimulate much liberality à l'américain unless the money went for regional or local projects. When private giving had so long been regarded as public vice, it was not easy to change the criterion of virtue.

The core of funding for the Caisse was always the pari-mutuel funds, and no source of funds is more certain than that based on gambling. It is too bad that the function of the Caisse was not expanded with a bigger share of the fruits of hope and greed. Pari-mutuel funds reached a billion francs in 1921. With a 10 or even 11 percent share, the state had the use of 100 million or more francs that were not incorporated into the official budget, although use of the funds was regulated by law, with a substantial part directed to welfare, hygiene, and agricultural improvements.[18] Only a small part of this lucrative rake-off could be siphoned to even obviously applied scientific research. Still, the existence of the Caisse meant that many scientists did get funds from a source that would otherwise have remained untapped. Indeed, it is a moot point whether parliament would have allowed the Caisse to come into existence if it had to depend on funds officially a part of the budget. A reluctance to incorporate in the budget permanent commitments to peripheral items like scientific research was characteristic of a majority of politicians, most of whom flattered themselves as guardians of the public purse. The Caisse was a simple, ingenious mechanism, funding useful if generally unspectacular work within the narrow boundaries of research on public health and disease control. Even if one argues, somewhat unfairly from a contemporary viewpoint, that the Caisse was a failure, it probably is one of those cases where it is better to have funded and failed than never to have funded at all. The road to success is paved with novel failures, alas! But

I have whiggishly argued for a much more favorable view of its importance in the history of French science funding.

Science, technology, and war: L'Office national des recherches scientifiques et industrielles et des inventions

French problems in failure and success: chemistry, optics, aviation

By the early twentieth century a considerable scientific research and teaching structure had been created in France. Two big problems had not been satisfactorily solved: how to fund basic research, and how to connect it with industrial and agricultural production. These problems were not uniquely French, of course. Nor were French efforts at solutions insignificant, as we saw in the case of the development of the institutes of applied science. How to achieve an efficient, fruitful, and lucrative input of science into industry had become a standard item in political rhetoric by the 1870s. At the annual meeting of the Association française pour l'avancement des sciences in 1874 the mayor of Lille (Catel-Beghin) was certainly the ideal politician to celebrate the marriage of science to industry, a union for which his city was acquiring a justly deserved fame. One of the problems in achieving the happy union characteristic of Lille became clear in the speech following the mayor's. Adolphe Wurtz, president of the association, one of France's leading chemists, began auspiciously enough with a Baconian salute to the aim and role of science in changing utopia into reality. True, most of the speech was an analysis of the exact sciences from the viewpoint of the atomic constitution of matter, but this might be viewed as quite salutary for chemistry. For Wurtz the important thing was the intellectual aim of science – understanding the order of the universe – a sort of footnote to the ultimate satisfaction the human mind derived from consideration of the first, unique, and universal cause, God.[19] There was some resistance in the scientific community to whoring for industry, as some might have viewed it. Perhaps it was just as well that there was a strong band of purists around, although they were always small in number. By the early twentieth century most scientists, especially in provincial faculties, were intimately connected with local industry and agriculture. And it had become clear that the chief problem was not the unwillingness of scientists to go to industry but the reluctance of industry to make an investment in science.

It is generally agreed that the most serious weakness in the linkage between science and industry in France was in chemistry, precisely the area where it was strongest in Germany. A good argument could be made that after 1860 French chemists increasingly neglected the practical aspect of their science. There was also the striking strength of inorganic as compared with organic chemistry in France; worse, Marcelin Berthelot, holder of the chair of organic chemistry at the Collège de France during most of the second half of the nineteenth century, opposed atomic theory, which became important for organic chemistry and the industries that were based on it.[20] At the turn of the century the Ministry of Commerce (Ministère du commerce, de l'industrie, des postes et des télégraphes) supported a study to find out exactly what the differences were between the French and German systems of relations between chemistry and industry. A long study by a scientist confirmed traditional opinion on the subject but added piquant details.[21] A serious input existed only in new industries in which chemists had a creative role at the beginning. Older manufacturing kept its traditional organization of work, following slavishly a factory routine based on safe empirical data, as in the case of the ceramic and glass industries. Ignorance of scientific explanations of the manufacturing processes made it difficult to avoid false maneuvers or to reduce the number of accidents. Some industrial groups wanted to change things in France. One of the leaders was the Chambre syndicale de la céramique et de la verrerie, the originator of the study (mostly on ceramics) done under the aegis of the Ministry of Commerce. But much of the trouble in France was found to lie in the educational system. Generally, studies by professors found the trouble to be in industry, which was deaf to their sound advice, and studies done by or, more usually the case, for industry blamed education and even the professors. No doubt neither one nor the other was so guilty as the accuser assumed.

Industrialists thought that higher education was very theoretical, to the point where professors were not encouraged to have relations with industry. In chemistry, students were not always directed toward practical research. Mundane, if difficult, issues, such as the composition of clay, were not studied in courses. For reasons that are not clear, and even less convincing, the French always believed the educational set-up in Germany – often meaning Prussia – to be better than that in France. One argument does make some sense. Granger, the author of the report for the Ministry of Commerce, in the course of his visits to eight schools, including those in Berlin, Aachen, and Hanover, was especially impressed by the extremely specialist nature of the German schools. Higher technical education was not scattered in different institutions, or at least an attempt was made to group as many types of study as possible under the same supervision. There was a tendency to create one type of school for higher industrial education embracing distinct sorts of study. This differed from the French model, according to which engineers for a certain service came from several establishments with different educational programs, for each French school had its own curriculum. Metallurgy, for example, was taught at both the Ecole nationale des mines and the Ecole centrale. German schools generally had five or six clearly separated divisions, with an autonomy unheard of in French schools. For the French interested in German technical education with an industrial orientation, the greatest fascination was the Königliche mechanisch-technische Versuchanstalt at Charlottenburg (Berlin), which had 2,425 students in the winter term of 1898–9, of whom 1,150 were in mechancial engineering, although chemistry had grown in an extraordinary fashion, far beyond the facilities available at any one institution in France. An amazing growth in industrially related chemistry in University institutes would change this comparison somewhat in favor of France, but the range and depth of organic chemistry remained strikingly superior in Germany.

At the same time that a few industrialists were getting worried about the absence of science in French industry, leading politicians were also making sympathetic noises. Léon Bourgeois made an intelligent speech lamenting the lack of testing laboratories open to industry.[22] One problem of significance for external commerce as well as for the practice of medicine was that only commercial thermometers were verified, whereas in Germany all thermometers used by medical doctors were tested, thus avoiding the not unusual irregularity in the graduation of French instruments. It was typical of the use of laboratories for industrial purposes that Charlottenburg had two laboratories of metrology.[23] But no recognition of a national problem requiring an immediate solution penetrated French political consciousness before the First World War. France had no national equivalent of the English National Physical Laboratory, the American Bureau of Standards or the Carnegie Institution, or the Physikalisch-Technische Reichsanstalt. Important changes of national significance did take place, but they were usually decisions taken by several universities with the collaboration of municipal and departmental authorities.

Several scientists attempted to get more science into industry during the last two decades of the nineteenth century. Two of the outstanding successes were the founding of the Ecole municipale de physique et de chimie industrielles of Paris and the founding of a scientific journal devoted exclusively to dyes and the dye industry. Scientists, industrialists, and politicians collaborated between 1878 and 1883, when the school opened in the old buildings of the Collège Rollin, to plan a school in which science directly relevant to industry would make up a program unique in French education. Paul Schützenberger and Charles Lauth were the chemists chiefly responsible for the maneuvers that eventually resulted in the decision by the city of Paris to fund the school. Schützenberger, Lauth, and Haller were the directors of the school between 1883 and 1923, when the physicist Paul Langevin took over and pushed the school more in the direction of basic research than had been the case previously.[24] Another means for the rapid injection of chemistry into industry was provided by Lauth, Ed. Grimaux, professor of chemistry at the Polytechnique, and Ch. Girard, head of the chemistry laboratory of the city of Paris, who took the initiative, with the support of Schützenberger, in establishing in 1897 the *Revue générale des matières colorantes et des industries qui s'y rattachent.* Things had moved far and fast since Lauth had been impressed enough with German achievements in chemistry shown at the exhibition of 1878 to ask, in vain, that the Ministry of Commerce create an Ecole nationale de chimie.

The career of Lauth provides a model of the top-level nineteenth-century French industrial-academic chemist. Born in Strasbourg in 1836, he began his ascent in the laboratory of Gerhardt. Arriving in Paris in 1862, he studied under Wurtz; becoming assistant to Persoz at the Conservatoire determined his career in the direction of the chemistry of colors and the techniques of dyeing and printing. He worked in Lyon, collaborated with Depouilly and then with Poirrier and Bardy in industrial experiments. He also worked with Grimaux on new dyeing techniques. As a result of his collaboration with the firm of Poirrier and Chappat, later called the Société anonyme de matières colorantes et produits chimiques de Saint-Denis, he became an officer in the company for the rest of his life. Between 1890 and 1904 he did experiments in Schützenberger's laboratory at the Collège de France. As administrator of the Manufacture nationale de porcelaine de Sèvres he invented, with Dutailly, a new porcelain that was baked at 1,300 degrees. On leaving Sèvres he founded an Ecole de céramique, which necessarily united scientific and artistic functions in producing students of use to industry.[25] The founding of the Ecole municipale, in the Alsatian tradition, and of the *Revue générale des matières colorantes* provided a glorious culmination to an extremely creative career in industrial chemistry and in educational administration, in the tradition of Berthelot, of Schützenberger, and of Pasteur.[26]

The French were also slow in developing a close connection between the science of optics and the manufacture of optical instruments. The curious reason for this may be that the French were extremely good at manufacturing a range of practical items like binoculars while remaining dangerously blind to the new world of scientific optics that had opened up. Success can be ultimately more stultifying than failure. Fortunately, the tradition of physical optics remained quite strong, as can be seen from the work on light by Foucault and Fizeau in experimental physics. It seems that the manufacture of binoculars was created in Paris about 1815 and made a great leap forward when J.-P. Lemière took out a French patent for the improvement of the Dutch double telescope. The manufacturers of glasses prospered in an extraordinary fashion in an artistic environment by collaborating with luxury crafts.[27] Paris became the great world center for the manufacture of a wide variety of binoculars ranging from articles of high luxury to cheap models. The appearance of prism models did not slow down French activity. J. Dubosq even took up the regular manufacture of the prism lens stereoscope. French manufacturers supplied glasses to the allied armies during the First World War. Yet a set of serious problems existed. Although the French and the English pursued work nearly identical with that done in optics in Jena, the Germans were quite superior in the manufacture of lenses by 1914. There was quite serious technical catching-up to do in the mass production of optical equipment. The tables had been turned since 1887, when one of the reasons advanced for the founding of the Physikalisch-Technische Reichsanstalt was German inferiority in precision engineering, including optics.

Of course there was a fairly strong French connection between optics and industry. Science required delicate and well-designed instruments, which industry alone could supply, but only if the specifications were furnished by physics. The interaction took place at several levels: in higher education for the formation of technical personnel, in professional education for the formation of workers, in research and testing in manufacture, and in an increasingly vital quality control and efficiency. Application of these general ideas was evident in the creation of establishments based on one branch of science and its related technology, such as in the founding of the Institut d'optique. In the optical industry scientist and industrialist, theoretician and practitioner, whether in the laboratory or in the workshop, collaborated closely. Yet, as one of France's leading optical physicists lamented in 1928, French optics did not have the same type of industrial-scientific relations that had ensured forty years of German success in this area.

It was not until 1920 that a modest private establishment was created in Paris to serve as an Ecole supérieure de l'Institut d'optique. In 1927 the institute moved into a new building, where the Chambre syndicale des constructeurs opticiens was also housed. A high-quality scientific journal was founded: the *Revue d'optique théorique et instrumentale*. Although private, the institute was connected with the state and the University of Paris, with an administrative council of forty-odd members made up of representatives from ministries and *grandes administrations*, of scientists, and of industrialists. There was a state subsidy.

There were several sections in the institute. First, the Ecole supérieure d'optique produced optical engineers for the cadres of the optical industry;

their high-level education was nearly all given by teachers from the University of Paris. Students in this program needed mathematics at least at the level of the certificate of general mathematics. Studies lasted one year but could be spread out over several years for people in industry. The faculty of sciences offered a certificate in applied optics at the licence level, which could be done at the same time as the optical engineering degree. Twenty students graduating annually were enough for industry. Most of the students who were state employees came from the navy, for which optics had a considerable importance. A section of the institute called the Ecole professionnelle trained annually sixteen youngsters, aged thirteen to fifteen, who had earned certificates of primary studies. After three years of general education, along with theoretical and practical optics, a student acquired the *diplôme d'ouvrier opticien*, which easily got him a job in the optical industry. Foremen and workers could take their own advanced night course. Many opticians also took a night course designed to help them fill medical prescriptions correctly. The institute covered a remarkable gamut of courses catering to a wide range of industrial, commercial, and medical needs.

Much of the instruction took place in the laboratory, which necessarily occupied a central role in the institute's programs. The laboratory also did testing and measuring for the optical industry. Charges for services were low, but numerous requests produced an income of 40,000 francs in 1927. Research was also done in the laboratory; so it was not a typical industrial laboratory doing analysis rather than research. Fear of industrial espionage did not exist – perhaps because there were so few industries – and neither did the issue of publishing research funded by an industry that did not want it published. All research results were published in the institute's *Revue d'optique*. Research was given the supporting service of a library that received a good many, but not all, of the specialized journals to which such an organization should have easy access.[28]

It is easy enough to pick out many specific examples of French success in integrating science and industry. One can take an old industry like watchmaking, raised to an entirely new level by basic science. When Voltaire established his watch factory at Ferney he proudly recommended the Enlightenment precision of his products: "bonnes et bien réglées" meant that the watches did not gain or lose more than twenty minutes per day. When watchmaking came to be more the province of the engineer than of the craftsman, the engineer himself depended on the foundations of scientific watchmaking laid by people like Phillips, who devoted his life to studying the mathematical laws regulating the oscillations of the spiral balance wheel.[29] Watchmaking at Besançon was no more exempt from the rule of theory than winemaking in Bordeaux.

312

The newer the industry, the more certain that science would be of key importance in its success, competitiveness and profitability. Georges Claude, a student of d'Arsonval's, was the creator of a giant industry in liquid air and low temperatures. Beginning with 10,000 francs in 1899, the company, Air liquide, was worth 11,000,000 francs by 1913. Based entirely on the purely theoretical data of thermodynamics, the industry's starting point was regarded as a moment of basic science, an experiment done in a laboratory for the first time by "the father of low temperatures," Louis Cailletet, in 1877. Claude could well afford to proclaim that theory is always right.

A similar situation soon developed in the electrical industry, whose increasingly powerful installations and more numerous customers required more and more scientific as well as practical activity on the part of engineers. The needs of the electrical industry led to legislation recommended by conferences and international congresses.[30] Measurement of internationally acceptable standards required precise controls that could be developed only in laboratories like the Laboratoire central d'électricité, Janet's pioneering prototype at the Ecole supérieure d'électricité. The grim prospect of devoting a lifetime to getting another decimal point might have led some theoretical physicists to consider more exciting careers as pianists or social scientists, but precision of measurement was what excited many scientists who worked in electricity. Janet believed that industry and science advance together in the construction of scales of measurement. "Those who know something of these delicate questions will appreciate the cost of the conquest of a decimal point in careful details, patient research, perseverance, and originality."[31] With the arrival of tediousness in science the scientist had finally caught up with novelists, who had long established boredom as an essential ingredient of their craft.

In no area was the collaboration of science, technology, and industry more vitally necessary than in aviation, a truly new creation of the twentieth century. An aggressive governmental policy in promoting aviation was able to take advantage of scientific and institutional resources to make France into one of the world's few leading air powers. As in the case of electricity in the Ecole supérieure d'électricité, aviation's success was in large part the product of brilliant specialization. Aviation also had the advantage of being regarded as important in national defense and in facilitating communications within the extensive French empire, especially for mail service in North Africa. For expertise in motors France could fall back on a group of first-rate engineers and scientists who had been studying internal combustion engines in industry, especially automobiles, for a generation. Not of least importance was French strength in mechanical engineering. The Ecole supérieure d'aéronautique et de construction mécanique was founded in Paris in 1909

313

by Commandant Roche with the support of the government, especially the ministry of war, and the support of leading scientists. Modeled on the new but distinguished Ecole supérieure d'électricité, it had a faculty from Mines, the Polytechnique, the Ecole centrale, Ponts et chaussées, and graduates from these and other top technical schools. In its first year the school had 120 students, of whom fifty were graduates of leading engineering schools and university institutes and sixteen were *licenciés ès sciences*. Paul Painlevé gave the course "mécanique de l'aviation." Marchis was also on the school's staff. Those who graduated as aeronautical engineers had a fine theoretical and practical education in everything relevant to aviation.

As seems typical of the French in their pursuit of science, in the case of aeronautical studies they looked abroad for a model to emulate. But this one was Russian, not German. In 1904 Dimitri Riabouchinsky spent 100,000 of his own roubles to build an institute of aerodynamics; it had an annual budget of 36,000 roubles. In 1906 it established a bulletin to publish the results of its experiments. At this time the greatest deficiency in aeronautics was lack of exact knowledge of the action of air on a solid body; that knowledge could come only from methodical and varied experiments in a laboratory. In 1911 Henri Deutsch de la Meurthe gave half a million francs of capital, with 15,000 additional francs annually for maintenance, in order to establish the Institut aérotechnique de l'Université de Paris. The faculty of sciences already had a chair of aviation, founded by Basil Zaharoff, given to Marchis, who had been a student and then Duhem's colleague at Bordeaux. His work was much more experimental and practical than that of Painlevé, whose equations gave the mathematician's "explanation" of the movement of an airplane. The institute soon established a bulletin and attracted the collaboration of some of the best known of the Sorbonne's scientists along with engineers and military men: Cailletet, Janet, Koenigs, Maurain (directeur), Marchis, Painlevé, Picard, and Urbain – mathematicians, physicists, and chemists. Armont de Gramont, duc de Guiche, a scientist well known for his work in aviation, established a laboratory, which published the results of its experiments from 1911 on. Eiffel's specially constructed laboratory carried out experiments on air resistance. Theses, journals, and books bore witness to the growing fascination of the scientists with the airplane, which would, Lecornu predicted in 1909, become as common as the automobile and the ocean liner.

Before the Great War the French enjoyed a period of consciousness of being in the forefront of one of the greatest technological innovations of modern times. Like Americans, the French took pride in the fabulous feats of their heroes: Louis Blériot, a graduate of the Centrale who built his own planes, comte Henry de la Vaulx, and comte Georges de Castillon de Saint-

Victor fired the nation's imagination. French nobility seems to have been as addicted to aviation as was the Red Baron. The *Almanach de l'action française* for 1934 shows the comte de Paris, resplendent in his aviator's costume, oozing modernity, standing next to his plane.[32]

Given a powerful fillip by the war, French aviation remained strong in the 1920s. The *Avion Bernard* won the world's speed record in 1924 with a dazzling 448.17 km/hr. Exhibitions had many French models. Foreign governments bought French products because of their high quality. Based on solid research in governmental services, schools, and a wide variety of laboratories in such areas as mechanics, chemistry, medicine, and meteorology, French aviation moved with confidence into an industrial phase, offering services to a growing civilian clientele on a basis of commercial profits. It was not easy to distinguish between civilian and military activities because the best airplanes could be easily transformed into engines of war.[33] This had the advantage of ensuring strong governmental support for avaiation. Scientific, industrial, national, colonial, and military advocates who were interested in promoting aviation gave this industry a character quite different from most other industries. Like much of modern activity, aviation was given its impetus by the triple stimulus of science, industry, and the military.[34]

Yet, a few years later, by 1930, the French were lamenting that they lagged in aviation research. Many foreign laboratories were recognized as better; the United States was carrying out experiments on large apparatuses. French laboratories had a shortage of personnel. Governmental laboratories were plagued by raids from private industries for technicians after an interministerial commission assimilated technicians into the bureaucracy as *fonctionnaires*.[35] Although by no means moribund, French aviation would not regain a leading place in the research sector until well after the Second World War.[36]

Fear of decline in science and technology seems as endemic in France as a consciousness of literary superiority; both emotions distort perceptions but are nonetheless anchored in substantial reality. One of the themes harped on by the glossy magazine *L'Air* (*Revue mensuelle de la ligue nationale populaire de l'aviation*), founded in 1919, was the imminent danger of France's falling behind both England and Germany in aviation. The Michelin brothers, who established a prize of half a million francs to encourage flying skills, issued an ominous warning: Postwar German planes could be transformed into bombers within an hour. French laboratory activity, which had fallen to practically nothing during the war, seemed unlikely to receive funding on a scale comparable to England's after the armistice. By 1922 France's experts on aviation were turning envious eyes on "the flourishing condition of

English aviation," in large part the result of the wisdom of the British government's early founding of a national advisory committee. Governmental support enabled the National Physical Laboratory and establishments of the Royal Air Force to carry out research beyond the capacity of private organizations. Complemented by university research, the input of these publicly supported institutions into British aviation produced great flying machines. The French *Services techniques* were far from receiving support comparable to that of their British counterparts.

During World War I the French aviation industry expanded considerably in order to supply planes to the allied armies: 50 percent of production went to the allied forces; 8,000 Bréguet planes alone were used during the war. With the armistice a brutal economic move by the government canceling credits placed the industry in "a grave crisis." The existence of an active Groupe parlementaire de l'aviation ensured that aviation continued to be funded. These political supporters of aviation included the marquis de Dion (Nantes), an industrialist, René Fonck (Vosges), Heurtaux, and Laurent Eynac, president of the editorial committee of *L'Air* and later air minister. While the French were worrying about the possible death of French aviation, the British air force was being reorganized with credits equivalent to 600 million francs. The French budget for aviation was finally established at 300 million francs for 1920. One of the problems throughout the 1920s was the absence of coordination and centralization in French aeronautical services: Military, maritime, and colonial aviation were parceled out to the ministeries of war, the navy, and the colonies. It was not until 1928 that the French hunger for centralization was partially satisfied with the establishment of the Ministère de l'air.

It might be argued that French aviation was still in excellent shape in the 1920s, among the best in the world. In February 1923 the Spanish Ministry of War organized a competition before buying several types of airplanes. A French machine won first place in each category of competition. The pessimism existing in French aviation circles derived to a large extent from the prevalent opinion that French research was inferior to that being done in several other countries, which would consequently be producing superior aircraft in the future. This argument was developed in *L'Air* in 1919 by the aeronautical engineer A. Toussaint, a firm believer in the necessity of the fertile coupling of scientific research and technology. Toussaint pointed out that the future of aviation was chiefly dependent on the scientific study of problems arising from aerial navigation. Construction itself was based on engineering, which used known science like the theory of elasticity and the resistance of materials, although there was also the issue of the conception and calculation of the proportions of forms, which demanded a deeper knowledge of the laws of aerodynamics. This knowledge could only be supplied through the design and study of new machines in laboratories of aerodynamics furnished in the Anglo-American style; this procedure would avoid expensive and dangerous experiments outside the laboratory. True, the science of aerodynamics had hardly begun to develop. Until *experimental* aerodynamics became more precise and organized, the scientific rigor of mathematics could not be applied to the complex phenomena of air resistance. As in late nineteenth-century winemaking, empiricism still played an excessively large role in the early twentieth-century aviation industry.

All Western countries serious about aviation had significant research institutions in aerodynamics: England had the National Physical Laboratory at Teddington and the air force establishment at Farnborough; Italy had laboratories in the central aeronautical institute of Rome and in the Turin Polytechnic; and Germany had the laboratory at Göttingen, famous even before the First World War for the precision of measurements done there. France did not lack laboratories doing research in aviation. It had pioneered in this area and, given the limited money available, as in the case of marine biology, had gone on to proliferate perhaps more laboratories than it should have.[37]

Among the pioneers were two private laboratories, those of Eiffel and Guiche. The German laboratory at Göttingen was founded in 1908–9, when Eiffel's laboratory of the Champs de Mars also opened. But Eiffel moved to Auteuil in 1910, after completing his experiments on the tower. At Auteuil he had two wind tunnels, one and two meters in diameter with wind speeds of 30 to 40 meters per second. Eiffel's first laboratory was a modest operation; the laboratory at Göttingen was also modest, until the war provided Ludwig Prandtl, the founder, with the resources to built a new, vaster facility. In all countries the war provided a strong stimulus to research in aviation. Colonel Lafay did some experiments during the war at the laboratory in the Ecole polytechnique. And the military laboratory built at Issy-les-Moulineaux by the Section technique de l'aéronautique militaire was of considerable importance. The duc de Guiche had carried out essentially high-level scientific experiments before the war. After 1914 his laboratory at Levallois-Perret became a shop for the construction of precision instruments and even airplane engines.[38] War blurred the distinction between public and private sectors. Of course, in aviation, as in most other sectors of French technological activity, the amount of research done in state laboratories was greater than that done in the private sector.

The initial effect of the war on France's greatest laboratory for aviation research was catastrophic. L'Institut aérotechnique de l'Université de Paris at

Saint-Cyr was used as a barrack until 1916, when it was totally redone and the technical section of aeronautics was put under Commandant Cagnot. When Maurain was director the laboratory did experiments on aircraft, especially wings and propellors. From the beginning there was also an emphasis on the practical and theoretical research needed to perfect all types of motors. In 1912 Henry Deutsch de la Meurthe gave the money to add a large aerodynamic section to the laboratory in order to increase its experimental potential. Between 1912 and 1914 Maurain obtained funds from the state to construct a wind tunnel of two meters in diameter with a maximum wind speed of 38 meters per second, although its final installation was interrupted by the war. In the same period Toussaint and Lepère developed a method of experimental research on airplanes in flight. The creation of the Section technique de l'aéronautique militaire in 1916 was a means of making the institute an effective instrument of the Services militaires de l'aéronautique. By 1918 the military had installed a station for testing avaiation motors in a cold, rarefied atmosphere. The maximum speed attainable in the wind tunnel was increased to 50 meters per second. During the war even nonmilitary activity was directly related to harnessing science for victory. Jean Perrin directed an Ecole d'écoute des avions that was based at the institute. It is clear that the institute emerged from the war transformed, strengthened, and considerably expanded in scope.

In the early 1920s it was reasonable to argue that L'Institut aérotechnique de Saint-Cyr, with its wide selection of technical equipment, its chemical laboratory, and its laboratory of general physics and electricity, was one of the best-equipped aeronautics laboratories in the world. It carried out joint experiments in 1926 with the Eiffel laboratory on projects for new planes. Also attached to the institute was the testing station at Villacoublay for planes in flight. Use of the institute by the Services techniques de l'aéronautique continued until 1933, when, as a result of a decision by Painlevé, then air minister, it was transferred to the Conservatoire national des arts et métiers in order to add a component of experimental research to the teaching of aeronautics at the Conservatoire. The transfer was based on a ten-year renewable contract with the University of Paris, which retained the right for students of the faculty of sciences, especially doctoral candidates, to do experiments at the institute. But the institute did not acquire giant wind tunnels, or tunnels producing variable densities of air, or high-speed tunnels. In the 1930s France did not possess one of the most modern aerodynamics laboratories.

As early as 1930 the head of the technical section of the Air Ministry, A. Caquot, published an admission of France's backwardness in failing to install expensive laboratory equipment for experimenting on large-scale machines. The situation in laboratories was made potentially disastrous by a shocking lack of qualified personnel, from workmen to engineers; this reduced research to perhaps one-third of what it should have been. And some weaknesses in industries vital for a strong aircraft industry made the French position more a cause for alarm than for hope. Production of aluminum alloys needed to be greatly increased. Quality control was poor in the steel industry, which produced large quantities of medium-quality steel, rather than the uniform, high-grade metal required for the manufacture of aircraft. Nor was the precision tool industry at all comparable to that of Germany or the United States; indeed, most of the precision tools required for plane construction were not even made in France. Engines, practically a monopoly of Gnôme-et-Rhône, were mediocre and high-priced. The aviation industries themselves, mired in a semi-artisanal state appropriate to a previous period of airplane manufacture, were numerous and financially weak.[39] By 1936 they were also struggling under the threat of nationalization, which was applied to the aircraft industry in a flexible way in the hope that research and innovation would not be tainted by bureaucratic blight. It was not until the end of the 1930s that there was widespread recognition of the superior air force developed in Germany and of the bad state of French aircraft production. By then the only quick way to increase the size of the French air force was to try to buy planes from the United States, which did not have any surplus to sell. And an unholy alliance of liberal financial circles in politics, builders of airplane engines, unions, and the Communist party opposed buying abroad.[40] The lack of facilities for large-scale experimentation had become a very minor issue by 1939. Saint-Exupéry provides a classic pessimistic view. "On ne fabrique pas un matériel en quinze jours. Ni même . . . la course aux armements ne pouvait être que perdante. Nous nous trouvions quarante millions d'agriculteurs face à quatre-vingts millions d'industriels!"[41] *Si non è vero* . . . It was clear that several scientific areas of national importance could not be left to the institutes of applied science, where regional interests were usually paramount. Naturally a big enough regional interest was of national interest, but this was the exception rather than the rule for the provincial faculties of science. In areas of science like organic chemistry and optics, no institute was deemed adequate by scientists and politicians. In areas like aviation and big physics (e.g. magnetism) only heavy state research support could be of much significance. There was a prevalent recognition of the need for the creation of a new national funding mechanism for this kind of science, but nothing much happened before the First World War.

The Office des inventions in war and peace

Although scientists quâ scientists have played a sporadic role in war since antiquity, it was not until the late nineteenth century, with the striking exception of the French Revolution, that they were formally integrated in the war machinery of the state. No one could be regarded as even a mild version of Dr. Strangelove. In 1870–1 Marcelin Berthelot became president of the scientific committee for the defense of Paris. A member of the War Ministry's commission on explosives after 1874, he became president of a new commission in 1878. Initially, the Commission d'examen des inventions intéressant les armées de terre et de mer (1887; extended to include the navy in 1894) had only a veterinarian representing science and medicine. The horse was still a key piece of war equipment. But five scientists were added a week after the new commission was born: Mascart, Moissan (replaced by Violle in 1907), Troost, Appell, and Boussinesq, all members of the Academy of Sciences.[42] Vieille (Ingénieur des poudres et salpêtres) was also on the commission; his work with Berthelot had "laid the foundations of a new scientific study of the mechanism of explosions."[43]

General scientific journals like the *Revue scientifique* carried articles in the 1880s on topics directly related to military issues: indirect targeting by the infantry, armaments, steel, rapid-fire weapons, and big guns. Scientists taught at schools under the Ministry of War (e.g., the Ecole polytechnique); they were on the admission committees of specialist schools such as the Ecole d'application d'artillerie navale. Alfred Ditte (physics, Sorbonne) did research on aluminum in the Laboratoire central de l'armement. In 1892 the minister of war, Freycinet, a scientist-mathematician, created a special commission to study possible uses of aluminum by the army. In 1893 Moissan, who had become a key member of the committee, was granted permission to work in the Section technique de l'artillerie on possible military applications of aluminum.[44] Army officers doing research sometimes got permission to publish in journals and to send notes to the Académie des sciences. Scientists and military men developed cozy relations during the Third Republic long before 1914.

The First World War brought scientists into making war and into defense on an unprecedented scale. And physicists were not left out during the finest hour of the chemists. Pierre Weiss and Aimé Cotton worked out a system of detecting the location of artillery through sound that was in extensive use by the time of the armistice. Weiss and Hadamard, as part of the physics section of the Direction des Inventions, set up by Painlevé in 1915, were responsible for establishing a liaison with the educational center for antiaircraft fire.[45] But it was the unexpected use of gas in April 1915 that led to the biggest mobilization of the French scientific community, especially chemists and medical researchers, who had to overcome rapidly France's total lack of preparation for a type of warfare that had been declared illegal by the European powers at the Hague on July 29, 1899.

A dangerous situation faced the French because of the superiority of the German chemical industry. Although France had perhaps one-tenth the number of chemists as Germany, there were a few factories producing the required organic chemicals, and they provided the skeleton for the vast production of explosives, gas, and other war matériel that was quickly set up. There were some serious problems in production. There was also a lack of well-trained chemists who could supervise all stages of the production of organics, an industry different from the manufacture of mineral or inorganic products, which could get by in those days with poorly trained chemists because of the predominant role of the engineer and the mechanic.[46] Respectable amounts of many vital chemicals were produced. Sulfuric and nitric acids, chlorine gas, chloride (bleaching), soda, phosphates, mineral salts, and superphosphates, for example, were produced by industries as good as those in metallurgy and electrochemistry. Among serious needs were industries for liquid chlorine and for bromine. There was also practically no production of potash. Germany had three factories producing liquid chlorine, France had none. But by August 1916 the French were able to reply equally to the Germans by the same poisonous means used by the enemy.[47] The army explicitly recognized their debt to industry and science for the Z companies (*Génie*), specially created to gas the enemy.[48]

Shortly after the first German use of gas in the West, on April 22, 1915, a committee came into existence for chemical warfare research. Three subcommissions were organized: first, a group presided over by Kling, head of the municipal laboratory of Paris, to study action at the front; second, a group, presided over by Charles Moureu of the Collège de France, to study possible aggressive action; and, third, a group presided over by Vincent, a professor and also an army medical inspector, to study protective measures. Organizations were created to do research and experiments on protective devices and to supervise the manufacture of toxic gases. The Section des produits agressifs was headed by General Perret with Moureu as vice president. Members were nearly all professors in French schools and faculties: Bertrand, Grignard, Job, Kling, Lebeau, Maquenne (replaced by Délépine), Simon, and Urbain. Members of the Section de protection were all medical doctors and pharmacologists. With the physiologist Emile Terroine as secretary and General Perret as president of both groups, a close liaison existed. Thirteen Parisian laboratories were turned into chemical research laboratories to work on toxic substances used in warfare. Experiments were carried out in several testing sites. Three other laboratories concentrated on protective devices and regulation of products manufactured for the Z companies and other troops.[49] The "success" of the gas warfare operation showed the great resourcefulness of the French scientific community and French industry. Integrated into the mechanisms of the state, the scientific community was totally at ease in collaborating with the army and with industry before and during the war. Many of the same people continued similar functions, suitably modified for peacetime, on a Commission des études et expériences chimiques in the 1920s.

The military use of science and technology was not new in World War I, but the nature and number of changes that took place seem to put previous relations in a kind of prehistoric age. Promiscuity was replaced by institutionalization, infrequency of consultation by regularity and constant demand, and dependence on acts of Promethean invention by endless improvement in existing technology. The technological surprise of World War I was not technical. The use of tanks, airplanes, and gas presented no fundamental innovation in science or in technology, although they were new parts of the increasingly efficient death-machine, representing considerable innovation in the utilization of existing technology, in planning, and in production.[50] Long-frustrated technicians suddenly found themselves solicited by desperate governments; blocked currents of reform found the dikes invitingly opened. In November 1915, when the mathematician Paul Painlevé was minister of education, the Direction des inventions intéressant la défense nationale was created as the first step in a movement of "scientific mobilization" for victory. The mathematician Émile Borel headed Painlevé's Service scientifique de la défense nationale; Borel would also be minister of the navy in the Painlevé cabinet of 1925. After a number of administrative metamorphoses and ministerial peregrinations, the direction jelled in 1922 as the Office national des recherches scientifiques et industrielles et des inventions under the Ministry of Education. As Blancpain points out, an important step had been taken: There now existed a small, autonomous, administrative but specialized organism responsible for the organization of applied research.[51] The Caisse des recherches scientifiques (1901), operating since 1903, was combined with the Office des inventions in 1922.

As early as 1887 the government had created the Commission d'examen des inventions intéressant les armées de terre et de mer, reorganized in 1894. The purpose of the commission was to tell the ministries of war and of the navy about the inventions submitted to them, especially those that might be used for national defense. Totally inadequate to satisfy the enormous demands put upon human ingenuity for death and destruction in wartime, the commission was replaced in August 1914 by the Commission supérieure chargée d'étudier et éventuellement d'expérimenter les inventions intéressant la défense nationale. In spite of its long name and noble intention, the Commission supérieure was unable to use its scientists to fulfill the most important part of its mission because it had no means of experimenting or of perfecting promising ideas. In November 1915 Paul Painlevé ended this absurdity by creating the Direction des inventions intéressant la défense nationale. Its purpose was to complete industrial mobilization by scientific mobilization. Without sulfuric acid and science, revolutionary *audace* would not lead to victory. To coordinate the activity of the technical services and to increase their productive capacity, the Direction des inventions was transferred in December 1916 to the Ministère de l'armement et des fabrications de guerre and elevated to a Sous-secrétariat d'état under J.-L. Breton, a scientist turned politician. In April 1917 the new unit grew in stature and in favor with the state as it became the Sous-secrétariat d'état des inventions, des études et des expériences techniques, bringing together all the technical offices and research organs of the Ministry of Armament. Breton continued to head the unit when it came under the authority of the Ministry of War in September 1917. Two months later the unit was reduced to the status of a board of directors. Breton stayed in charge without a salary or personal subsidy. In spite of these administrative gyrations, the technical services remained together in this unit during the war.

After the armistice the danger was that the Office would be abolished in a fit of economy. Fortunately Breton knew his way around well enough in politics to keep his organization alive, provisionally as the Direction des recherches scientifiques et industrielles et des inventions (decree of April 14, 1919). Then in May 1919 the government put forward legislation to create an Office national des recherches scientifiques et industrielles et des inventions, which would absorb the services of the board of directors. Although the chamber passed the law in July 1919, the Senate took years to approve it. When the Office national was officially created on December 29, 1922, Breton, a master of survivorship, was in charge. He had kept the board of directors in operation, thus ensuring continuing contacts with scientists, industrialists, and inventors. He finally had a genuine peacetime organization.

Undoubtedly, prevalent knowledge of the useful, widely advertised secret work done by Breton and his organization during the war contributed to his bureaucratic survival. The Commission supérieure des inventions had the daunting task of examining 44,976 proposals before passing on 1,654 of the 1,958 proposals seriously studied to the different sections of the Sous-secrétariat d'état et de la direction des inventions. Only 781 proposals were passed on to the technical services as inventions having practical applications.

Like the Bergsonian model brain, the commission's chief function was to forget. Breton advertised himself as the creator of the assault tank and the person who introduced it into the army. Certainly a great deal of important work was done by the organization, especially in artillery, automatic weaponry, transport, and the ballistics of various types of missiles.[52]

It is clear that most powerful politicians thought that France needed governmental stimulus in technological innovation in peace as well as in war; indeed, in technology and in science was not peace a continuation of war by other means? The consensus was that the French had a genius for making inventions but were inept in manufacturing and in selling them. The roots of this mythology – grounded, perhaps, in an elusive reality – went back to at least the mid-nineteenth century. This sin of commercial inferiority was expiated in a boast of a nearly genetic superiority in French inventive skill. Analytical skill produced innovation, imagination constantly exploited this skill in work, French taste led to care in the details of manufacture, and France's geography and ethnic makeup placed its people at an intellectual crossroads where they could play the role of universal adapters. Just in case that was not enough, France also had a superb system of education. And what could be more logical than to have the state complete and possibly coordinate the supportive action of private organizations and ministries who aided inventors? Intellect complemented by institutions has always been a French prescription for success.

There was a clear recognition of the eternal problem of an absence of contact between science and industry, or at least this was the common opinion expressed in editorials in scientific journals, mostly by people who had little to do with industry. It would be easy enough to cite a list of people like Haller and Le Chatelier who were top scientists closely connected with industry, but they were leaders of a special chorus complaining about the lack of science *in* industry. And there was the model of the fanciful German paradise, where close cooperation between industry and science produced rapid industrial development and great wealth. It is true that before the Great War French inventors did not get much help from the few official organizations that existed. Although frugal French industrialists increased profits by buying foreign patents rather than supporting costly scientific research in their own operations, there was a patents office, the Office national de la propriété industrielle. For tests, analyses, and reports, there was the Laboratoire d'essais du conservatoire des arts et métiers. The Ministry of War had its Commission supérieure des inventions intéressant les armées de terre et de mer to signal anything of potential use in the gadgets advanced by inventors as great contributions to the progress of humanity. Clearly the condition of the French inventor, not so desperate as the

Cassandras cried, could be easily improved.

The Marin commission made a laundry list of services that could be provided by a governmental agency attached to the national office responsible for scientific, industrial, and agricultural research and inventions. Inventors should have the means to put their inventions into practical working condition. Inventors should know what others have done; given the absence of governmental research on patents, this would be a particularly useful service in France. Technical personnel should be provided for detailed work that inventors could not do. Industrialists and constructors should be persuaded to give inventions a practical test. Inventors should be protected against the theft of their ideas as well as have their capital and other investments assured of protection from the unscrupulous entrepreneur. Bureaucratic babysitting of the inventive class appealed to several powerful commissions and to the chamber, which then asked the government to clone yet another commission to study the issues and to recommend a permanent organization to promote scientific, industrial, and agricultural research.

Influenced by the ancient but powerful and notorious apostolic diatribe written by Le Chatelier in 1916 on the state of scientific research in industrial laboratories, the ministries of commerce, education, industrial reconstruction (formerly Ministry of Armament), and agriculture established an extraparliamentary commission to design the new organization – a case of new presbyter being old priest writ large if there has ever been one.[53] This so-called Clémentel-Painlevé commission was certainly loaded with power and brains. Clémentel, Painlevé, Millerand, Picard, Borel, Breton, Maurain, Appell, Haller, Lucien Poincaré, Violle, Citroën, Renault, and Janet were in the galaxy of politicians, scientists, and industrialists on the commission. Most of the work was done by a subcommission presided over by Painlevé, France's leading savant-turned-politician.[54] Painlevé was only one of a number of republican scientists who went into politics, but he was certainly the most powerful and successful of this group after the meteoric rise and fall of Paul Bert. Like Freycinet, he rose to the pinnacle of power. Justification for the new office was put in terms that would catch the attention of even jaded French politicians: economic competition. Science would play the same role in peacetime economic struggles that it had played in the war. Since 1914 Germany had perfected a powerful scientific and technical organization that would provide in an industrial victory the ransom of its military defeat. Both England and the United States had taken action to harness scientific resources in the service of national production. In England a Privy Council committee had a budget of the equivalent of twenty-five million francs for five years, and in the United States an appeal had been made to the National Research Council. A French centralized research organ, founded with the general aim of developing and coordinating scientific research, especially that applicable to industry and to agriculture, would assure the undertaking of studies requested by governmental agencies and would help in the completion of serious projects proposed by inventors.[55] Public funds would be used to create a service for technical information, promoting research, or cooperating with industrial and agricultural groups on projects of mutual interest. Much of the later ideology of scientific research, expressed at greater length in the charter of the founding of the CNRS, can be found in the justification proclaimed for the establishment of the Office national des recherches scientifiques, industrielles et agricoles et des inventions. Breton and his minions were quite keen on exploiting the siren call of the mystery-laden word *invention*.

Invention summed up one of Breton's key ideas – perhaps his *idée fixe* and a fatal flaw in his grandiose scheme for funding science in France. Because progress in modern society requires constant improvement in its industrial plant and labor, there must be an industrial organization of invention itself. How this should be done was Breton's second basic idea concerning the function of the Office: The encouragement of inventions along with their protection and diffusion would require a state service. In the end the Office never did move seriously or creatively beyond its role of assuring the execution of the studies requested by public agencies. These were not insignificant, but they did not fulfill the dreams of many of France's most creative scientists, led by Jean Perrin. This failure would lead the political allies of Perrin to abolish the Office itself in 1938. A program limited to solving problems for the army and industry appeared very dull fare to most scientists, and given the technological conservatism of both *patrons*, was perhaps doomed from the start, in spite of Breton's noble long-term view of putting great emphasis on basic research.

The budgetary home of the Office was in the Ministry of Public Education, but it was run by a national council, presided over by the minister and designated by a curious congeries of parliament, scientific organizations, scientific and technical societies, public agencies, industrial and agricultural groups, and workers' associations. This was science in the public interest with a vengeance. Although no scientist could be forced to do something he did not want to do – the freedom of the marketplace was a sacred dogma in scientific research – the Office could call on any scientist in a publicly funded laboratory to do something it deemed vital for the exploitation of the country's scientific resources. And what savant could resist the moral authority of a request in the public or national interest? Because there was no central funding mechanism for industrial and agricultural research, a great deal of duplication and waste could be avoided if the Office were able to establish fiscal control and technical centralization for the different laboratories. A reform of the legislation on patents was also given as one of the Office's noble aims. It was reasonable that the old Direction des inventions became the nucleus of the new Office. The future was under way at last.

In administrative theory, where payoffs are frequently meager, the link of the Office with basic research was to be assured by putting the Caisse des recherches scientifiques under it. In personal terms the ostensible imperialism of the Office was reassuring, for Breton had supported Audiffred in the project that had produced the Caisse forty years earlier. As Minister of Hygiene, Breton reestablished the annual subsidy to the Caisee from the parimutuel funds, which amounted to only 150,000 francs in 1914 but was suppressed anyway in February 1914. Breton got the subsidy up to 1.25 million francs, and it never dropped to below half a million thereafter. Although the Office was authorized to accept bequests, they were never an important source of income – the two most important being 15,000 and 10,000 francs. The French tax structure has not promoted the establishment of foundations or encouraged acts of ostensible generosity on the part of modern Maecenases. A minor source of revenue was found in income from contracts with inventors who had been subsidized by the Office. It seems that by 1939 the Office expected large amounts from the rubber vulcanization process of Dufour and Leduc and from work in metallurgy (aluminum) by Séailles. By 1938 a considerable amount of money was being derived from patents exploited abroad. Still, all of these sources of income were insufficient for the operation of the Office.

The two main sources of income were state funds and an exhibition permanently organized by Breton in the Grand Palais. Breton made a strenuous effort to increase the amount of revenue generated by the Office itself, hoping to continue his scheme of reducing the state subsidy by 200,000 francs annually. In 1932 state subsidies were nearly 3.5 million francs, or 67 percent of its financial resources. Its increasing ability to generate funds probably saved the Office from impotence as the depression began to take its toll, with the government reducing its funding of frills like science and technology.

Year	Total income of Office (francs)	State subsidy	
		Amount (francs)	Percentage
1928	2,567,766	2,045,710	80
1931	5,451,353	3,651,572	66
1934	4,055,607	2,395,000	60
1937	4,807,600	1,993,400	41

Subsidies by ministries varied from year to year, but defense and public works were frequent clients for the services of the Office. Between 1925 and 1935 ministerial departments gave annual subsidies ranging from a minimum of 90,000 to a maximum of 492,500 francs, and averaging less than a quarter million francs annually over the eleven-year period.[56] Some of the subsidies were big: In 1931 the Ministry of Defense gave the Office nearly a million francs for research relating to national defense; the normal subsidy of the Ministry of War was also a quarter-million francs in 1937. On the eve of its abolition the Office was about to receive another subsidy of 400,000 francs from the Ministry of War. The ministries of air and the navy did not use the services of the Office very much, although air gave a subsidy of 53,000 francs in 1938. In the legislature the financial status of the Office was always a bit shaky, in spite of the astute fiscal management that produced an annual budget surplus, something the heads of agencies were proud of in this frugal period in fiscal history. Breton even asked parliament to reduce his budget by 200,000 francs annually after 1935; the request was not denied.

Breton's finest achievement was the creation of the Salon des arts ménagers, although the snobbism inherent in elite science required that any activity reeking of kitchen technology not be welcomed into the fold, or even be permitted to erect a shrine to scientific popularization in Perrin's Palais de la découverte. In 1923 Breton came to the conclusion that the shortage of domestic servants would be remedied by the widespread use of kitchen equipment, which at that time was made mostly in Germany and the United States. Equally repugnant to enemies of allying science to commerce was the fact that Breton got 10,000 francs from the Chambres syndicales des grands magasins et des galeries et des bazars to offer as prize money to French inventors who could improve family organization. Anyone who knows anything about French kitchens must laud the worthy aim behind this noble failure. Over the years many items vital to modern kitchen progress had their day in the Salon: aluminum, paint, rubber, and even the mail. Fish, milk, butter, cheese, and wines were displayed by ministries. The florists came. Contests were held for the best household. Contracts were made with firms in architecture and in construction materials. In 1927 the monthly *Art ménager* appeared; André-J.-L. Breton had taken the initiative in collaboration with the Salon, of which Paul Breton was the commissary general. A family enterprise.

By 1938 the budget of the Salon was up to 6,953,000 francs with the number of annual visitors at over half a million. On the strength of the samples provided, especially in the sections consecrated to wine and cheese tasting, a substantial amount of business was done. The success was so striking that it was possible to get a statement from Charles Maurain, dean of the Paris faculty of sciences and a member of the Academy of Sciences, that, from a certain point of view, the Salon was an exposition of the applications of science, whose salutary and fruitful role in improving the conditions of life was made clear to the public. Politicians and journalists were more extravagant in their praise of what Louis Lumière called household science. Its success did not make it any more acceptable to scientists as a funding scheme.

Given its limited resources, the Office was involved in an incredible array of activities – perhaps too many – ranging from the most mundane of technical problems to some of the most interesting challenges in extending scientific knowledge. It was also involved in the use of radio in education, or at least it tried to promote its use in spite of the apathy of the Ministry of Education.

First let us glance at the Office's technological tentacles. The Office's love of committees was revoltingly modern but, given the vastness of its scientific and technological ambitions, inevitable. The technical committees of the Office with their presidents were the Comité de mécanique (Auclair), the Comité technique de physique (Cotton), the Comité technique de photographie et de cinématographie (Comandon), the Comité de chimie (Copaux), the Comité de navigation et génie (André Broca), the Comité technique d'aviation (Toussaint), the Comité de biologie (Lapicque), and the Comité d'hygiène (Pottevin). There actually existed within the Office a Commission supérieure des inventions, contrived in August 1914 although its origins went well back into the nineteenth century. Between 1887 and 1913 the Commission d'examen des inventions intéressant les armées de terre et de mer led an unexciting existence looking at various ideas for infernal machines of war and contraptions with the potential to kill enemies more efficiently. In November 1915 Painlevé put the Commission supérieure in the newly hatched Direction des inventions. After the war the role of the commission was extended to allow it to examine inventions of possible interest to industry and the economy as well as defense. Items for examination came from the Directeur de l'office. By 1927 the commission had over sixty members representing the ministries of education, war, the navy, agriculture, commerce, colonies, public works, etc. War, education, and the navy had the most representatives. There were ten members representing the Académie des sciences. One of the new functions assigned to it by the interministerial decree of 1927 was to give advice on the coordination of France's scientific research and on the use of its scientific resources. France had a bureaucratic infrastructure of scientific research before its language had the jargon.

In energy-poor France it was inevitable that the Office have some connection with research on various forms of energy. It began research in 1922 for the Ministry of War, with a grant of 100,000 francs from the Direction des pétroles et essences. The Office then teamed up with the Office national des combustibles liquides to help this organization spend its money by implementing their programs of research. Both were interested in carrying out relevant tests. In this area Breton was "following the bucks," for in 1923 the Direction des pétroles et essences (Ministry of Commerce) gave him a grant of 750,000 francs to create the needed facilities. Testing of non-petroleum-burning truck motors for the Ministry of War began in 1927 in the new buildings of the Station des recherches et expériences techniques, as the new installation was called. Within the station's 2,000 square meters there existed splendid facilities for experimental work in general mechanics, electricity, hydraulics, materials testing, as well as in the testing of thermal engines. A bureaucracy inevitably metastisized in the Station for administrative and managerial matters. Also in 1928 the Office national des combustibles liquides established its own Service des essais for a very specialized research program, which was needed in addition to the work done under the aegis of the Office national des recherches et inventions. The quantity of French scientific and technical institutions was matched by a bureaucratic web whose complexity had its toxicity limited only by the parsimony of parliament in funding its growth.

The Station des recherches et expériences provided technical and administrative services for the scientific achievement of which Breton's office was proudest, the big electromagnet (120 tons) of the Academy of Sciences. Here the Office, suspected of technical taint in the circles of highbrow science, could claim direct contact with science on the distant frontier of first-class research, or at least with scientists who thought they were doing it. As early as 1912 a group of French physicists who wanted big, powerful magnetic fields was dreaming of its construction. Aimé Cotton asked the council of the faculty of sciences (Paris) to build a powerful magnet. Both the faculty council and the council of the University of Paris agreed to set aside 50,000 francs from the Commercy fund for the project. The Academy of Sciences set up a commission to study the proposal of Cotton and Weiss to build a big magnet as well as a competing proposal put forward by Henri Deslandres and Alfred Pérot to build several small magnets. Deslandres and Pérot were not in the same league as Cotton and Weiss, but they were good physicists. It was difficult to turn down their project, which did have experimental promise; so the commission gave them 50,000 francs but recommended that the big magnet should also be built. Paul Appell's proposition was that a big laboratory for magnetic research would be paid for by the Academy of Sciences and administered by the University of Paris. Estimated cost of the project was 200,000 francs, which could be cut by one-third if the amount of copper were reduced. Nothing happened until after the Great War, although the ever generous Prince Bonaparte was ready to fund the project.[57]

In 1920 Madame Boas de Jouvenel and the organist and composer Charles M. Widor founded an organization with the optimistic name of Bienvenue française for the purpose of promoting intellectual and spiritual exchanges between nations, the lees of the euphoria of victory. One standard type of exchange was scientific, but it was brought to the attention of the organization that French laboratories were in such a bad state that they were unlikely to attract anyone. By happy chance the centenary of Pasteur's birth was at hand, and what better source of sentimental but useful exploitation than the savior of small children from the ravage of rabies to prise open tight French purses? A national subscription for French laboratories easily got the support of the Academy of Sciences, the obvious manager of any money that might be collected. A suitable strophe was put out by the academy on the effects of the war, the need for France's scientific genius to shine on the world once more, and the threat of a defeated but powerful scientific enemy who still wanted to be the world's leader in civilization. The Germans could not be trusted to accept the natural subjection of *Kultur* to *culture française*. The fund-raising drive was a great success. "La science française comptera encore des jours glorieux."[58] Or as glorious as they could be with another thirteen million francs, the amount collected.

The construction of a big magnet was not possible until 1923, when the public powers authorized this elegant begging for scientific research. A million francs of credit from the *journée Pasteur* was administered by the Academy of Sciences, which decided to follow Aimé Cotton's recommendation that it built a powerful electromagnet. Cotton, professor of physics at the Sorbonne, a member of the Academy of Sciences, was president of the Comité technique de physique of Breton's Office. The project was an augur of future physics. The installation required close collaboration with engineers and technicians; Cotton's colleague was Mabbonx, an engineer who was also a member of the Comité technique de physique. It turned out that the organization best qualified to handle the installation, especially the technical difficulties, was Breton's Office. Because of all the experimental facilities and other resources existing there, Bellevue was the obvious choice of a site for the instrument that could finally produce a field well above 70,000 gauss.

Some foreigners came to work at Bellevue, but the best-known experiments were done by Rosenblum and by Louis Leprince-Ringuet. The work on magnetism that later brought Néel a Nobel prize had a connection with Bellevue. Néel was a researcher in Pierre Weiss's Institut de physique de Strasbourg (Laboratoire de magnétisme). Weiss and Cotton had been at the Normale together and during the war developed the Cotton–Weiss acoustical method for determining enemy gun emplacements. This work was done in the Office des inventions. The Bellevue magnet differed from the one constructed by Kapitza at Cambridge in creating permanent, strong, and extended fields. So the English physicists were also part of the international elite of scientists who made a pilgrimage to Bellevue.

The magnet also contributed to the cementing of the old Dutch connection in French science. Huygens is the patron saint of that connection, but the romance did not end in the seventeenth century and was still full of sedate passion in the nineteenth and early twentieth centuries. In 1866, after an existence of 114 years, the Dutch Society of Sciences (Haarlem) decided to publish a journal to counteract the ignorance outside the Netherlands of the work of Dutch scientists. Publishing it in Dutch would do little to overcome the bad effects of isolation; the annunciation of Dutch scientific research to the world was to be in French through the *Archives néerlandaises des sciences exactes et naturelles*. Zeeman and Lorentz continued the tradition. Both were given honorary doctorates by the University of Paris. Lorentz was invited several times by the Société française de physique to talk about his work; two notable occasions were the celebrations of the fiftieth anniversary of the society and the centenary of Fresnel's birth. The polyglot nature of Dutch education frequently ensured a mastery of French. "Lorentz aimait la France et parlait notre langue avec une rare perfection." Zeeman, who often went to France and wrote some of his scientific work in French, became very interested in the Bellevue magnet, thereby returning the compliment Cotton and Weiss had paid him when they had worked on Zeeman's effect in 1906–7.[59]

After 1929 other scientific enterprises spun off the magnet, especially a laboratory of magnetism and magneto-optics, complemented by a laboratory for the study of low temperatures. Perhaps the most famous laboratory in cryogenics was in Leiden, where the studies of the properties of bodies at low temperatures by Heike Kamerlingh Onnes (Nobel prize in physics, 1913) gave rise to a unique center to which people came from all over the world. Kamerlingh Onnes was also interested in the mundane matters of food preservation, refrigerated transport, and ice production.[60] France was not in the forefront of research on refrigeration. A worried business group, the Syndicat général de l'industrie frigorifique de France, got the government to establish a Comité technique du froid for the study of all aspects of refrigeration and for the supervision of experiments and tests in a Centre expérimental du froid to be organized by Breton's Office. The low-temperature laboratory, constructed according to specifications supplied by Cotton and Georges Claude, was partly the result of business anxiety over deficient French technology. As Breton proudly announced, "a research series in pure science" was done jointly by the two laboratories on such matters as the optical and magneto-optical properties of bodies at low temperatures, dielectric properties, and superconductibility. Applied work was done mostly at the Station expérimental du froid in Meudon. Insufficient funds, chiefly the result of parliamentary lack of interest, especially after 1936, and the rise in industrial prices, prevented the Station from carrying out requests for experiments by the Ministry of Air, the refrigeration business organization, and a company engaged in transporation under refrigeration. Again Breton had made an important move in science and technology. But it was not enough to prevent some people from arguing that the Institut international du froid should be moved out of France because the French had not contributed much to technical progress in refrigeration. (The first international congress on refrigeration had been held in Paris in 1908.)

Some other fairly important work was done on technical matters by Breton's Office. Both government and private industry had enough interest in saving money and in safety to pay for many tests on paints and varnishes. In 1933 a Comité de recherches et expériences techniques for paints and varnishes was created through the collaboration of the ministries of war and of education. Not even fire prevention escaped the tentacles of the Office: A technical committee was created by interministerial action in 1929. Engineers found much to do in the laboratory of the Commission des barrages et charpentes, which got a subsidy of 190,000 francs in 1929; the grant was down to 15,000 in 1936. Ponts et chaussées had more interest than anyone in this operation, but some of the basic questions on the resistance of materials and on hydroelectric construction, for example, were close enough to the scientific interests of faculties to justify the presence on the commission of Charles Camichel of the faculty of sciences in Toulouse. Most of the activity of Breton's Office would have merged quite well with the interests and research programs of provincial universities. Joseph Auclair, president of the Comité de mécanique, directed much of the experimentation in the Laboratoire de barrages et charpentes. His report on ten years of experimentation by the Office appeared in *Recherches et inventions* in

December 1926, the month he died. The disappearance of a powerful political supporter coincided with the triumph of the Popular Front, which followed the advice of scientific gurus having no use for the Office. Its days were numbered.

Breton complained of the failure of the Caisse nationale de la recherche scientifique, established in 1935, to give any financial help to the Office for work on inventions and applied research, although it did pay for a small part of the cost incurred by the office in running the big electromagnet of the Academy of Sciences. In light of the existence of a state organization (Centre National de la recherche scientifique appliquée, established in 1938) for the funding of applied research, Breton's complaint was not entirely justified, but, more important, it shows that he did not understand that the scientists who got the new Caisse founded wanted an entirely different type of science funded. Breton was also wounded by the refusal of the Palais de la découverte to give the Office any exhibition space. This seemed rank ingratitude after the Office had given the Palais valuable electrical supplies and had been only partially paid for some work done in its own workshops – in spite of the many millions at the disposal of the Palais, Breton enviously noted. The word *invention* was conspicuously absent from the program of the Palais and the decree creating the Centre national de la recherche scientifique appliquée. This hostility was partly that between the tinker and the thinker in science, but it is also true that Breton's enterprise was not a useful means of funding science; worse, it was increasingly unsuccessful in getting direct state funding for its work. Why then should Perrin and company not keep their distance from an organization that was remote from their aims? Why should they tie the destiny of science to an organization whose scientific activity was increasingly peripheral to functions of science? No reason at all, especially when Breton was a sick old man whose colleagues and friends were no longer powerful in politics.

On May 10, 1938, Breton learned that the Ministry of Education was preparing a decree to abolish his most important work, the Office national des recherches scientifiques et industrielles et des inventions. Energetic protests, unanswerable and unanswered objections, whining pleas, literary tears, and predictions of disastrous results did not move Jean Zay, radical-socialist, minister of education (1936–39), to reconsider a decision whose brutal suddenness shocked Breton. Emile Borel, a family friend of the Perrins, mathematics tutor for Francis, and president of the administrative council of the Office, tried in vain to get input by the Office's council into the framing of the decree. Maurain was eventually asked for advice. Zay simply wrote Breton that his apprehensions would be shown unfounded when the

Centre national de la recherche scientifique appliquée was created in a reorganization and extension of the Office des inventions. A commission of the Office indicated errors, omissions, and dangers in the decree-law of May 24, 1938. Modifications were recommended by the council and sent to Zay. (The scientists on the council were Borel, Lacroix, Cotton, Lapique, Laugier, Lecornu, Maurain, Perrin, and Picard.) Perrin was unable to defend before the commission the decree that he had hatched. All to no avail. The death of the Office preceded the birth of the national center of applied scientific research. Politicians had other things to worry about in 1938. Breton retired with an annual pension of 36,000 francs. He died in 1940, not far from his beloved magnet in Bellevue.

The Office des inventions had experienced some specific failures. Two in particular were a source of bitter chagrin for Breton: the failure to establish a central supply depot for laboratories and the failure to get free delivery of German scientific equipment in large quantities when there was an excellent opportunity. The first failure was probably just a case of bureaucratic inertia and political perversity. The second, a case of high-level stupidity, was more serious.

According to a decree of 1932, one of the essential functions of the Office was to establish a liaison between public services and laboratories while making sure that the laboratories had the proper equipment. In 1926 Breton pointed out to the Ministry of Education the advantages of having a central supply for laboratories. Wholesale buying would be cheaper. The Office already had a considerable amount of matériel, acquired when the Americans and the English liquidated their war stocks. The ministry agreed to subsidize this venture in 1926. French governmental services like war, the navy, air, and telecommunications sent lists of surplus supplies that they could sell cheaply. A good deal of out-of-date matériel could be used for demonstrations in teaching laboratories. Many items were modified and sold to laboratories. All organizations agreed on the usefulness of this service provided by the Office. The issue was studied *ad nauseam*, and recommendations were made by scientists and industrialists, but the Office expired before the "magasin central des laboratoires" could be established. Ten years is but an instant in the life of bureaucracies.

The affair of the German laboratory was part of an imbroglio, involving the ministries of education and of finance, which ultimately resulted in the assessment of a charge of 200,000 francs on the budget of the Office by the Ministry of Finance for matériel that should have been given to it free of charge. With the reduction of state subsidies this was a serious financial blow. In 1922 the minister of education told officials in higher education that it was possible to get laboratory materials and equipment as part of the war indemnity scheme. A list of supplies worth 20,000 francs was drawn up. Nothing happened. Changes in prices led the minister to ask for a new list nearly two years later. In July 1927 a third list was asked for. A financial ballet went on for another two years until in May 1929 an ultimate fantasy resulted in the establishment of nearly 130,000 francs of credit for technical establishments and other services of the Ministry of Education. Contracts were signed with German industrialists for 13,000 francs, an omen of the eventual illegal suppression of 85 percent of the credits given to the Ministry of Education. Meanwhile Borel, Jean Perrin, and Marie Curie wrote a letter to the minister of finance emphasizing the importance of getting 60 million francs' worth of German equipment in order to end the material poverty of French laboratories. The substitution of the Young plan for the Dawes plan caused great confusion, or gave the occasion for a sly maneuver in finance, resulting in a reduction of 85 percent in credits. (The Ministry of Public Works got over 50,000 additional francs of credit; it then took 70 million francs' worth of German paving stones that could have been supplied by French quarries.) A chance had been lost for greatly improving French laboratories, and a large number of scientists had wasted their time preparing useless lists. This was high-class bungling at the top political level, in a confused time, it is true, but the Office should not have been billed 200,000 francs for the fiasco.

These failures were not typical. The Office could usually boast of its successes, even if many were quite modest. An assessment of success done by Breton inevitably listed inventions for which the Office had been chiefly responsible or to which it had made some contribution in providing facilities or matériel. The war was the golden age of invention but peacetime was by no means sterile in innovation. Inventions with considerable commercial potential sometimes provided valuable resources for the Office. Among the significant discoveries for which the Office took credit were the sonar of Langevin and Chilowsky, Chauveau's automatic rural telephone, Legendre's process for preserving cereals (500,000 quintals of humid cereal saved each year), Panis's artificial respiration apparatus and the mask of Legendre and Nicloux, Dufour's high-frequency induction oven, Jean Breton's apparatuses for developing industrial photographic paper, Fonbrune's micro-manipulator, manufactured under licence by Carl Zeiss, Séailles's aluminum process, and the new processes for the vulcanization of rubber developed by Dufour and Leduc. Not a bad record; one or two of these activities could be classified as scientific by even a high priest of *Wissenschaft*. And when the Office was about to be terminated, Breton could quote some of France's best scientists on the services his institution had rendered to science. Picard, Lacroix, Cotton, Calmette, Viola, Lapicque, Hadamard, Janet, Joubin,

Sabatier, Lecornu, Weiss, Louis de Broglie, Cartan, and Richet were among those willing to sing its public praises.

Breton flattered himself with the belief that because the Office des inventions was at the crossroads of science and technology, it provided a unique input into the French economy and national defense. The Office provided not only a rare interaction between science and engineering; it also provided a cross-section of scientists and engineers with quite unusual training, making them capable of tackling any problem. So it could carry out a set of experiments on corrosion, or on paint, or on the effect of fire on steel structures for the Office technique pour l'utilisation de l'acier. Studies on concrete arches were done for the Comité français des grands barrages. Other satisfied clients: the Société du gaz de Paris, the Etablissements de Saint-Chamond-Granat, the Syndicat général de l'industrie frigorifique, the Chambre syndicate des couleurs et des vernis, and the Société des Aciéries de Longwy. Most governmental ministries also called on the Office for technical studies that could be best done at the "Centre national de Bellevue." All the services of the three ministers in national defense testified to the importance of Breton's Office. It seemed that the only enemies of the Office were Zay and Perrin! And even its enemies recognized its utility.

Perhaps the most important tangible legacy of the Office was the giant scientific and technical complex at Bellevue: workshops, laboratories, testing rooms, and storerooms made up a small scientific city equal to handling any problem. Among its resources: high-tension electricity, thirty-six converter groups of all kinds, high-pressure steam, compressed air, powerful machine tools, many specially arranged platforms, a giant hydraulic installation, powerful testing machines, a mercury hydrostatic pressure installation, lifts, refrigeration down to $-60°C$, a low-temperature laboratory, a laboratory for making all kinds of rubber, and an acoustical laboratory. Curie, Joliot, and Cotton with his students could do scientific work at Bellevue impossible to do anywhere else in France. Big science was born in Bellevue, with Breton as its midwife, and although he might not have recognized the Bellevue monster twenty years after the Office was abolished, he would not have been surprised or displeased by its reorganization and growth, except, no doubt, the dropping of the word invention from titles.

In relations of the Office with the armed forces, Breton had used to great advantage the connections he had forged during the war. Large subsidies came to the Office from the military, which was unpleasantly surprised by its abolition. Relations with new organizations were not easily established, partly because of the hostility of the armed forces toward the leftist Front populaire. In fact, those administering military research refused to submit to the authority of the CNRSA, and, adding injury to insult, stayed illegally in their offices for several weeks.[61] Two months after the Office disappeared, the three ministers responsible for national defense created an Institut de la recherche scientifique appliquée à la défense nationale. Breton hoped that the Bellevue complex would be utilized by the new institute of military research in order to avoid the floundering that had hampered military research in the early period of World War I. But the great generation of politicians who ran France during the war had gone to meet Napoleon's marshals in Valhalla, and the politicians of the late 1930s were incapable of creating a military counterforce to the *Wehrmacht*. The relations between science, technology, and the military would be slowly reestablished during the Fourth Republic, in spite of considerable resistance on the part of leftist scientists, and effectively consolidated in the Fifth Republic, where Breton's dream of a well-functioning troika of science-technology, industry, and defense would be as much a part of the French scene as it is of the rest of the Western world.

The Office des inventions prefigured post-World War II science in some ways: interaction of science and technology, cooperation of physicists and engineers, big funding (or a hope for it), connections with many governmental departments, and the attempt at forging profitable relations with industry and the military. No military–industrial–scientific complex existed, but the desire of the scientists had been aroused. When Bellevue developed after World War II, many of the old characteristics of the Office persisted, but there was a striking shift in the direction of science – discovery took precedence over invention – Perrin, not Breton, had triumphed. It is only fair to point out that Breton believed in the necessity of a long-range governmental policy encouraging basic research in the laboratories of the universities and the *grandes écoles*.[62] In the glossy pamphlet put out in 1958 on the Bellevue laboratories the frontispiece is a photograph of Aimé Cotton (1869–1951), a suitable icon for Bellevue. A clearly different emphasis separated the 1950s from the interwar period. Twenty-two different laboratories (1958) formed a group, including laboratories for basic research and laboratories specially oriented toward applied research with the mission of establishing a liaison between scientific research and industry. Physics still predominated: magnetism, low temperature, high pressure, X-rays, and also high tension (in a laboratory created in 1939 for the study of defense matters). A Centre d'analyses et de techniques physico-chimiques included a service for testing paints. The Laboratoire de physique du froid carried out numerous tests for services such as the railways (SNCF), Electricité de France, and the Services des subsistances de l'armée, as well as work for the government and the refrigeration industry. Clients of the Laboratoire des rayons X, which necessarily had close relations with other CNRS laboratories, included the Institut du pétrole, the Institut de recherches de sidéurgie, the Institut textil de France, the Institut de Caoutchouc, and the CEA.[63] This was the stuff of which Breton's dreams were made.

The failures of the Office des inventions were incapacitating in some sectors. One of the Office's main programs was a failure: Direct investment in industrial research did not lure conservative, protectionist industrialists to it. Complex bureaucratic rules did not help the Office move beyond low-level research activity in its collaboration with industry. It just did what industry asked it to do, meaning, as Blancpain argues, that in this case the Office functioned as a factory laboratory. Auclair and Breton wanted direct industrial support for a "bureau d'études" at Bellevue, but the top bureaucrats of higher education feared loss of control over a part of the Office. No doubt in this case it would have been an advantage for the Office to have been under the Ministry of Commerce and industry rather than education, but because most laboratory work in this period was university connected, this was not feasible and probably undesirable. The Bellevue laboratories were also too primitive or lacked equipment for some key experimental areas. One notorious example was in aeronautics, where research was also retarded by the refusal of the Air Ministry to grant a subsidy to the Office's engineer specializing in this subject. New high technology was beyond the Office's capabilities. The Office also remained a small operation under the Malthusian fiscal restraints of the 1930s. In 1930 it could pay only eleven engineers and had only forty people in administration for the entire Office.[64] A truly primitive prototype of the giant research complexes funded by the CNRS and the CEA.

9

THE DENOUEMENT OF
THE 1930S
A NEW SCIENTIFIC FUNCTION FOR
THE STATE

"For the Kingdom of God *is* not in word, but in power."
– 1 Corinthians 4:20

T HE changing organization of French scientific research in the 1930s
might easily be interpreted as a sign of confusion and vacillation, in
tune with national politics. Nothing could be more mistaken. A
strong sense of purpose animated the scientific community, whose devotion
to the research imperative along with its persistent political lobbying had
already produced by 1930 the foundations of state organization and funding
of research. This basic fact is somewhat obscured by the birth of several
different organizations with similar names during the 1930s; fortunately for
the memory of posterity, they were all gobbled up by the Centre national de
la recherche scientifique, the mega-organization created in 1939. Breton's
Office des inventions, which had absorbed the Caisse des recherches
scientifiques (1901–22), so adept at survival in the 1920s, was replaced in
1938 by the Centre national de la recherche scientifique appliquée, an
ephemeral creature that expired in 1941. In 1930 parliament created two
funds, one for letters and one for science. The Caisse nationale des sciences
for the support of scientific research used a system of short, renewable
contracts rather than create a permanent group of researchers. There was a
novelty here. The state had set a new direction by granting research
scholarships, for it now pushed the creation of a corps of researchers, rather
than just pay for research materials. Nearly three years later the government
agreed to the creation of the Conseil supérieur de la recherche, representing
both science and letters, made up of bigwigs in all disciplines who could
make recommendations to the government through the minister of education, chairman of the council. In 1935, as a result of governmental insistence
on the consolidation of organizations in order to save money through the
elimination of bureaucratic duplication, the Caisse nationale de la recherche
scientifique was created to manage the funds of the old Caisse des recherches
scientifiques (part of the Office des inventions) and the Caisse nationale des
sciences (1930), along with the credits based on the *taxe d'apprentissage* and
other revenues, all amounting to seventeen million francs. The creation of a

Service central de la recherche scientifique in 1936 established a liaison between the Conseil and the Caisse. The Service central prepared the budget of the Caisse according to the recommendations of the Conseil and distributed funds to researchers and laboratories, subsidized publications, and gave aid to needy scientists and their families. In spite of the economic depression, governments supported a modest and growing scientific research program. Without the generally friendly attitude of parliament, especially the Chamber of Deputies, toward science, it would not have been possible for a small, dynamic group of leading scientists to have changed the basic organization and funding of science in a few years.

After the First World War a new general organization came into existence: the Confédération des sociétés scientifiques françaises (1919), a creation of the old scientific societies, which were both extremely worried by obvious threats to scientific research and hopeful for the economic revitalization of France through science, an idea encouraged by the widespread belief, especially among scientists, that science had been a key factor in winning the war.[1] This new organization, a paper army twenty thousand strong, was greatly concerned about the effect of war losses on French science. A legitimate worry: Death and injury eliminated 40 percent of the Sorbonne students who had gone to war; about half of the 161 students of the classes of 1911–13 of the Ecole normale disappeared. "The carnage deprived France of a generation of scientists."[2] Postwar inflation made it difficult for scientific societies to continue their main contribution to science, the publication of their journals. Some folded while many were reduced in size. Combined with the decrepitude and poverty of many laboratories, these factors were probably responsible for the fall in the number of students going into science.

The reorganization of French science in the 1930s was the culmination of a widely based movement of reform coming from the scientific community, it is true, but it also enjoyed the general support of the creative cultural elite. The French penchant for melodramatic statement was soon satisfied by Maurice Barrès's useful propagandistic phrase "la grande misère des laboratoires." In parliament and in the press, Barrès, primed by Charles Moreu, evoked the proper pathos for the situation, just as he had done before the war with equal brilliance for the old religion in "la grande pitié des églises de France." Henry Bernstein gave 60,000 francs, the income from the first staging of his play *Judith* (1922), to the big scientific organizations. A Comité d'aide à la recherche scientifique made up of scientists and industrialists collected two million for laboratories in 1922. A general gift of twelve million francs to the Sorbonne was made in 1923 from the estate of the Marquise Arconati-Visconti. In 1923 Criqui contributed 100,000

341

francs, his share in the profits from a boxing match. In 1924, the year in which Paul Pousson of the Paris faculty of pharmacy left his fortune of several hundred thousand francs to science, the Léonard Rosenthal Foundation for the Advancement of Science was born. Parliament voted Marie Curie a national pension of 40,000 francs.[3] An ostensibly catastrophic situation was reversed by the alarmist campaign of the 1920s, followed by a striking expansionist scientific movement in the 1930s. The threat of decline – Perrin warned of the fall of France to the status of a third-rate scientific power – stimulated a sort of Toynbeean response that eventually reversed the danger of scientific eclipse by an eternally renascent Germany.

The growth of scientific organizations, both major and minor, accelerated in the 1930s, a period of creative reordering unequaled in the history of French science since the 1870s and 1880s. Some of these organizational efforts are more important historically for the tendencies they represented than for their actual accomplishments. In 1934 some excitement was generated by an organization with the alluring name of La Cité des sciences. Copying the Centre Marcelin Berthelot of the Maison de chimie, it tried to coordinate the activities of scientific societies, facilitate the documentation and diffusion of scientific knowledge, organize centers of specialized documentation like the Maison de chimie, emphasize documentation for the application of science in improving the human condition, and develop international scientific relations. The umbrella of the Cité des sciences would cover the three federations grouped in the Confédération des sociétés scientifiques françaises. The chemists had seven societies with over 10,000 members, the physicists eight societies with over 14,000 members, and the natural scientists fifty-four societies with over 27,000 members. These seventy societies grouped a motley army of 50,000 researchers.[4] Many believed that organizing them was the necessary prelude to the restoration of scientific glory for the French nation.

It is conventional for reformers to puff up the importance of their reforms by describing in dramatic terms the conditions existing before their great work began. Perrin enjoyed regaling audiences with descriptions of the decline into which science had fallen in France compared with other countries after the Great War, then to be saved from the dragon of despair by his friend Borel's "sou des laboratoires," added to the apprenticeship tax. It is not surprising that Perrin saw a turn-around begin with an institution in whose founding he played a large role. Perrin argued that the signal for revival came from private initiative in 1926, when Edmond de Rothschild, who had talked to Claude Bernard about life and was still interested in it, gave fifty million francs to found an Institut de biologie physico-chimique.

342

(In 1926 the franc sank to 10 percent of its prewar value but soon rose steadily to 20 percent as a result of Prime Minister Raymond Poincaré's program of austerity.) Mounted on the hard sciences, this institute, devoted solely to research, like the Kaiser-Wilhelm-Gesellschaft, Perrin noted, set out in pursuit of the mechanism of life, an idea then popular in biology. The University of Paris gave part of the land, close to the Laboratoire de chimie physique et de radioactivité, whose construction had been the result of Léon Blum's parliamentary initiative in 1922. Perrin played a large role in getting the Rothschild institute started. Along with two physiologists, Pierre Girard and André Mayer, and two chemists, André Job and Georges Urbain, he also ran the institute.

Weekly discussions at the institute led to the more ambitious scheme of a "service national de recherches" on the model of the institute. A hierarchy of research positions equivalent to University posts would be given to refugees from higher education. A Conseil supérieur de la recherche scientifique would make decisions on personnel and funds. French scientific production could easily be doubled if a generous parliament provided forty million francs. Perrin's eloquence convinced his rhetorical friend Herriot to support his scheme. In 1930 Herriot asked the Chamber of Deputies for twenty million for scientific research from national defense funds. He got five million, enough to establish the Service national de la recherche scientifique. The Service was integrated into the Caisse nationale des sciences, whose functions up to then were limited to giving handouts to needy scientists and their families. With the support of Painlevé, Herriot, Blum, Borel, and Hippolyte Duclos (deputy of the Haute-Garonne and reporter on the education budget) parliament had made a significant commitment to scientific research and approved an organization to encourage it. Scientists and their friends were showing themselves to be quite good in politics – indeed, a few scientists and professors had become deputies – but they were also lucky to be operating in a period when most politicians were willing to believe that national defense required state-supported scientific research.

The struggle for funds required a sustained political effort by Perrin, who wisely engaged the dying Marie Curie to accompany him on begging visits to the parliamentary satrapy. It was evident that the efforts paid off, for the budget of 1934 included eight million francs for researchers and eight million for the subsidy to laboratories in higher education, although due to a bureaucratic mix-up only a total of twelve million was actually disbursed. During the first cabinet of Perrin's friend Léon Blum in 1936 a Sous-secrétariat de la recherche was established, with Irène Joliot-Curie, a reluctant political virgin, pushed into service. In Blum's second and more futile ministry of 1938 Perrin gallantly took the job, permitting Irène to flee back to her laboratory. Both politics and science probably profited from the new arrangement. In 1937 Perrin was given ten million for researchers, of which one-fifth, instead of one-seventh as previously, went for what the French quaintly designate "les sciences humaines," and an extraordinary grant of sixteen million for laboratories ("crédits des grands travaux"), equally divided between Paris and the provinces. This generosity caused Perrin not to request more than eight million for the regular subsidy to laboratories. A corps of technical laboratory aides cost another two million francs (three million to begin), part of the French "fight against intellectual unemployment." Another two million went for publication; the same amount was budgeted to reward scientists whose discoveries brought honor and wealth to France. So under Vincent Auriol as Blum's minister of finance the budget for research went to more than twenty-six million francs, an increase of eleven million over the preceding budget. The research budget for 1938 rose to thirty-one million francs. One excessive estimate puts the budget of the Caisse for 1939, when the research share from the apprenticeship tax was close to ten million francs, at the splendid total of ninety-nine million francs. By the end of the 1930s it was clear, as I. I. Rabi once said (1963), that "there is something like a Parkinson's law that scientific activity will grow to meet any set budget and find it to be grossly inadequate."

After World War II scientists could easily win the sympathetic ear of politicians and frequently get generous funding for science, especially in big physics and in military-related research. The chief reason for this funding was that in the 1940s science pulled a rabbit out of a nuclear hat and promised increasingly prolific litters if funding were increased. But the funding of science has come to depend on non-rabbit factors. The editor of *Science* laments that "it's been a long time since science pulled a rabbit out of a hat." Scientists working in the 1930s, having no rabbits to show or even promise to politicians, had to use their wits in convincing sceptical custodians of shrinking public purses to fund the fulfillment of their remote research dreams. Of course sly intimations of potential bunnies were not unknown, even in the oppressively virtuous environment in which French scientists boasted of working. Not that France was the only country in which scientists' souls were being tempered by a little poverty: The total annual laboratory budget for Columbia's department of physics was about $10,000 in the early 1930s. In praising the great successes obtained in the anisotropic world of French science by full-time researchers subsidized for six years by the Caisse, Jean Perrin singled out for special mention the work of Frédéric Joliot and Irène Joliot-Curie (married to Joliot), their discovery of artificial radiation and the finding of new radioactive elements by a method that was then used by other researchers to discover "two or three hundred fleeting isotopes of the stable elements of the old chemistry." The glorious results of the creation of the Service national de la recherche scientifique were the transformation of chemistry, probable biological applications of the elements discovered, a Nobel prize for the Joliots, and great prestige for France. Perrin believed in the old rule that pure research produces more applications than applied research.

Throughout the reforms of the 1930s the scientific community worried about its autonomy and especially its freedom of research. The growing interest of the ministeries of national defense in the military uses of science increased anxiety among scientists, who finally concluded that their salvation lay in a dual system of applied and basic research with the latter dominant. At the center of the debate over the organization of research was the Conseil supérieur de la recherche scientifique. The parliamentary Commission des offices recommended that all financial agencies supporting scientific research be grouped in one service under the Ministry of Education but with a separate research budget. Another parliamentary commission did not agree: The report of the Commission des économies emphasized the importance of following the French tradition of separating applied and pure research. This tradition was perhaps more a figment of some scientific minds than a working reality of the faculties, but a powerful figment nevertheless and certainly a potent weapon against the dangerous simpletons who could not see the delicate interaction between pure and applied science. The government, following the positions favored by the professors, created the Caisse nationale de la recherche scientifique (1935). In spite of increased funding of research after the Popular Front came to power in 1936, Perrin judged the situation of science to be still politically unstable. While politicians were calling for the organization of team research and the elaboration of a science policy, Perrin succeeded in getting his Service central de la recherche scientifique created as a permanent organ of science policy. But the integration of the Service central and the Caisse nationale de la recherche scientifique, a sort of administrative and autonomous creature of the Ministry of Education, made it difficult to establish the autonomy of scientific research or an independent science policy. It is not surprising that given an opportunity, the Conseil supérieur de la recherche scientifique proposed revamping the whole research system.

Assaulted by the threat of economic disaster, which was aggravated by the absence of systematic and sustained technological innovation in French industry, and the danger of war with Nazi Germany, French politicians were more than usually receptive to promises of salvation from science and technology. The scientific community was not slow in showing its patriotism by proposing a reorganization of scientific research to assure the optimum rate of scientific production in France. In an atmosphere charged with threats of war and rumors of economic reform, it was to be expected that the first priority would be the reorganization of applied research, including the forging of a mechanism for getting a quick input into applied from basic research – this was the major obsession of the day. In shocking politicians and scientists into serious collaboration, Hitler played a major role in the reorganization of science in France.

Throughout the 1930s a clearly growing interaction existed between science and military strength, a link that had been forged in the First World War but neglected during the 1920s, when the German menace appeared to be more scientific and industrial than military, although such distinctions soon became trivial. Herriot secured the first millions for the Caisse nationale des sciences from the funds for the Maginot line ("le milliard de la ligne Maginot"), a bottomless fiscal pit. This decision was so politically acceptable and the amount so small within the budget that the minister of finance could make the generous gesture of temporarily abandoning the Scrooge-like role required by the job and increase the total research credits by half. Perrin purred with satisfaction: The decision deserved the future gratitude of the motherland. "La Recherche [capitalized, like German nouns] équivaut à une sorte d'armement très peu couteux, très efficace, et irremplaçable." Perrin was not wrong for the future, when a later generation would harvest the dragon's teeth of scientific input into the military state. The Conseil supérieur de la défense thought that the national research council in Fascist Italy was a useful model to imitate in milking science for military uses. For the solutions of its problems in high technology the Ministry of Air had started contacts in 1929 with University laboratories; this was not a surprising decision in light of the historic role the University of Paris had played in aeronautical research. After 1930 the Ministry of War formalized its own system of liaison with basic research. The whole trend was consecrated in the *décret-loi* of 1938 creating the CNRSA: The first aim given in the definition of its mission was to facilitate scientific work relevant to national defense by establishing connections between all the research services of various ministries and ten private organizations. A military–industrial complex was recognized in legislation long after the womb of reality was quickened.

One of the traditional strategies used by French scientists to get research money from governments had always been to cry wolf: general French decline coupled with German superiority, including substantial evidence from one scientific area of economic importance identified as being in very bad condition, usually organic chemistry. In the 1920s and the 1930s greater emphasis was put on certain branches of physics, increasingly seen as a science of great economic and military potential. This view was assiduously cultivated by physicists. In 1920, when Emile Borel succeeded the venerable Boussinesq in the chair of mathematical physics and the theory of probability at the Sorbonne, it was generally recognized that theoretical physics had not kept up with German developments for about a generation. The explanation of this scientific scandal is partly historical, for the kinetic theory of gases and Maxwell's electromagnetic theory had never fully penetrated French scientific education to the degree justified by their importance. Even worse, relativity and quantum theory were later systematically ignored, with the powerful exception of Langevin at the Collège de France. It was this dangerous and deplorable situation that stimulated Borel to create the Institut Henri Poincaré with gifts from the Rockefeller foundation and from Edmond de Rothschild. Borel, head of the institute, was assisted in its operation by Charles Maurain, Perrin, and Langevin. Foreigners were imported to challenge the minds of French scientists long mired in the rut of traditional physics: Einstein, C. G. Darwin, Max Born, Dirac, R. A. Millikan, and R. von Miles gave lectures that were also published in the institute's *Annales*, edited by L. Brillouin, L. de Broglie, and M. Fréchet, also apostles of the new physics.[5] So French physicists hoped to join the mainstream of the movement in which the University of Berlin was the "world's foremost center for research in physics," and the *Zeitschrift für Physik* (1920) accepted as "the semiofficial journal for . . . main-line [atomic] physicists" in Copenhagen, Munich, and then Göttingen.[6] Practical difficulties accompanied theoretical recalcitrance to the new physics. It is no surprise to find that few provincial faculties had laboratories where students could specialize in studying X-rays and particle physics. Things had changed considerably by the 1930s. In 1934 even the small faculty of sciences in Besançon opened a well-equipped laboratory to introduce students to the experimental side of the new physics. Of course Maurice de Broglie's private laboratory had been doing basic experimental work in this area throughout the 1920s. And by the late 1930s Fred Joliot's empire of nuclear physics was firmly established at the Collège de France and Ivry. The generation of the Curie circle was moving effectively to replace the scientific gerontocracy, now faced by a two-pronged offensive on intellectual and institutional fronts.

The structure of scientific research in Germany was still a much venerated icon in France in the 1930s; indeed, some alarmists though that the French scientific lag was worse than ever. What features of the German research organization did the French scientific community look upon with envy? A primary French concern was the crass matter of funding. Albert Ranc, the leading Gallic accountant in science funding, estimated the per-capita research expenses of France to be only one-fifth those of Germany's annual equivalent of 200 million francs (1930). Another persistent complaint of the French scientists was that no opportunity for a career devoted exclusively to research really existed in France. Even the advantage of light teaching loads was negated by salaries so low that they encouraged professors to take extra jobs. The Kaiser-Wilhelm-Gesellschaft's laboratories had an annual budget equivalent to 42 million francs, half paid by governments. Thus the laboratory of physical chemistry of the KWG could spend the equivalent of 1.4 million francs on matériel and working personnel compared to the 200,000 francs given to a similar laboratory of the University of Paris. Industries like Siemens and I. G. Farben also had research laboratories, which were still not typical of French industries. The fall of the mark brought into existence the *Notgemeinschaft*. This organization, devoted to saving laboratories from the revolutionary ravage visited on the middle classes by the government's allowing the mark to inflate into nothing, gave funds equal to 60 million francs a year to university research laboratories. Ranc did not find it surprising that German scientific production, like even that of the English, was greater and better than that of the French. He saw only one remedy for the French disease, a typical French cure, the creation of a state-supported research organization, namely the Service national de la recherche scientifique proposed by Perrin and put before parliament by Herriot. The security of France required nothing less, the advance of science needed nothing more.[7]

The interaction of scientists and politicians in the 1930s led to a substantial increase in the range of French scientific activity through both the funding of scientists devoted exclusively to research and through the establishment of new laboratories and programs as well as the regular funding of existing ones. Among the specialized laboratories founded were Georges Urbain's Laboratoire de gros traitements chimiques, known for its exotic chemical activities like the extraction of precious metals by semi-industrial procedures; Joliot's Laboratoire de synthèse atomique (Ivry), a "real factory for the production of new elements," run on funds provided by the Caisse nationale de la recherche scientifique; Aimé Cotton's Laboratoire des basses températures at Bellevue; and the Service d'astrophysique with one station in Paris and another in Haute Provence. Some construction, that for astrophysics and Urbain's laboratory, for example, was done with money from the funds for Grands Travaux. Funds of the Caisse could also be used to put a laboratory on a regular fiscal footing, as in the case of the Laboratoire de magnétisme of Pierre Weiss at the University of Strasbourg. Experimental medicine was elevated to the status of a research section of the Service national in 1937. The name of the section honored Claude Bernard. When the physiologist Henri Laugier was made "Chef du service central de la recherche scientifique," Perrin praised him as a man who knew what "la Recherche conquérante" was all about. Power was the name of the game. Bernard would have been pleased with the new probes into nature's secrets, although the big payoff would be in physics, not physiology.

The study of the organization of applied research seems to have occupied nearly as many people as the research itself. Study begat study, report begat report. The professors wanted to protect science from too much contamination with application, the politicians wanted applications, and the Conseil supérieur de la recherche scientifique wanted to please both groups by maintaining a wall of separation while shortening the time lapse between discovery and application. To facilitate the transfer process, the Conseil recommended that the Office des inventions, administratively isolated, be replaced by another organization, one closer to the needs of industry. An ancient refrain, it is true, but on May 24, 1938, the Centre national de la recherche scientifique appliquée was created by a *décret-loi*, which was itself an application of a law passed a month earlier enabling the government to take all necessary measures for the purposes of national defense and economic and financial recovery. All of the research institutes, largely created by the scientific left, were held together, at least in theory, by the capitalistic invention of interlocking directorates, reinforced by the CNRSA Haut comité de coordination, presided over by Perrin and including Irène Curie and Henri Laugier. The result: more committees, more studies, a credit of thirteen million francs added to the pot taken from the defunct Office des inventions, and a labyrinthine system of administration, but no real science policy. Yet, as most of the professors of science were quick to recognize and often deplore, a new structure of scientific research had been born. Policy would not tarry long.

Professors are a notoriously discontented lot, tropistically critical, like the intellectuals of which they are an extremely specialized subspecies. French professors exhibit in exaggerated form most of the quarrelsome features of their group. It would be naïve to assume that the reforms of the 1930s elicited universal support from the scientific community, especially in the faculties, who felt threatened by the rise of a research system outside the control of the old University administrative mechanism. To function successfully the new system had to depend on the faculties, and an interfacing of the personnel and functions of both was unavoidable. It seemed bad policy and worse morality to a professor in a provincial faculty of sciences for the government to pour money into a new research organization while faculties were starving. State aid to higher education had become insignificant except for a great deal of sometimes unneeded construction, a further drain on already strained University budgets. In

349

1938 most faculties could no longer afford to keep up their subscriptions to foreign periodicals and could hardly pay their heating bills, while the state gave fifty million to the Service central de la recherche scientifique. Was there a danger that the University would lose its research role, reverting to an exclusively teaching function? Some thought that such a fate might be deserved. Had not the faculties presided over the fall of France as a scientific power from the group of first rank to that of fourth or even fifth on the world scene? This was the opinion of Jean Delsarte, mathematician at Nancy. Raoul Anthony, professor at the Muséum national d'histoire naturelle, did not agree with this assessment because he saw the decline in France as part of a quasi-universal decadence into which science had fallen since the war.[8] This comforting idea of sinking into a slothful decline with both friend and foe was definitely the view of a minority. But even those who agreed on the uniqueness of the French decline disagreed on its causes.

Delsarte identified the disease of the provincial faculties by referring to four sore points. There were too few students, and too many of these were mediocre, although by the late 1930s some better students, attracted by the possibility of new and interesting careers, were showing up in chemistry and the natural sciences. The condition of the teachers was as alarming as that of the students. Chosen for the wrong reasons by local faculty councils, professors were often as mediocre as the students. Many professors being poorly paid, or just greedy, at least in medicine, pharmacy, science, and law, took extra jobs rather than do poorly rewarded research. Perhaps worst of all for a professor in a provincial faculty was the eternal domination of the University of Paris in the university system. With only 5,000 less than the provinces' 30,000 students, 30 percent of the 676 *fonctionnaires* in the system, and its immense prestige, the Sorbonne aroused the appetite of most provincial professors for a chair in Paris.

People in Paris, even at the Muséum, would disagree with the form if not the matter of Delsarte's opinions, none of which were new, but Anthony drew attention to a new disease undetected by Delsarte. This was the disorganization of University programs, permitting a student to get a licence and then a doctorate in the natural sciences, for example, without doing basic natural science. It was possible to get a licence by simply presenting a specified number of certificates from a large group. Anthony called for a return to the old licence based on three basic certificates with no equivalences; this old rigor would end the dangerous fantasies existing in scientific education. It would also end a lot of teaching jobs in the new areas that had spun off from the core curriculum. Those who want to return to the basics are always generous enough to sacrifice other people's jobs along with peripheral intellectual pursuits. The proliferation of certificates had been a concession to specialization, but the change had the potential paradox of producing graduates ignorant of the basics of the field that the degree was supposed to guarantee. It seems that the French had made an early discovery of the pedagogical principle that the more choice a student has, the less basic knowledge he is likely to acquire. Fortunately the changing definition of basic knowledge makes the whole debate an exercise in harmless frivolity, while science keeps on getting done, expanding in content and depth without much influence from changes engineered by the arbiters of educational fashions.

The growth of the research budget from the Caisse's five million francs in 1930 to the fifty-three million of the Service less than a decade later inevitably led to criticism of the way funds were distributed. Delsarte aired his views in the *Revue scientifique*, a forum that gave them the patina of disgruntled orthodoxy.[9] Delsarte complained that the distribution of research funds was not made public knowledge through the normal method of publication in the *Journal officiel*. No doubt as the administrative mechanism started to function properly, such shocking irregularities would disappear. Most important for Delsarte was the establishment of a coherent doctrine on two research matters. First, the moral problem. In the case of the professor who did research, there was always a benefit for the state because whether or not he made a discovery, the pursuit of research gave a certain luster to the University and even improved the quality of teaching. In the case of the person who did only research, a problem could easily arise if no one understood what the researcher was doing. Cauchy did not understand Galois. Essentially, one had to take the word of the researcher, whose conscience became the first guarantee of the state. Delsarte did not think that research organizations had given enough attention to this issue. It is doubtful that French scientists were immoral enough to concoct patchwork mice, but Delsarte had identified a key problem for future mega-research organizations.

Naturally the pontiffs of research funding would argue that regulation of the quality of work was the function of the Conseil supérieur de la recherche. For Delsarte that was the second problem of the new organization. In spite of the high competence of its members, the council would be reduced in the end to judging researchers by the wretched criterion of quantity of publication. How could it be any different in the new tower of Babel created by modern science where one specialist could speak only to his fellow specialist? Since he admitted that only the researcher could control his own research, with his conscience as his guide, Delsarte had no way out of the basic difficulty of judging the worth of researchers. He was happy to have discussed the problem and to be able to reject Perrin's solution of a judicious mixture of peer review and patronage.

Perrin stood firm in the scientific credo of a line of French scientists stretching back to the Second Empire, perhaps even to the Enlightenment, but his *ersatz* religion was less overtly anti-Catholic than the secular faith of his fathers had been. His basic religious dogma was that research and discovery would enable mankind to fulfill all its dreams in the areas of creativity and emotional longing. "Recherche et découverte constituent le seul moyen, pour l'humanité, de réaliser et de dépasser ses vieux rêves dans la puissance et la liberté, dans l'Art et la Beauté, dans la Fraternité; c'est là notre espérance et notre idéal; au sens le plus élevé du mot, c'est notre religion." Delsarte was quick to point out that religion implied morality. This meant that scientific progress would bring moral progress; worse, it meant that scientists were beings of higher morality. All of this sounds like a rehash of the late nineteenth-century debate over the so-called bankruptcy of science, although in 1939 the French were spared this intellectual atrocity. Delsarte rejected Perrin's implied belief in scientistic *Uebermenschen*, who would lead mankind along the path of the progress of the human mind to a final stage where no alienation frustrated man's creative impulses. Perrin's synthesis of Enlightenment scientism, Littréan aesthetics, revolutionary camaraderie, all tinged by the Marxian hope of liberation from *Entfremdung*, could not be more than a secular religion based on science. Delsarte argued that because it lacked the essential ideas of the act of spiritual growth toward perfection and of sacrifice, Perrin's concoction could not provide the basis for a real religion. Since the times of Comte and of Berthelot it had become clear that scientific progress did not necessarily bring moral progress. The role of science in war had stripped away the last ambiguities. Everyone understood the key role science had come to play in national defense, but no thoughtful person could also accept uncritically a naïve faith in the moral value of science.

The creation of a research structure outside the control of the University's dispensers of patronage produced jealousy, resentment, and an attempt by the University's mandarins to influence the distribution of funds. Emmanuel Dubois, dean of the faculty of sciences in Clermont-Ferrand, thought the new research effort fraught with danger and full of waste. He found it shocking that a student who had just managed to pass his licence might be given a research grant on the recommendation of an influential professor of good faith but ignorant of the student's past. Coordination of the granting of funds with rectors, deans, and *chefs de service* could avoid this problem. Even more galling to Dubois was the fact that a young researcher, who could spend 30,000 francs without difficulty, had credits at his easy disposal far superior to those of senior professors. And a professor who got a grant from the Caisse risked running afoul of new laws against double employment (*cumul*), in spite of vehement protests by the universities and in spite of the contrary intent of parliament, all ignored by the perversely zealous Administration des finances. Nor should foreigners be given research funds when good French scientists were in desperate need of laboratory equipment. Meanwhile, the Palais de la découverte, a useful creation in the exhibition, had sneaked into permanence at an annual cost of 8 million francs, the total cost of matériel for all the faculties of the University of Paris. So far, Dubois concluded, much had been spent for little. He thought that the future would be better because of reforms establishing closer cooperation between the traditional mandarins of higher education and the new gurus of research. This interfacing of structures came about to some extent because of the loud protests of the faculties. Dubois smugly concluded that the new research structure could only improve because of these changes.[10]

At the end of the 1930s, as Blancpain emphasizes, a deep transformation of the scientific function of the state had taken place. The creation and organization of the Centre national de la recherche scientifique (October 1939) was the culmination of the process. A parliamentary Comité de réorganisation administrative recommended, as an economy move, that the services of pure research of the ministry of education be combined with the CNRSA. (Emile Borel was president of the CNRSA's administrative council; Henri Laugel was the general secretary; and Longchambon, dean of the faculty of sciences at Lyon, was the director.) Combination would be a matter of coordination and finance. For the committee there was not a clear distinction between basic science and its applications. If politicians and bureaucrats had lost all hope of salvation through the stimulating prospect of directing scientific research from administrative heights, they had not abandoned hope of avoiding damnation through suggesting a general direction for scientists to take. Although the Centre nationale de la recherche scientifique appliquée was absorbed by the CNRS, basic and applied research remained separate sections until the suppression of the applied section on March 25, 1941. For basic research, of course, the transformation from Caisse nationale to Centre nationale was little more than a change of name, the settling of "the last layer of a phenomenon of institutional sedimentation," consecrating the hold of the University on the research empire that was now a safe part of the state bureaucracy.

NOTES

The following abbreviations are used in the notes:

AN   Archives nationales
DSB   *Dictionary of Scientific Biography*
ED   *Enquêtes et documents*
PP   *La Philosophie positive*
RIE   *Revue internationale de l'enseignement*
RGSPA   *Revue générale des sciences pures et appliquées*
RO   *Revue occidentale*
RQS   *Revue des questions scientifiques*
RS   *Revue scientifique* (*Revue des cours scientifiques* up to 1870)

INTRODUCTION

1 François Guizot, *Historical Essays and Lectures*, ed. Stanley Mellon (Chicago, 1972), pp. 141–3 (quoting Mill). Ernst Robert Curtius, *The Civilization of France: An Introduction* trans. Olive Wyon (New York, 1962), pp. 92–6 (quoting Hugo).
2 François Furet, preface to Terry Shinn, *Savoir scientifique et pouvoir social: L'Ecole polytechnique, 1794–1914* (Paris, 1980), p. 5.
3 *Le Monde*, September 17, 1982. Actual budgets are different from plans: The total spent in 1984 was just over 48 billion francs.
4 *Le Nouvel observateur*, May 22, 1982.
5 Charles Coulston Gillispie, *Science and Polity in France at the End of the Old Regime* (Princeton, N.J., 1980).
6 Charles Coulston Gillispie, *The Edge of Objectivity: An Essay in the History of Scientific Ideas* (Princeton, N.J., 1960).
7 Michel Serres, *Rome: Le Livre des fondations* (Paris, 1983), p. 27.
8 David Edge, "Is There Too Much Sociology of Science?" *Isis*, 74 (1983), 252.
9 George Weisz, *The Emergence of Modern Universities in France, 1863–1914* (Princeton, N.J., 1983), p. 8.
10 Derrida's axioms contrived by E. D. Hirsch, Jr., in the *London Review of Books*, 5 (July 21 to August 3, 1983), 17–18.

11 Clive Trebilcock, *The Industrialization of the Continental Powers, 1780–1914* (London, 1981), chs. 1 and 3. A brilliant book, witty and well written, rare in economic history.

12 John H. Weiss, *The Making of Technological Man: The Social Origins of French Engineering Education* (Cambridge, Mass., 1982), p. 25. For the limited nature of French industrialization in this period, see Trebilcock, *The Industrialization of the Continental Powers*, pp. 143–50.

13 On the universities, Weisz, *The Emergence of Modern Universities*, is now the basic work; see also Theodore Zeldin, *France, 1848–1945* (Oxford, 1973–7), vol. 2, *Intellect, Taste and Anxiety*, pp. 316–45. The best survey of the faculties of science is Terry Shinn, "The French Science Faculty System 1808–1914: Institutional Change and Research Potential in Mathematics and the Physical Sciences," *Historical Studies in the Physical Sciences*, 10 (1979), 271–332. See also H. W. Paul, "The Issue of Decline in Nineteenth Century French Science," *French Historical Studies*, 7 (1972), 416–40.

14 Weisz, *The Emergence of Modern Universities*, ch. 8, 270–314.

15 *Ibid.*, pp. 10, 86, 271–2, and for a full treatment of intellectual, including scientific, factors with their political ramifications, see Harry W. Paul, *The Edge of Contingency: French Catholic Reaction to Scientific Change from Darwin to Duhem* (Gainesville, Fla., 1979); "The Debate over the Bankruptcy of Science in 1895," *French Historical Studies*, 5 (1968), 299–327; "The Crucifix and the Crucible: Catholic Scientists in the Third Republic," *Catholic Historical Review*, 58 (1972), 195–219; and "Scholarship versus Ideology: The Chair of the General History of Science at the Collège de France, 1892–1913," *Isis*, 67 (1976), 376–97.

16 Duruy, cited in Weisz, *The Emergence of Modern Universities*, p. 91.

17 C. Rod Day, "Education for the Industrial World: Technical and Modern Instruction in France under the Third Republic, 1870–1914," in Robert Fox and George Weisz, eds., *The Organization of Science and Technology in France, 1808–1914* (Cambridge, 1980), pp. 127–54.

18 See Peter Lundgreen, "The Organization of Science and Technology in France: A German Perspective," in Fox and Weisz, eds., *The Organization of Science and Technology*, p. 332.

19 Thomas S. Kuhn, *The Essential Tension: Selected Studies in Scientific Tradition and Change* (Chicago, 1977), pp. 141–7.

20 Michael Sanderson, *The Universities and British Industry, 1850–1970* (London, 1972), pp. 119–20.

21 See the enthusiastic preface in Isidore Pierre, *Chimie agricole* (6th ed.; Paris, 1882), vol. 2, pp. v–vi.

22 See Jeremy Hardie, "A Confucian Kind of Capitalism," *Times Literary Supplement* (July 9, 1982), p. 745, a review of Michio Morishima, *Why Has Japan "Succeeded"?* (Cambridge, 1982).

23 Paul Verlaine, "Pierrot Gamin," *Oeuvres poétiques complètes* (Paris, 1962), p. 562.

24 Kuhn, *The Essential Tension*, p. 187 n. 11.
25 Weisz, *The Emergence of Modern Universities*, p. 196.
26 Victor Karady, "Educational Qualifications and University Careers in Science in Nineteenth-Century France," in Fox and Weisz, eds., *The Organization of Science and Technology*, pp. 95–124.
27 Mary Jo Nye, *Molecular Reality: A Perspective on the Scientific Work of Jean Perrin* (London, 1972), pp. 157–65.
28 Kuhn, *The Essential Tension*, pp. 60, 220.
29 Paul Verlaine, "Art poétique," *Oeuvres poétiques complètes*, p. 327.
30 For the campaign of the 1930s, see Spencer R. Weart, *Scientists in Power* (Cambridge, Mass., 1979), pp. 26–36; for the post–World War II period, see H. W. Paul and T. W. Shinn, "The Structure and State of Science in France," *Contemporary French Civilization*, 6 (1981–2), 153–94, especially pp. 165–8.

CHAPTER 1. FROM SECOND EMPIRE TO THIRD REPUBLIC

1 Georges Dupeux, *La Société française, 1789–1960* (Paris, 1964), p. 168.
2 François Crouzet, "French Economic Growth in the 19th Century," *History*, 59 (June 1974); Gordon Wright, *France in Modern Times*, 3rd ed. (New York, 1981), ch. 21.
3 David S. Landes, *The Unbound Prometheus: Technological Change and Industrial Development in Western Europe from 1750 to the Present* (Cambridge, 1969), pp. 11, 32 ff.; but see François Caron, *An Economic History of Modern France*, trans. Barbara Bray (New York, 1979). The debate between economic historians over the performance of the French economy since the early nineteenth century is amusingly treated by Clive Trebilcock, *The Industrialization of the Continental Powers, 1780–1914* (London, 1981), ch. 3.
4 Wright, *France*, p. 288.
5 Theodore Zeldin, *France, 1848–1945* (Oxford, 1973–7), vol. 1, *Ambition, Love and Politics*, pp. 577, 620.
6 Dupeux, *Société*, p. 170; Zeldin, *Ambition*, pp. 94 ff.; but see Rod Day, "The Making of Mechanical Engineers in France: The Ecoles d'arts et métiers, 1803–1914," *French Historical Studies*, 10 (1978), 439–60, who pointed out that the graduates moved into engineering, supervisory, and managerial positions during the early stages of the industrial revolution and that "over the three generations from 1830 to 1914, the families of graduates remained in French industry, business and transport."
7 Antoine Prost, *Histoire de l'enseignement on France, 1800–1967* (Paris, 1968), pp. 32, 45; but see Fritz K. Ringer, *Education and Society in Modern Europe* (Bloomington, Ind., 1979), ch. 3, especially pp. 135, 140. For the number of degrees, see Paul Gerbod, *La Condition universitaire en France au XIXe siècle* (Paris, 1965), p. 637, and Prost, *Histoire*, p. 243.
8 On the social origins of students, see George Weisz, *The Emergence of Modern Universities in France, 1863–1914* (Princeton, N.J., 1983), pp. 23–7; Terry

Shinn, *Savoir scientifique et pouvoir social: L'Ecole polytechnique, 1794–1914* (Paris, 1980), pp. 185–223; and John H. Weiss, *The Making of Technological Man: The Social Origins of French Engineering Education* (Cambridge, Mass., 1982), pp. 70–8. Craig Zwerling, "The Emergence of the Ecole normale supérieure as a Centre of Scientific Education in the Nineteenth Century," in Robert Fox and George Weisz, eds., *The Organization of Science and Technology in France, 1808–1914* (Cambridge, 1980), pp. 50 ff., states that the children of the working class made up 11 percent of the science students at the Normale in the nineteenth century and jumped to 23 percent between 1904 and 1914, and see the refinements of Robert J. Smith, *The Ecole normale supérieure and the Third Republic* (Albany, N.Y., 1982), ch. 3, especially p. 34. Authors differ on the percentages of class composition of schools according to what groups are included in the definition of different classes.

9 Zeldin, *Ambition*, pp. 22 ff; Weisz, *The Emergence of Modern Universities*, p. 9; and James Q. Graham, *French Legislators, 1871–1940: Biographical Data* (Inter-university Consortium for Political and Social Research, Ann Arbor, Mich., 1983).

10 Zeldin, *Ambition*, p. 483.

11 John H. Weiss, "*Professeurs* and Proletarians: A Social Profile of Two Generations of French Science Teachers," AHA paper, 1974.

12 Jean Lhomme, *La Grande bourgeoisie au pouvoir (1830–1880)* (Paris, 1960).

13 Gabriel Compayré, *Histoire critique des doctrines de l'éducation en France depuis le XVIe siècle* (Paris, 1885); AN, F[17] 13072.

14 "Rapport présenté au Conseil supérieur sur un projet de Décret relatif au Doctorat ès Sciences, par M. G. Darboux, membre du Conseil," in A. de Beauchamp, *Recueil des lois et règlements sur l'enseignement supérieur, 1889–1898* (Paris, 1880–8), vol. 5, pp. 787–90. Candidates for the doctorate would be required to do one of the following three groups of certificates:

I	II	III
Calcul différentiel et intégral	Physique générale	Zoologie ou physiologie générale
Mécanique rationelle	Chimie générale	Botanique
3e certificat au choix du candidat	3e certificat au choix du candidat	Géologie ou minéralogie

15 G. Rayet, *Histoire de la Faculté des sciences de Bordeaux, 1838–1894* (Bordeaux, 1898), pp. 9–12.

16 Prost, *Histoire*, p. 27.

17 *Séance annuelle des facultés*, Lille, 1856–76.

18 Ibid.

19 AN, F[17] 13178 (Rennes). Lallemand later became dean at Poitiers.

20 Ibid.

357

21 Prost, *Histoire*, pp. 224–8.

22 "Discours prononcé par le Ministre . . . à la distribution des prix du concours général (12 août 1852)," AN, 246 AP 19 (Fonds Fortoul).

23 "Nouvelle organisation académique," 246 AP 19. See "Rapport à l'Empereur . . .," *Moniteur*, February 15, 1856.

24 AN, F^{17} 13157 (Fortoul to rectors, July 22, 1855, and March 7, 1856). Nicole Hulin, "A propos de l'enseignement scientifique: Une réforme de l'enseignement secondaire sous le Second Empire, la 'bifurcation' (1852–1864)," *Revue d'histoire des sciences*, 14 (1982), 217–45.

25 AN, F^{17} 12958 (Conseil impérial de l'instruction publique, session of Dec. 9, 1867).

26 Circular of Nov. 8, 1882, and letter of Duvaux, minister, to Rectors. *ED*, 4 (1883), 3–5.

27 Decrees cited in de Beauchamp, *Recueil* . . . , IV, 203–22 and V, 404–5.

28 AN, 246 AP 19.

29 See the "Epilogue" of C. Stewart Gillmor, *Coulomb and the Evolution of Physics and Engineering in Eighteenth-Century France* (Princeton, N.J., 1971).

30 AN, F^{17} 13071–2.

31 Ibid.

32 AN, F^{17} 13175. See also the note for the minister from the director of higher education (ler bureau), AN F^{17} 13157 (1873). An *arrêté* by Fortoul (1852) stipulated that the three committees of higher education, secondary education, and primary education meet each Friday under the presidency of the minister or the vice president of the Conseil supérieur. The Comité des inspecteurs généraux de l'enseignement supérieur – in 1861, Dumas, Nisard, Ravaisson, Brongniart, Lafferrière, Denonvilliers, Giraud (all inspectors); Artaud, vice rector of the Academy of Paris; Petit, head of the first division in the ministry; Dumaige, head of the second division; and Mourier, secretary of the committee – met every two weeks.

33 AN, F^{17} 13175.

34 AN, F^{17} 13178 (Toulouse). See the accusations of Duhem against Berthelot: Pierre Duhem, "Thermochimie à propos d'un livre récent de M. Marcelin Berthelot," *RQS*, 42 (1897), 361–92.

35 AN, F^{17} 13178.

36 AN, F^{17} 13178 (Strasbourg).

37 AN, F^{17} 13177 (Marseille).

38 AN, F^{17} 13177 (Lyon).

39 AN, F^{17} 13177 (Montpellier).

40 AN, F^{17} 13176.

41 AN, F^{17} 13158 (Bordeaux). Marcelin Berthelot, "La Théorie atomique," *RS*, 16 (1875), 442–47. Low-temperature work supported atomic theory; see Raoul Pictet, "Essai d'une méthode générale de synthèse chimique," *RS*, 56 (1895), 257–72. See n. 19, ch. 7.

42 AN, F^{17} 13157 (Documents généraux pour les programmes des cours, 1855–6).

43 See George Weisz, "The French Universities and Education for the New Professions, 1885–1914: An Episode in French University Reform," *Minerva*, 17 (1979), 98–128, and Terry Shinn, "The French Science Faculty System, 1808–1914: Institutional Change and Research Potential in Mathematics and the Physical Sciences," *Historical Studies in the Physical Sciences*, 10 (1979), 271–332.
44 D. S. L. Cardwell, *The Organisation of Science in England* (rev. ed., London, 1972), p. 142.
45 AN, F^{17} 13178 (Strasbourg); Berthelot, letter of September 22, 1876, AN F^{17} 13157.
46 AN, F^{17} 13178 (Rennes).
47 Marseille complained bitterly about the absence of a *jardin des plantes*, especially needed for students in medicine and pharmacy. AN, F^{17} 13177 (Marseille, 1868, rector's report).
48 AN, F^{17} 13177 (Nancy).
49 AN, F^{17} 13177 (Lyon), reports of the dean in 1863–4.
50 Prost, *Histoire*, p. 224.
51 *Statistique de l'enseignement supérieur, 1878–88* (Paris, 1889), p. xxviii.
52 On French consciousness of foreign learning, see Claude Diegon, *La Crise allemande de la pensée française (1870–1914)* (Paris, 1959), ch. 7, and Harry W. Paul, *The Sorcerer's Apprentice: The French Scientist's Image of German Science, 1840–1919* (Gainesville, Fla., 1972), ch. 1.
53 Prost, *Histoire*, pp. 236–7.
54 *Statistique*, p. xxvii.
55 Louis Liard, "Exposé des motifs," in ibid., pp. 113–23.
56 Weisz, *The Emergence of Modern Universities*, p. 318.
57 Shinn, "The French Science Faculty System," pp. 305, 312; Weisz, *The Emergence of Modern Universities*, pp. 170–4.
58 See Fox and Weisz, eds., *The Organization of Science and Technology in France*, passim, especially the tables on pp. 12, 328–9; Table 4 in Chapter 4 herein; and Weisz, *The Emergence of Modern Universities*, 225–69.
59 Rayet, *Histoire de la Faculté des sciences de Bordeaux*, p. 6.
60 The quotation is an accusation against J. D. Bernal by Michael Polanyi, *Personal Knowledge* (1958; New York, 1964), p. ix.
61 Louis Crié, "L'Enseignement de la botanique dans les lycées," *RS* 27 (1881), 653–8. For the whole debate on Bert's reforms, see G. Bonnier, H. de Lacaze-Duthiers, and G. Pouchet in *RS*, 29 (1882), 252–69, 334–8, 648–56, and letters by Pouchet and Bonnier, 306–8, 375–6. Bert and Lacaze-Duthiers slugged it out over zoology in *RS*, 30, (1882), 49–51, 66–74, 124, and passim! On Bert's role in the anticlerical crusade, see Jan Goldstein, "The Hysteria Diagnosis and the Politics of Anticlericalism in Late-Nineteenth-Century France," *Journal of Modern History*, 54 (1982), 209–39, especially 230–3.
62 Steven Lukes, *Emile Durkheim: His Life and Work: A Historical and Critical Study* (New York, 1972), pp. 386–7.

63 Jerome S. Bruner, *Towards a History of Instruction* (Cambridge, Mass., 1966), p. 22.
64 Lukes, *Dürkheim*, p. 388.
65 *Statistique*, pp. 3–4.
66 AN, F^{17} 12968.
67 AN, F^{17} 12958-12962-12963-12970-12974-13641. The scientists on the Conseil in 1880 were Berthelot, Institut, Collège de France; Bertrand, Secrétaire perpétuel de l'Académie des sciences; Burta, professeur, Ecole centrale; Chatin, Institut, directeur de l'Ecole supérieure de pharmacie; Fremy, Institut, directeur du Muséum; Gavarret, Faculté de médecine, inspecteur général de l'enseignement supérieur; Godard, Directeur de l'Ecole Monge; Hervé-Mangon, Institut, Conservatoire des arts et métiers; Hacquier, Professeur au Collège de Vitry-le-François, Laussedat, Directeur d'études, Ecole polytechnique; Lespiault, Doyen de la Faculté des sciences, Bordeaux; Moitessier, Doyen de la Faculté de médecine, Montpellier; Risler, Directeur de l'Institut agronomique; Saint-Claire Deville, Ecole normale supérieure; Vintéjoux, Lycée Saint Louis; Voigt, Lycée de Lyon; Vulpian, Institut, doyen de la Faculté de médecine, Paris. As a result of the elections to the Conseil in November 1923, the following scientists represented their institutions: Picard, Academy of Sciences; d'Arsonval, Collège de France; Lacroix, Muséum; Koenigs and Petit, Faculty of Sciences, Paris; General Gossot, Polytechnique; Gabelle, Conservatoire des arts et métiers; Guillet, Ecole centrale; Wery, Institut agronomique. Lycées, collèges, and various areas of primary education also had representation, as did the faculty of medicine and the Ecoles supérieures de pharmacie and the *facultés mixtes* (medicine and pharmacy).
68 AN, F^{17} 12968.
69 *Statistique*, pp. 113–23.
70 R. Steven Turner, "The Growth of Professorial Research in Prussia, 1818 to 1848 – Causes and Context," *Historical Studies in the Physical Sciences*, 3 (1971), 137–82.
71 AN, F^{17} 12958 (*Conseil impérial de l'instruction publique*, session of December 9, 1867).
72 Ministère de l'instruction publique, *Rapport à l'Empereur à l'appui des deux projets de décret relatifs aux laboratoires d'enseignement et de recherches et à la création d'une Ecole pratique des hautes études*, by V. Duruy, Le Ministre secrétaire d'état au département de l'instruction publique (Paris, 1868). See also, Duruy, *L'Administration de l'instruction publique de 1863 à 1869* (Paris, 1870), pp. 674–739.
73 Fremy fumed: "N'oublions pas que les affreux Prussiens construisent des *Palais* pour la chimie." AN, F^{17} 13614.
74 "Ce titre ne qualifie donc pas un établissement spécial, mais un système d'études, dont les décrets du 31 juillet 1868 ont réglé le fonctionnement, en instituant des commissions de surveillance et un conseil supérieur." AN, F^{17} 13614.
75 On the Ecole pratique des hautes études and for the plan to add another section (social sciences), which originally Duruy specified as "Une cinquième section

NOTES TO PP. 49–54

... formée pour les études juridiques," see Terry Nichols Clark, *Prophets and Patrons: The French University and the Emergence of the Social Sciences* (Cambridge, Mass., 1973), pp. 42–51.

76 The *Bulletin des sciences mathématiques et astronomiques* was funded to publish work done at the Ecole pratique.

77 Clark, *Prophets and Patrons*, p. 43.

78 Guy Lazerges, "Une Ecole de physique au XXe siècle: Le Laboratoire Cavendish et l'école de Cambridge," *RGSPA*, 37 (1926), 5–15.

79 *Rapport sur l'Ecole pratique des hautes études*, 1872–2, 1882–3, 1893–4 (Paris: Imprimerie nationale). "By 1874 the Ecole normale had replaced the Ecole polytechnique as the leading centre of scientific education in France." For how this mutation came about, largely induced by Louis Pasteur as administrator of the school, see Craig Zwerling, "The Emergence of the Ecole normale supérieure as a Centre of Scientific Education in the Nineteenth Century," in Fox and Weisz, eds., *The Organization of Science and Technology*, pp. 31–60.

80 *La Science française* (new ed., 2 vols., Paris, 1933), vol. 1, p. 145. For the basis of a different, more favorable view, see Ronald Newburgh, "Fresnel Drag and the Principle of Relativity, *Isis*, 65 (1974), 379–86, dealing with the importance of the work of Fresnel, Fizeau, Mascart, Potier, and Jamin.

81 Ibid., p. 146. For a vew of Duhem's work as excessively French-centered and ignoring the modern thermodynamics of Nernst and Planck, see a review by Walter Mecklenburg of books in thermodynamics, *Scientia*, 15 (1914), 267–72.

82 Lawrence Badash, "The Completeness of Nineteenth-Century Science," *Isis*, 63 (1972), 48–58.

83 There is a highly speculative article by Marjorie Malley, "The Discovery of Atomic Transmutation: Scientific Styles and Philosophies in France and Britain," *Isis*, 70 (1979), 213–23, which argues that "Curie's positivism did not facilitate progress in radioactivity" and "made him excessively cautious," predisposing "him to an unproductive thermodynamics model," all of which ensured that Rutherford and the British carried off most of the honors.

84 Paul, *Sorcerer's Apprentice*, pp. 59–60.

85 See the article of Jean Rosmorduc, "Aimé Cotton (1869–1951): Le Rationalisme et l'expérience," *La Pensée*, no. 165 (1972), 112–26.

86 Mary Jo Nye, *Molecular Reality: A Perspective on the Scientific Work of Jean Perrin* (London, 1972).

87 On the role of the doctorate in the sciences, see Paul, "The Issue of Decline in Nineteenth-Century French Science," *French Historical Studies*, 8 (1972), 427–9. For the later evolution of the municipal Ecole supérieure toward the basic sciences, see Terry Shinn, "Des sciences industrielles aux sciences fonda-mentales: La Mutation de l'Ecole supérieure de physique et de chimie (1882–1970)," *Revue française de sociologie*, 22 (1981), 167–82.

88 Darboux, "Rapport," in Beauchamp, *Recueil*, V, 787 ff.

89 "Le sujet traité est tout à fait élémentaire, il fait partie des premières applications des cours ordinaires d'analyse; fouillé dans tous les sens depuis une vingtaine

361

d'années par des géomètres habiles, il est aujourd'hui complètement épuisé et tout à fait impropre à donner lieu à des recherches vraiment originales." Issaly tried again in 1887 with another thesis, but Paul Appell's report, also signed by Picard, rejected the thesis because it added nothing fundamental to the theory of surfaces in spite of its elegant formulas. *Archives de l'Académie de Paris, AJ* 16, carton 27 (AN, temporary classification).

90 "Les collections allemandes à la galérie d'anatomie comparée au Muséum," *RS*, 25 (1880), 669–708. For some ecstatic remarks on the high status of botanical research (circa 1884) at the Muséum under Philippe van Tieghem, see A. Lacroix, "Un Grand botaniste français: Henri Lecomte (1856–1938)," *RIE*, 93 (1939), 83–97.

91 Quotations are from Camille Limoges, "The Development of the Muséum d'histoire naturelle of Paris, c. 1800–1914," in Fox and Weisz, eds., *The Organization of Science and Technology in France*, pp. 211–40. See also RS, 33 (1884), 289–90 and 344–5, dealing with Fremy's pamphlet on the deplorable state of the Muséum. More material is in *RGSPA*, 11 (1900), 662, 667–8, and 819–20, and in *RS*, 63e année (1925), 1–17, which contains part of Alfred Lacroix's lecture to the Academy of Sciences on A. Milne Edwards. For a moment of optimism, in 1898, when the galleries of anatomy, paleontology, and anthropology were opened, see *RGSPA*, 9 (1898), 558–60.

92 Charles Maurain, *Répertoire de laboratoires français* (Paris, 1922), and Charles Maurain and A. Pacaud, *La Faculté des sciences de l'Université de Paris de 1906 à 1940* (Paris, 1940).

93 Paul Lemoine, "Le Muséum national d'histoire naturelle et son développement en 1934," *RS*, 73e année (1935), 299–307.

94 Shinn, *Savoir scientifique*.

95 Prost, *Histoire*, pp. 469–70; Shinn, "The French Faculty System."

96 The German situation was similar. See Russell McCormmach, "On Academic Scientists in Wilhelmian Germany," *Daedalus* (Summer 1974), 157–71. On education and society see Fritz Ringer, *Education and Society in Modern Europe*, and Robert Fox's perceptive review of it in *Minerva*, 18 (1980), 164–70. Further studies on France are in a fat tome edited by Donald N. Baker and Patrick J. Harrigan, *The Making of Frenchmen: Current Directions in the History of Education in France, 1679–1979*, a double issue of *Historical Reflections*, 7 (1980).

CHAPTER 2. POSITIVISM IN BIOLOGY

1 Dorothy L. Sayers, *Strong Poison* (London, 1930), p. 32.

2 Their adoption of Darwinism set English physiologists "apart most clearly from their European colleagues." Gerald L. Geison, *Michael Foster and the Cambridge School of Physiology* (Princeton, N.J., 1978), pp. 334–6.

3 For a perceptive study of the fusing of the political and social in the cognitive and institutional birth of physiology in Germany, see Everett Mendelsohn,

"Revolution and Reduction," in Y. Elkana, ed., *The Interaction between Science and Philosophy* (Atlantic Highlands, N.J., 1974), pp. 407–26.

4 M. D. Grmek, *Raisonnement expérimental et recherches toxicologiques chez Claude Bernard* (Geneva, 1973), p. 280. See pp. 280–5 for the intriguing story of how Bernard used the publication process to "hide," for a period, his most interesting discovery on the galvanization of animals given curare.

5 First article of the "Règlement de la Société de Biologie," *Comptes rendus des séances et mémoires de la Société de Biologie*, 2 (1850), v.

6 See Georges Canguilhem, "La Philosophie biologique d'Auguste Comte et son influence en France au XIXe siècle," *Etudes d'histoire et de philosophie des sciences* (Paris, 1970), pp. 61–98.

7 F. Papillon, *Histoire de la philosophe moderne*, t. 1, p. 300, cited by E. Gley, *Essais de philosophie et d'histoire de la biologie* (Paris, 1900), p. 186. "Rien, a-t-on dit, ne retirera du tissu de la science les fils d'or que la main du philosophe y a introduits."

8 "Sur la direction que se sont proposée en se réunissant les membres fondateurs de la Société de Biologie pour répondre au titre qu'ils ont choisi," *Comptes rendus des séances et mémoires de la Société de Biologie*, 1re année, 1849 (Paris, 1850), read to the society on June 7, 1848. See also W. M. Simon, *European Positivism in the Nineteenth Century* (Ithaca, N.Y., 1963), ch. 4. In the first year of *La Philosophie positive* (1867) Robin wrote three articles essentially expanding the basic Comtean ideas he had set forth in his programmatic speech to the Société de biologie in 1849. "De la biologie. Son objet et son but, ses relations avec les autres sciences, la nature et l'étendue du champ de ses recherches, ses moyens d'investigation," *PP*, 1 (1867), 78–101, 212–32, 392–412.

9 Robin, *PP*, 1 (1867), 81 and 212.

10 Robin, *PP*, 1 (1867), 229; *RS*, 42 (1888), 435, a review of Letellier's thesis, *Etude de la fonction urinaire chez les mollusques acéphales*; and A. Gautier's banquet speech when he replaced Chevreul at the Institut, reprinted in *RS*, 44 (1889), 76–80.

11 "A côté de la géométrie, l'architecture; à l'astronomie répond l'art nautique; à la physique et la chimie répondent les arts industriels . . ." Robin, "Sur la direction," p. vii.

12 " . . . l'art medical . . . emprunte . . . à l'histoire naturelle. . . . Ainsi, pour l'étude des médicaments, il est à chaque instant obligé de puiser ses renseignements dans la *taxonomie végétale* et les autres parties de l'histoire des plantes. Dans l'étude des parasites, c'est encore à l'histoire des plantes. Dans l'étude des parasites, c'est encore à l'histoire naturelle des animaux et des végétaux que le médecin doit recourir, puisque les maladies qu'ils causent ne sont que le résultat des moeurs de ces êtres ou de plusieurs de leurs phénomènes vitaux." Robin, "Sur la direction," p. x.

13 L. Lévy-Bruhl, *The Philosophy of Auguste Comte* (1900; London, 1903), book 2, ch. 4.

14 Charles Letourneau, "Variabilité des êtres organisés," *PP*, 3 (1868), 173–205, 333–72; 4 (1869), 5–30.

15 E. Jourdy, "Darwin et Agassiz," *PP*, 8 (1872), 24-58.

16 G. Wyrouboff, *PP*, 22 (1879), 23-49. Review of A. Lefèvre, *Les Philosophes* and *La Philosophie* (1879), published in the Reinwald series Bibliothèque des sciences contemporaines.

17 E. Littré, "De la philosophie positive," *PP*, 13 (1874), 161-2.

18 G. Wyrouboff, "De l'individu dans le règne inorganique," *PP*, 4 (1869), 52-70.

19 G. Wyrouboff, "De l'espèce et de la classification en zoologie," *PP*, 5 (1869), 121-38.

20 E. Jourdy, "Darwin et Agassiz," 59-87.

21 André Sanson, "Les Expériences de Darwin sur les pigeons" (translation of Darwin's work by J. J. Moulinié in 1868: *De la variation des animaux et des plantes sous l'action de la domestication*), *PP*, 10 (1873), 64-95.

22 E. Littré, "De quelques questions soulevées à propos du transformisme," *PP*, 15 (1875), 441-8; a review of A. de Quatrefages, *L'Espèce humaine* (2nd ed., 1877), *PP*, 20 (1878), 161-9; and "L'Hypothèse de la génération spontanée et celle du transformisme . . . ," *PP*, 22 (1879), 165-79. A. Proost, *Revue des questions scientifiques*, 3 (1878), 157, distinguished between *Darwinisme* ("qui a la prétention d'expliquer le pourquoi, de formuler la loi de l'évolution organique") and *transformisme* ("qui suppose le fait sans prétendre l'expliquer"). See also Mary Jo Nye, "The Boutroux Circle and Poincaré's Conventionalism," *Journal of the History of Ideas*, 40 (Jan.–March 1979), 107-120, and Harry W. Paul, *The Edge of Contingency: French Catholic Reaction to Scientific Change from Darwin to Duhem* (Gainesville, Fla., 1979).

23 *PP*, 15, (1875), 445-6; *PP*, 25 (1880), 125.

24 Yvette Conry, *L'Introduction du Darwinisme en France au XIXe siècle* (Paris, 1974), pp. 415-20. For Darwin's use of functional analysis, see Michael T. Ghiselin, *The Triumph of the Darwinian Method* (Berkeley, Cal., 1969), ch. 6.

25 It was probably used earlier than 1800. See Jacques Roger, "Chimie et biologie: Des 'molécules organiques' de Buffon à la 'physico-chimie' de Lamarck," *History and Philosophy of the Life Sciences*, 1 (1979), 43. This journal, edited by M. D. Grmek, is one of the "pubblicazioni della stazione zoologica di Napoli," section II.

26 Raoul Morgue, *La Philosophie biologique d'Auguste Comte* (Lyon, 1909), pp. 8-10 and 57. This work was originally published in the *Archives d'anthropologie criminelle et de médecine légale* in Oct., Nov., and Dec. 1909.

27 Morgue, *La Philosophie biologique*, p. 19.

28 Ibid., pp. 19-21; see n. 1, p. 21, for an interesting example of Comte's being "right" and Bernard's being "wrong": Cf. "Bernard enseigne que la structure ne révèla pas la fonction."

29 Ibid., pp. 22-23, 26.

30 Ibid., pp. 29-31.

31 Frederick Lawrence Holmes, *Claude Bernard and Animal Chemistry: The*

Emergence of a Scientist (Cambridge, Mass., 1974), p. 374, citing Canguilhem.

32 Holmes, *Bernard*, p. 374.

33 Charles Robin, *Du microscope et des injections suivi d'une classification des sciences fondamentales, de celle de la biologie et de l'anatomie en particulier* (Paris, 1849). See also his *Tableaux d'anatomie* (Paris, 1851).

34 M. D. Grmek, "Robin," *DSB*, XI, 491–2.

35 Robin, *Du microscope*, p. 129. The definitions are italicized in Robin's text.

36 Robin, *Du microscope*, p. 129, citing Henri-Marie Ducrotay de Blainville, *De l'organisation des animaux, principes d'anatomie comparée* (Paris, 1822), Introduction, and Auguste Comte, *Cours de philosophie positive* (3rd ed., Paris, 1860), vol. 3, p. 205.

37 Ch. Robin, *Traité du microscope: Son mode d'emploi; ses applications à l'étude des injections; à l'anatomie humaine et comparée; à la pathologie médico-chirurgicale; à l'histoire naturelle animale et végétale; et à l'économie agricole* (Paris, 1871). Robin displayed among his various titles at that time that of "Professeur d'histoire à la Faculté de médecine de Paris."

38 E. Littré, *Dictionnaire de la langue française*, I, 348; "ce mot [biologie] crée par un naturaliste allemand, Treviranus, a été employé pour la première fois, par Lamarck, dans son *Hydrologie*, 1802, et dans son *Discours d'ouverture sur la question de l'espèce* (1803)." For correction, see n. 25 of this chapter. See *DSB*, XIII, 460, on Treviranus.

39 Littré broke down physiology further: *Physiologie générale*: "Science qui, sans traiter d'une espèce vivante déterminée, traite d'une manière philosophique et abstraite des phénomènes de la vie;" *Physiologie spéciale*: "Science qui prend pour objet d'étude une espèce vivante distincte, et décrit le mécanisme de la vie dans cette seule espèce."

40 Porfirio Parra, "Biologie et physiologie," *RO*, 20 (1900), 317–33.

41 Parra also found the same confusion in Bernard's *Leçons sur les phénomènes de la vie communs aux animaux et aux végétaux*, 2 (1879).

42 E. Littré, "Du déterminisme de Claude Bernard," *PP*, 21 (1878), 5–11.

43 Parra, "Biologie et physiologie," pp. 321 ff.

44 Ibid.

45 Ibid., pp. 330 ff.

46 Mathias Duval, "Claude Bernard," *PP*, 20 (1878), 424–44.

47 E. Gley, "Influence du positivisme; développement des sciences biologiques en France," *Annales internationales d'histoire* (Paris, 1900), 164–70; D. G. Charlton, *Positivist Thought in France during the Second Empire, 1852–1870* (Oxford, 1959); Reino Virtanen, *Claude Bernard and His Place in the History of Ideas* (Lincoln, Neb., 1960). For Canguilhem, Grmek, and Holmes, see the works already cited. Annie Petit, "D'Auguste Comte à Claude Bernard: Un Positivisme déplacé," *Romantisme*, nos. 21–2 (1978), 45–62.

48 Claude Bernard, *Philosophie*, "Manuscrit inédit, texte publié et présenté par Jacques Chevalier" (Paris, 1954), pp. 25–43, especially p. 37. But see *RS*, 6

(1868–9), 136, for Bernard's recognition of Bacon's elevation of the experimental method to the status of a true scientific philosophy. Bernard was then a member of the Royal Society.

49 Ibid., pp. 32, 42–3.

50 Ibid., p. 26.

51 Joseph Schiller, *Claude Bernard et les problèmes scientifiques de son temps* (Paris, 1967), ch. 12, "Claude Bernard et la métaphysique."

52 Judgment by a leading contemporary biologist, Joseph S. Fruton, in Holmes, *Bernard and Animal Chemistry*, p. xii.

53 Letter by Charles Richet on "La Métaphysique de Claude Bernard d'après M. Letourneau," in *RS*, 24 (1879), 303–4.

54 Pierre Lamy (professeur au Collège colonial et climatique de France, docteur en Sorbonne), . . . *Claude Bernard, le naturalisme et le positivisme* (Paris, 1928), and *Claude Bernard et le matérialisme* (Paris, Alcan's Bibliothèque de philosophie contemporaine, 1939).

55 Georges Canguilhem, *L'Idée de médecine expérimentale selon Claude Bernard* (Paris, 1965), p. 10. Frank Manuel, *The Prophets of Paris* (1962; New York, 1965), pp. 276–7, points out that Vico and Turgot have modern claims to the paternity of this version of an old idea.

56 C. H. [Dr. Constant Hillemand?], "Auguste Comte et Claude Bernard," *La Revue positive internationale*, 14, no. 2 (18 Homère 126–15 Feb. 1914).

57 Bernard, *Introduction à l'étude de la médecine expérimentale*, p. 71, cited in Canguilhem, *L'Idée*, p. 10.

58 Bernard to Littré (Dec. 10, 1867), referring to his *Rapport sur les progrès de la physiologie générale* and to a forthcoming reprinting of an article in the *Revue des deux mondes*, cited in Stanislas Aquarone, *The Life and Works of Emile Littré, 1801–1881* (Leyden, 1958), pp. 185–6.

59 Bernard, *Principes de médecine expérimentale*, p. 51, cited in Canguilhem, *L'Idée*, p. 10, who emphasizes that Bernard used the word "fonder," not "créer" – proof of Bernard's modesty, no doubt.

60 Canguilhem, *L'Idée*, p. 16. Bernard, *Leçons de pathologie expérimentale* (2d ed., 1880), p. 466.

61 Bernard, "L'Évolution de la médecine scientifique," *RS*, 11 (1873), 900–1.

62 For the six chief aspects of biological comparison, see Robin, *PP* (1867), 396–97.

63 Ibid., p. 223.

64 Grmek, "Robin," pp. 491–2.

65 Ch. Robin and F. Verdeil, *Traité de chimie anatomique et physiologique, normale et pathologique* (3 vol., Paris, 1852–3).

66 Grmek, "Robin," p. 492.

67 Claude Bernard, *Rapport sur les progrès et la marche de la physiologie générale en France* (Paris, 1867), pp. 1–4. See John E. Lesch, *Science and Medicine in France: The Emergence of Experimental Physiology, 1790–1855* (Cambridge, Mass., 1984), for the continuities between Magendie and Bernard.

68 Bernard, *Rapport*, p. 131.

69 Ibid., p. 141.
70 Ibid., pp. 5, 132–3.
71 Ibid., p. 136.
72 Ibid., pp. 137–8.
73 Ibid., p. 136; Gabriel Gohau, "Alfred Giard," *Revue de synthèse*, 3e série, nos. 95–6 (1979), 395–8.
74 E. Littré, "Les Trois philosophies," *PP*, 1 (1867), 22.
75 On Wyrouboff, see Paul, "Scholarship versus Ideology: The Chair of the General History of Science at the Collège de France, 1892–1913," *Isis*, 67 (1976), 384–7.
76 Wyrouboff, "Qu'est-ce que la géologie?," *PP*, 1 (1867), 31–3, 50.
77 Ibid., p. 50.
78 Wyrouboff, "Le Certain et le probable," *PP*, 1 (1867), 167–82.
79 C. C. Gillispie, *The Edge of Objectivity: An Essay in the History of Scientific Ideas* (Princeton, N.J., 1960), p. 147.
80 Lévy-Bruhl, *The Philosophy of Auguste Comte*, p. 6, and Robin, "De la biologie," p. 407.
81 Wyrouboff, "Le Certain," p. 177.
82 Saint-John Perse, *Amers*.
83 Robin, "De la biologie," *passim*.
84 Saint-John Perse, *Amers*.

CHAPTER 3. BIOLOGY IN THE UNIVERSITY

1 AN, F^{17} 13157.
2 Plea from Rennes for the "dédoublement de la chaire," *ED*, 30 (1887–8), 167.
3 Claude Bernard, "Médecine expérimentale: Histoire de la chaire de médecine au Collège de France," *RS*, 11 (1873), 772.
4 "Collège de France – réorganisation administrative," *RS*, 11 (1873), 765–6, and Claude Bernard, *Principes de médecine expérimentale* (Paris, 1947), p. 215.
5 A. Dastre, "La Chaire de physiologie de la Sorbonne," *RS*, 40 (1887), 737–43, 779–88; "La Chaire de physiologie à la faculté de médecine de Paris," *RS*, 41 (1888), 500; and R. Dubois, "La Physiologie générale" (Lyon), *RS*, 51 (1893), 617–23. See also M. D. Grmek, "Claude Bernard," *DSB*, II, 24–34.
6 S. Arloing, *L'Enseignement de la physiologie dans les facultés des sciences: Leçon inaugurale faite à la faculté des sciences de Lyon* (Lyon, 1884).
7 Charles Richet, "L'Enseignement de la physiologie," *RS*, 48 (1891), 335–8.
8 E. Ray Lankester, "Les Etudes biologiques en Angleterre, en France et en Allemagne," *RS*, 32 (1883), 481–9.
9 See William Coleman, *The Interpretation of Animal Form* (New York, 1967).
10 Paul Lawrence Farber, *The Emergence of Ornithology as a Scientific Discipline: 1760–1850* (London, 1982), ch. 8, modifies the argument of Michel

Foucault, *The Order of Things* (1966; New York, 1970), ch. 5, concerning the nineteenth-century replacement of natural history by biology.

11 Claude Bernard, *Notes pour le rapport sur les progrès de la physiologie*, manuscrit inédit présenté et commenté par M. D. Grmek (Documents inédits du Collège de France, 2, Paris, 1979), passim.

12 Ibid., p. 63.

13 G. Canguilhem, "Techniques et problèmes de la physiologie au XIXe siècle," in René Taton, ed., *La Science contemporaine*, vol. 3, *Le XIXe siècle* (Paris, 1961), vol. 1, 480.

14 Bernard, *Notes*, p. 56.

15 Ibid., comment by Grmek, p. 17. For an analysis of Bernard's battle to establish experimental physiology as an autonomous discipline, see William Coleman, "The Cognitive Basis of the Discipline: Claude Bernard on Physiology," *Isis*, 76 (1985), 49–70; Coleman shows the "interaction of cognitive and social parameters in discipline formation," without wallowing in the mystifications of *Wissenssoziologie*.

16 Bernard, *Notes*, Grmek's comments, pp. 63–4. See J.-J.-V. Coste, *De l'observation et de l'expérimentation en physiologie, du laboratoire* (Paris, 1860).

17 Cf. "the bold sexual imagery" of the Baconian experimental method: Carolyn Merchant, *The Death of Nature: Women, Ecology and the Scientific Revolution* (New York, 1980), pp. 171–2. In an unpublished paper, Annie Petit draws attention to "l'érotique scientifique" in Claude Bernard's "net désir de pénétration," but Petit is amused by "the ambiguity of certain texts" rather than inclined to indulge in Merchant's type of phallic analysis.

18 Bernard, *Notes*, Grmek's comments, p. 59; Bernard, *Notes*, pp. 55–6. Bernard's thesis for his doctorate in the natural sciences was "*Recherches sur une nouvelle fonction du foie considéré comme organe producteur de matière sucrée chez l'homme et les animaux* (Paris, 1853). Milne Edwards was president of the jury, Dumas and de Jussieu the examiners. The thesis was dedicated to Magendie and Rayer.

19 Camille Limoges, "The Development of the Muséum d'histoire naturelle of Paris, c. 1800–1914," in Robert Fox and George Weisz, eds., *The Organization of Science and Technology in France, 1808–1914* (Cambridge, 1980).

20 H. Lacaze-Duthiers, *The Study of Zoology*, in *The Interpretation of Animal Form* . . . , trans. and intro. by William Coleman (*The Sources of Sciences*, no. 15, New York, 1967).

21 Alfred Giard, "Morphologie générale," *RS*, 75 (1905), 129–36.

22 Maurice Caullery, "La Méthode et les critères de la morphologie," *Rivista di scienze*, IV, no. 7, and *Revue des idées*, November 1908, no. 59.

23 An amiable description of Comtism by the dyspeptic Carlyle: "Poor 'Comtism,' ghastliest of algebraic spectralities."

24 P. B. and J. S. Medawar, *The Life Sciences: Current Ideas of Biology* (New York, 1977), p. 7.

25 Le Dr J. Joyeux-Laffuie (professeur à la faculté des sciences, directeur du laboratoire maritime de Luc-sur-mer), *Discours sur l'étude des animaux marins et*

l'utilité des laboratoires maritimes, prononcé le 3 novembre 1888 à la rentrée solonnelle des facultés . . . Caen.

26 René Sand, "Les Laboratoires maritimes de zoologie," *Revue de l'Université de Bruxelles,* 3 (1897–8), 23–47, 121–51, 203–35, 689–96.

27 Ernest van den Broek, "Une Visite à la Station zoologique et à l'Aquarium de Naples," *Bulletin scientifique du Nord et des pays voisins,* 14 (1882), 240–54; Alfred Giard, "La Direction des recherches biologiques en France," *Bulletin scientifique de la France et de la Belgique,* 27 (1895), 432–58; and A. Buisseret, "Les Stations zoologiques des bords de la mer," *RQS,* 25 (1889), 42–75; 446–70. There are some discrepancies in the figures given by the different articles.

28 The date of founding given in *Laboratoires de biologie marine des côtes de France* (Publication du Centre national de documentation pédagogique, 1955) differs from other dates I have found: 1853, 1857, 1859, and 1860; Lacaze-Duthiers gives 1860, the date Coste got governmental support. Coste comes and goes, with his initials, in Eleanor Clark's titillating account of *The Oysters of Locmariaquer* (Chicago, 1978; originally published in 1959).

29 Coste, "De l'observation et de l'expérience," cited by Ch. Robin and G. Pouchet in "Rapports adressés à M. le Ministre de l'Instruction Publique sur les laboratoires maritimes," *ED,* 13 (1884), 38–42.

30 G. P. [Georges Pouchet], "Rapport sur le fonctionnement du Laboratoire de Concarneau en 1887, et sur la sardine," *ED,* 25 (1887–8), 5–44. Marcel Baudouin, "L'Industrie de la sardine en Vendée," *RS,* 41 (1888), 651–60, 683–92; *RS,* 75 (1905), 715–22, and Fabre-Domerge (Directeur-adjoint du Laboratoire de zoologie maritime à Concarneau), "Le Régime de la sardine et la détermination des lois qui la régissent," *RGSPA,* 7 (1896), 429–32, who points out that the well-being of the maritime population of the littoral between Sables-d'Olonne and Brest depended on the sardine.

31 Other work done at Concarneau included Giard's research on parasitic crustaceans, Chabry's doctoral thesis *Contribution à l'embryologie normale et tératologique des ascidies simples* (*Journal de l'anatomie,* 1887), and an important document for French fauna: Bonnier's *Catalogue des crustacés malacostracés recueillis dans la baie de Concarneau,* published in the *Bulletin scientifique du départment du Nord* (1887) and as a separate monograph by O. Doin. Chevreux and Th. Barrois also did work on the crustaceans of Concarneau – a very studied group.

32 Henri de Lacaze-Duthiers, "Laboratoire de zoologie expérimentale de Roscoff," *ED,* 13 (1884), 1–14; Giard, "La Direction des recherches biologiques," pp. 449–50; and Toby A. Appel, "Lacaze-Duthiers," *DSB,* VII, 545–46. See also articles on Roscoff by Lacaze-Duthiers in *RS,* 27 (1881), 673–80; *RS,* 55 (1895), 161–70, 226–31, which also includes material on Banyuls-sur-Mer; and by H. Fol, "Le Laboratoire de Roscoff en 1883," *RS,* 31 (1883), 417–22. Lacaze-Duthiers's comments on students are in "Les Sciences accessoires de la médecine dans les facultés des sciences et les stations maritimes," *RS,* 53 (1894), 546.

33 See *ED*, 88 (1904–1905), 19, and *RSGPA*, 16 (1905), 546.
34 See Lacaze-Duthiers, *ED*, 13 (1884), 14–32; René Sand, "Les Laboratoires maritimes," 145–50; Lacaze-Duthiers, "Etablissement zoologique de Banyuls-sur-Mer, laboratoire Arago," *RS*, 28 (1881), 705–15; and Henry de Varigny, "Le Laboratoire de zoologie expérimentale de Banyuls-su-Mer," *RS*, 35 (1885), 371–74. De Varigny also wrote two short pieces on Cette and Concarneau, *RS*, 37 (1886), 593–96, 629–30.
35 Lacaze-Duthiers, *ED*, 13 (1884), 25.
36 *RS*, 42 (1888), 161–73, 198–212, and H. de Lacaze-Duthiers, *Le Monde de la mer et ses laboratoires*, a thirty-nine-page pamphlet published in 1888 by the Association Française pour l'Avancement des Sciences.
37 Alfred Giard, speech on the laboratory of marine zoology at Wimereux to the Association Française pour l'Avancement des Sciences, Congrès de Lille, 1874, pp. 12–13.
38 See "La Station zoologique de Wimereux," *RIE*, 23 (1892), 502–12.
39 See N. Ensch and L. L. Querton, "La Station zoologique de Wimereux," *Revue de l'Université de Bruxelles*, 1 (1895–6), 306.
40 *Laboratoires de biologie marine des côtes de France*, published by the Centre national de documentation pédagogique, 1955. Clermont-Ferrand had the first fresh-water laboratory; see Ph. Glangeaud, "Un Laboratoire biologique dans les volcans d'Auvergne," *RGSPA*, 11 (1900), 626–30.
41 A. Ménégaux, "Visite aux laboratoires maritimes français de la Méditerranée," *Bulletin de l'Institut général psychologique*, 1908–10, and Alfred Binet, "La Vie psychique des micro-organismes," *Revue philosophique de la France de l'étranger*, 24 (1887), 449–89. See Yves Delage, *Comment pensent les bêtes* (Paris, 1911), extract of the *Bulletin*, Institut général psychologique, no. 1, 11e année, 1911.
42 Maurice Caullery, "L'Enseignement à la station zoologique de Wimereux," *La Revue de l'enseignement des sciences*, 1 (1907), 329–38. On the explicit pedagogical aim of Marion's laboratory, see A. Ménégaux, *Le Laboratoire maritime d'Endoume, à Marseille*, extract from the *Bulletin de l'Institut général psychologique*, 4 (1910).
43 For details on the appointment of Giard to the chair of the "Evolution des êtres organisés," see Harry W. Paul, *The Edge of Contingency: French Catholic Reaction to Scientific Change from Darwin to Duhem* (Gainesville, Fla., 1979), pp. 56–9, and Marc Viré, "La Création de la chaire 'Evolution des êtres organisés' à la Sorbonne en 1888," *Revue de Synthèse*, 3e série, nos. 95–6, July–Dec. 1979, 377–91.
44 AN, F¹⁷ 25793.
45 AN, F¹⁷ 25793, and Andrée Tétry, "Alfred Giard," *DSB*, V, 385–6.
46 Félix Le Dantec, *Lamarkiens et Darwiniens: Discussion de quelques théories sur la formation des espèces* (Paris, 1899).
47 AN, F¹⁷ 25672 and 25750. For a friendly assessment of Delage see Andrée Tétry, "Yves Delage," *DSB*, IV, 11–13.

48 Alfred Giard, "De l'influence néfaste des prix de l'Académie," *Bulletin scientifique du Nord*, 10 (1878), 56-62, 214-19.

49 Alfred Giard, "Fragments biologiques," *Bulletin scientifique de la France*, 20 (1889), 167-170. Yves Delage's memoir was a "thèse de la Faculté des sciences de Paris" (no. 452): *Contribution à l'étude de l'appareil circulatoire des crustacés édriophthalmes marins* (1881), 173 pages with thirty plates. Delage's thesis was awarded "boules blanches pour toutes les épreuves." AN, F^{17} 25750; there is also some material on Delage in F^{17} 25672. The review of the memoir in the RS, 27 (1881), 665-7, is a friendly, descriptive-analytical account.

50 E.-L. Bouvier, "Sur la circulation de l'écrevisse," *Comptes rendus de la Société de Biologie*, 17 (1889), also published by Giard in his *Bulletin scientifique* . . . , 19 (1888).

51 Yves Delage, "Réponse à M. Giard," *Revue biologique du Nord de la France*, 2nd year, no. 1 (Oct. 1889), 99-101. Published in Lille, this journal was edited by Théodore Barrois and R. Moniez, professors of natural history in the medical faculty at Lille, and Paul Hallez, professor of zoology in the faculty of sciences at Lille.

52 Alfred Giard, "La Direction des recherches biologiques," p. 447, and "Sur deux types nouveaux d'ascothoracida," *Bulletin scientifique de la France* . . . 23 (1891), 98-9, a note on the work of Fowler and Norman. Giard worked twenty-three years on his *Monographie des rhizocéphales*.

53 Yves Delage, *La Structure du protoplasma et les théories sur l'hérédité et les grands problèmes de la biologie générale* (Paris, 1895).

54 Jean-Louis Fischer, "Yves Delage (1854-1920): L'Épigenèse néo-lamarckienne contre la prédétermination weismannienne," *Revue de synthèse*, 3e série, nos. 95-96 (1979), 443-61.

55 Fischer, "Delage" pp. 448-61; Anne Diara, "Le Transformisme de Félix Le Dantec," *Revue de synthèse*, 3e série, nos. 95-6 (1979), 407-22; Kathleen Wellman, "Félix le Dantec et le néo Lamarckisme," ibid., pp. 423-42.

56 "Causerie bibliographique," *RS*, 56 (1895), 449-501.

57 Michael T. Ghiselin, *The Triumph of the Darwinian Method*, (Berkeley, Cal., 1969), pp. 128-9. For the discussion in the academy, see *RS*, 7, no. 34 (1870).

58 "La Continuité et la discontinuité dans les sciences et dans l'esprit," *RS*, 15 (1875), 828-31, reprinted in Jules Tannery, *Sciences et Philosophie* (Paris, 1911 and 1924).

59 Alfred Giard, "Les Faux principes biologiques et leurs conséquences en taxonomie," *RS*, 17 (1876), and "La Castration parasitaire et son influence sur les caractères extérieurs du sexe mâle chez les crustacés décapodes," *Bulletin scientifique du département du Nord*, 18 (1887), 1-28, along with F. Houssay's review of Remy Perrier (brother of Edmond), *Éléments d'anatomie comparée*.

60 Alfred Giard, "L'Evolution dans les sciences biologiques," *Bulletin scientifique de la France et de la Belgique*, 41 (1907), reprinted in *RS*, 76 (1905), 193-205.

61 Lacaze-Duthiers had written *Une Monographie anatomique des galles*. See J. Darboux and C. Hovard, *Catalogue systématique des zoocécidies de l'Europe et du bassin méditerranéen* in *Bull. sci.*, vol. 34 bis, 1901.

62 Quoted with the proper appreciation by Emile Picard, *L'Evolution des idées sur la lumière et l'oeuvre d'Albert Michelson* (Paris, 1935), p. 22.

63 Gohau, "Giard," pp. 398–406.

64 *RS*, 4 (1885), 499–501.

65 Maurice Caullery, "Les Études zoologiques. Critiques et desiderata," *RS*, 63 (1925), pp. 801–3. Cf. Paul Janet, "Sur la réforme de la licence ès sciences," *RS*, 61 (1923), pp. 2–7.

66 A reference to Delage's book, *La Structure du protoplasma*.

CHAPTER 4. THE INDUSTRIAL CONNECTION OF UNIVERSITY SCIENCE

1 See *ED*, 58 (1895), 12.

2 *ED*, 100 (1910), 201.

3 R. de Forcrand (directeur de l'Institut de chimie de l'Université de Montpellier), "L'Enseignement supérieur professionnel," *RIE*, 37 (1899), 2, 34.

4 For the old régime see Charles Coulston Gillispie, *Science and Polity in France at the End of the Old Regime* (Princeton, N.J., 1980), especially part 3 on applications, chs. 5 and 6.

5 The first number of the *Bulletin de la Société* . . . , 1, (1874) 64–72, carried Pasteur's "Etude sur la bière; nouveau procédé de fabrication pour la rendre inaltèrable." In 1904 Henry Le Chatelier's *Revue de métallurgie* took over the publication of memoirs done for the previous ten years in the bulletin.

6 Robert Fox, "The *Savant* Confronts His Peers: Scientific Societies in France, 1815–1914," in Robert Fox and George Weisz, eds., *The Organization of Science and Technology in France, 1808–1914* (Cambridge, 1980), p. 257.

7 Ibid. *Bulletin de la Société industrielle de Mulhausen*, 1 (1828) and 40 (1870). See Robert Fox, "Presidential Address: Science, Industry, and the Social Order in Mulhouse, 1798–1871," *British Journal for the History of Science*, 17 (1984), 127–168.

8 By 1873 the title of the Society was Société industrielle et agricole d'Angers et du département de Maine-et-Loire. See *Bulletin* of the Society for 1830 and for 1873.

9 *Bulletin de la Société scientifique industrielle de Marseille*, 1872; *Annales de la Société d'agriculture, science et industrie de Lyon*, 7th series, 1 (1893).

10 *Société chimique du Nord de la France*, 1re année (1891); *Bulletin de la Société industrielle du Nord de la France*, 40e année, 1912.

11 Charles Camichel, *Le Laboratoire scientifique et l'usine* (Toulouse, 1902), a sixteen-page pamphlet.

12 Louis Houllevigue, *Du laboratoire à l'usine* (Paris, 1904), preface. The book deals with machines, gas engines, transportation and distribution of energy, the

industrial Alps, electrochemistry, incandescent light, the science and applications of refrigeration, and molecules, ions, and corpuscles.

13 Terry Shinn, "The French Science Faculty System, 1808–1914: Institutional Change and Research Potential in Mathematics and the Physical Sciences," *Historical Studies in the Physical Sciences*, 10 (1979), 271–332.

14 A. Calmette, *Ce que Pasteur dut à Lille et ce que Lille doit à Pasteur* (Lille, 1910), p. 6. Unless otherwise indicated, the source for Lille is *Université impériale: Académie de Douai (Rentrée solenelle de la Faculté des lettres de Douai, de la Faculté des sciences et de l'Ecole préparatoire de médecine de Lille)*, 1856–76. The name of this gathering was later changed to the *Séance annuelle des facultés*.

15 *ED*, 108 (1914), 243–44, 303–4; George Weisz, *The Emergence of Modern Universities in France, 1863–1914* (Princeton, N.J., 1983), pp. 237–41.

16 AN, F^{17} 12987.

17 *RGSPA*, 8 (1897), 568–9; *ED*, 88 (1906), 205.

18 R. de Forcrand, "Enseignement supérieur professionnel," 27–28.

19 Louis Barbillion, "La Formation initiale de l'ingénieur: Enseignement secondaire ou enseignement technique, *RIE*, 85 (1931), 175–82.

20 *RIE*, 66 (1913), 84–90; Weisz, *The Emergence of Modern Universities*, pp. 177–86; J. Gosselet, "Des modifications à apporter aux examens de licence pour développer le nombre des étudiants," *RIE*, 23 (1892), 113–25.

21 *ED*, 69 (1898), p. 186; v 85, (1904), 236.

22 See the call to action by Emile Boirac, rector of the Academy of Grenoble, on the occasion of the founding of the Société pour le développement de l'enseignement technique à Grenoble; in *RIE*, 38 (1900), 38–50.

23 "Grenoble," *ED*, 106 (1913). Brenier was the president of the Grenoble chamber of commerce.

24 *ED*, 100 (1910), 46–70.

25 "L'Institut électrotechnique de Grenoble," *RIE*, 52 (1906), 432–40, 538–44; R. Gosse, "L'Institut polytechnique de Grenoble," *RGSPA*, 46 (1935), 520–9, one of a series of articles in the journal on "Les grands centres de recherches scientifiques."

26 Charles Kindleberger, "The Postwar Resurgence of the French Economy," in Stanley Hoffmann et al., *In Search of France* (New York, 1965), p. 119, and Léon Guillet, *La Métallurgie et les mines* (Paris, n.d.).

27 David Landes, *The Unbound Prometheus: Technological Change and Industrial Development in Western Europe from 1750 to the Present* (Cambridge, 1969), pp. 269–75.

28 Jacob Schmookler, *Invention and Economic Growth* (Cambridge, Mass., 1966), and *Patents, Invention, and Economic Change: Data and Selected Essays* (Cambridge, Mass., 1972); for a critique, see Nathan Rosenberg, "Science, Invention and Economic Growth," *Economic Journal*, 84 (1974), 90–108, and *Technology and American Economic Growth* (New York, 1972).

29 See also E. Bichat, "L'Enseignement des sciences appliquées à la faculté des sciences de Nancy," *RIE*, 35 (1898), 299–307.

30 Ibid., and A. Buisine, "Les Laboratoires de chimie appliquée . . . Lille," *RIE*, 34 (1897), 11–15, and Henry Le Chatelier, "Mémoires," *Revue de métallurgie*, 5e année, 1908, p. 114.

31 Theodore Zeldin, *France 1848–1945* (Oxford, 1973–7); vol. 2, *Intellect, Taste and Anxiety*, pp. 291–302.

32 *ED*, 82 (1904), 210–11; 86 (1905), 169–70; 95 (1908), 160–1. For Toulouse, see *ED*, 108 (1914).

33 *RIE*, 37 (1899), 490–3.

34 *ED*, 108 (1914), 174.

35 AN, F^{17} 12983, 13663–4.

36 AN, F^{17} 13663.

37 See Maurice Daumas, ed., *Histoire générale des techniques* (Paris, 1978), vol. 4, *Les Techniques de la civilisation industrielle*. "Energie et matériaux," 337–66, and A. Monmerqué, "Les Distributions d'énergie électrique," *RS*, 63e année, (1925), 387–406.

38 "Expériences de transport entre Grenoble et Vizille" (*Rapport de la commission municipale, 27 septembre 1883, Comptes rendus*). See *Notices sur les travaux scientifiques de M. Marcel Deprez* (Paris, 1883).

39 L., "L'Enseignement de l'électricité en France et en Allemagne," *RGSPA*, 39 (1928), 594–5.

40 AN, F^{17} 13662–3.

41 Shinn, "The French Science Faculty System," pp. 310–15.

42 AN, F^{17} 13622–66.

43 Louis Barbillion, "L'Enseignement technique supérieur universitaire et le décret du 30 juillet 1920," *RS*, 61e année (1923), 193–97. To avoid any doubt on Barbillion's status, his titles are given: "Professeur à l'Université, directeur de l'Institut polytechnique de Grenoble."

44 G. Koenigs, "Ingénieurs-docteurs," *RS*, 61e année (1923), 701–4. Koenigs was professor at the faculty of sciences and at the Conservatoire nationale des arts et métiers.

45 Louis Barbillion, "Le Protectionnisme étranger et l'enseignement technique supérieur français," *RIE*, 84 (1930), 62–65.

46 Louis Barbillion, "Sur l'enseignement de la physique dans les écoles d'ingénieurs," *RIE*, 88 (1934), 229–34.

47 Ibid.

48 This latent conflict between physicists and mathematicians was defused by the mathematician Robert d'Adhémar, "Physiciens et mathématiciens," *RGSPA*, 32 (1921), 273–5, where he reviews the Preface to Bouasse's *Résistance des matériaux* dealing with "De l'utilité des mathématiques pour la formation de l'esprit."

49 Louis Barbillion, "Sur la formation mathématique de l'ingénieur," *RIE*, 90 (1936), 193–9.

50 Shinn, "The French Science Faculty System," pp. 37–29.

51 P. Genvresse, "La Chimie appliquée à l'Université de Besançon," *RIE*, 35 (1898), 32–34.

52 See reports on Paris and Lyon in *ED*, 33 (1887), and on Bordeaux and Marseille in *ED*, 30 (1888).

53 Decree of July 31, 1893, instituting "enseignement préparatoire au certificat d'études physiques, chimiques et naturelles" in the faculties of science. Admission was based on the *diplôme de bachelier*; anyone seventeen years old or less, if approved by the faculty and possessing a *brevet supérieur de l'enseignement primaire* or a *certificat d'études primaires supérieures* was also eligible. The PCN certificate was given after four trimesters; see RIE, 25 (1893), 161–9. For the impact of the PCN program on the faculties see ED, from 1893 to 1907, *passim*. See George Weisz, "Reform and Conflict in French Medical Education, 1870–1914," in Fox and Weisz, eds., *The Organization of Science and Technology*, pp. 82–4. In 1934 the PCN was replaced by the PCB, "physique, chimie, biologie animale et biologie végétale." Nature was subjected to biology.

54 *ED*, 82 (1904), 143–4.

55 Information on Clermont from reports in *ED*, 69 (1898), 80 (1903), 92 (1907), and 108 (1914).

56 *ED*, 104 (1912)

57 "Rapport de M. Petit-Dutaillis sur la situation de l'enseignement supérieur à Lille en 1904–05," *RIE*, 51 (1906), 238–49.

58 *ED*, 92 (1907), 190–1; 95 (1908), 165.

59 *ED*, 98 (1909), 258–9; 100 (1910), 297.

60 A. Léger, "Un Laboratoire de chimie industrielle à Lyon," *Lyon scientifique et industriel* (1882), pp. 361–6; Pierre Papon, *Le Pouvoir et la science en France* (Paris, 1978), pp. 32–3; and Paul M. Hohenberg, *Chemicals in Western Europe: 1850–1914* (Chicago, 1967), pp. 35–38.

61 Genvresse, "Besançon," pp. 33–4, and Léo Vignon, "Rapport," cited in *RIE*, 73 (1919), 66.

62 A. Cornu, "L'Ecole polytechnique: Le But de son enseignement: L'Esprit qui doit inspirer ses programmes," *RGSPA*, 7 (1896), 902.

63 Note by "La Direction" in Pierre Weiss, "Les Nouveaux laboratoires techniques de l'Ecole polytechnique de Zurich et ceux de nos facultés," *RGSPA*, 10 (1899), 58.

64 Henry Le Chatelier, *Science et industrie* (Paris, 1925), ch. 10. See also *ED*, 106 (1913), section on the Sorbonne.

65 "Rapport fait au nom de la Société de l'enseignement supérieur sur le développement à donner dans les universités à l'enseignement technique supérieur," by Le Chatelier (secretary), in *RIE*, 58 (1909), 45–54.

66 D. Gernez, "L'École de chimie industrielle annexée à la faculté des sciences de Lyon," *RIE*, 30 (1895), 39–42.

67 *ED*, 97 (1909), 255, and 101 (1911), 348.

68 *ED*, 106 (1913), section on Nancy.

69 Mary Jo Nye, "The Scientific Periphery in France: The Faculty of Sciences at Toulouse (1880–1930)," *Minerva*, 13 (1975), 374–403: and "Nonconformity and Creativity: A Study of Paul Sabatier, Chemical Theory, and the French Scientific Community," *Isis*, 68 (1977), 375–91.

70 Pierre Duhem, "Usines et laboratoires," *Revue philomathique de Bordeaux et du Sud-Ouest*, 9 (1899), 385–400.

71 Bichat, "Enseignement," p. 307, and Paul, *Sorcerer's Apprentice*, p. 63. On "the difference between science and technology," see Derek de Solla Price, *Science since Babylon*, 2nd ed. (New Haven, 1975), ch. 6. Price deliberately avoids the obfuscating term "applied science." I follow French usage of the period in using it throughout this chapter. Nearly all the problems Price analyzes were recognized by the French at the end of the nineteenth century. See also Jacques Ellul, *The Technological Society* (New York, 1964), pp. 7–11.

72 Paul Forman, John L. Heilbron, and Spencer Weart, "Physics *circa* 1900: Personnel, Funding, and Productivity of the Academic Establishments," *Historical Studies in the Physical Sciences*, 5 (1975), 1–185 (90–114), on "new plant." See Paul Appell, "L'Enseignement supérieur des sciences," in Alfred Croiset [et al.], *Enseignement et démocratie, leçons professées à l'Ecole des hautes études sociales* (Paris, 1905), p. 47, for complaints similar to Duhem's.

73 J. Gosselet, "L'Enseignement des sciences appliquées dans les universités," *RIE*, 37 (1899), 97–107.

74 Hanotaux, "Discours inaugural du premier Congrès de la Houille Blanche (1902)," cited in Louis Barbillion, "Instituts techniques universitaires et grandes écoles," *RIE*, 82 (1928).

75 See David Landes, "Religion and Enterprise: The Case of the French Textile Industry," and Maurice Lévy-Leboyer, "Innovation and Business Strategies in Nineteenth- and Twentieth-Century France" in Edward C. Carter, II, Robert Forster, and Joseph N. Moody, eds., *Enterprise and Entrepreneurs in Nineteenth- and Twentieth-Century France* (Baltimore, 1976), pp. 41–86 and 87–135. See also C. R. Day, "The Making of Mechanical Engineers in France: The Ecoles d'arts et métiers, 1803–1914," *French Historical Studies*, 10 (1978), 439–60, for a favorable view of the role of the *gadzarts* in French industry; Patrick Fridenson, "Une Industrie nouvelle: L'Automobile en France jusqu'en 1914," *Revue d'histoire moderne et contemporaine*, 19 (1972), 557–78; and James M. Laux, *In First Gear: The French Automobile Industry to 1914* (Montreal, 1976).

76 Terry Shinn, "Des corps de l'état au secteur industriel: Genèse de la profession d'ingénieur, 1750–1920," *Revue française de sociologie*, 19 (1978), 39–71.

77 Weisz, *The Emergence of Modern Universities*, pp. 218–22, and Paul, *Sorcerer's Apprentice*, pp. 20–8.

78 Albert Turpain, "Le Problème téléphonique actuel en France," *RGSPA*, 19 (1908), 978–82. For a favorable view of the methods and science of the Polytechnique, see Henry Le Chatelier, "L'Enseignement technique dans ses rapports avec l'enseignement universitaire," *Revue de métallurgie*, 5e année, (1908), 794–5.

79 Terry Shinn, "The Genesis of French Industrial Research, 1880–1940," *Social Science Information*, 19 (1980), 607–40.

80 J. Delsarte, "De l'enseignement supérieur en France et spécialement des facultés des sciences," *RS*, 77e année (1939), pp. 371–6.

81   G. Ribaud, "Sur l'organisation de la recherche technique," *RS*, 77e année (1939), pp. 371–6.

CHAPTER 5. SCIENCE IN AGRICULTURE

1   Eugen Weber, *Peasants into Frenchmen: The Modernization of Rural France, 1870–1914* (Stanford, Cal., 1976), p. 356.
2   *Comptes rendus des travaux du congrès international des directeurs des stations agronomiques* (2 vols., Paris, 1881, 1891). The *Annales de la science agronomique française et étrangère*, up to vol. 10 by 1891, were published under the patronage of the Ministry of Agriculture. The volumes contain many translations; the editorial secretary of the *Annales* was L. Grandeau's son, Henry, who was also assistant director of the Station agronomique de l'Est.
3   *Comptes rendus... congrès international*, vols. 1 and 2.
4   Jean Guichard, "L'Ecole nationale d'agriculture de Grignon," in *Centenaire de Grignon* (Paris, 1926), pp. 7–22.
5   George Wery, "L'Institut national agronomique," in *Centenaire de Grignon*, pp. 45–56. From 1848 to 1852 the head of the institute was the comte de Gasparin, "le célèbre agronome"; from 1876 to 1902 the heads were Eugène Tisserand and Eugène Risler. In 1902 Paul Regnard, who had worked with Paul Bert for eighteen years, succeeded Risler. See *Le Cinquantenaire de l'Institut national agronomique* (Paris, 1929), and *L'Institut agronomique et son enseignement, 1876–1926* (Paris, 1926).
6   "Un Siècle d'enseignement agronomique au Conservatoire national des arts et métiers," in *Centenaire de Grignon*, pp. 67–80.
7   M. Le Rouzic and M. L. Esnault, "L'Ecole de Grand-Jouan-Rennes," in *Centenaire de Grignon*, pp. 23–44.
8   M. F. Duchein, "L'Ecole nationale d'agriculture de Montpellier," in *Centenaire de Grignon*, pp. 57–62. See bibliography in George Ordish, *The Great Wine Blight* (London, 1972), pp. 221–33; the charming monument to Foëx is number 11 of the illustrations after p. 98, and a German contrast is given. On phylloxera and the new wine industry, see Leo A. Loubère, *The Red and the White: A History of Wine in France and Italy in the Nineteenth Century* (Albany, N.Y., 1978), chs. 8 and 9. The *Revue de viticulture* started publication in 1894 under the editorship of P. Viala, head of the Laboratoire de recherches viticoles of the Ecole nationale d'agriculture in Montpellier and then professor of viticulture at the Institut national agronomique in Paris, and L. Ravaz, head of the Station viticole de Cognac. With its office in Paris, the journal's collaborators included Armand Gautier (chemistry, Paris faculty of medicine), Barbier (chemistry, Lyon), Flahaut (botany, Montpellier), Gayon, Henneguy, Marion, Raulin (head of the Station agronomique de Lyon, industrial chemistry), and Paul Sabatier. *Revue de viticulture* ("Organe de l'agriculture des régions viticoles"), 1 (1894).
9   AN, F^{17} 17071 (Comité des inspecteurs généraux de l'enseignement supérieur, Jan. 30, 1861).

10 M. Nicolas, "Les Instituts agricoles des facultés," in *Centenaire de Grignon*, pp. 199–212.

11 Grandeau, "Sur un nouveau procédé de titrage des liqueurs aidimétriques ou alcalimétriques," cited in L. Blaringhem, "Louis Grandeau (1834–1911) et les progrès de l'agriculture française à la fin du XIXe siècle," *RS*, 5e série, 51e année (1913), 523–8. Also AN, F¹⁷ 13177 (Nancy).

12 Boussingault's *Traité d'économie rurale* (2 vols., Paris, 1844) was rewritten as *Agronomie, chimie agricole et physiologie*, whose volumes began appearing in 1860; a third edition of eight volumes appeared between 1886 and 1891.

13 Richard P. Aulie, "Jean-Baptiste-Joseph-Dieudonné Boussingault," *DSB*, II, 356–7, and the article in *La Grande encyclopédie*, vii, 839–40.

14 Blaringhem, "Louis Grandeau," p. 525.

15 *La Fumure des champs et des jardins: Instruction pratique sur l'emploi des engrais commerciaux: Nitrates, phosphates, sels potassiques* (Paris, 1893; 6th ed., 1897).

16 See also L. Grandeau and Auguste Laugel, *Revue des sciences et de l'industrie pour la France et l'étranger* (2 vols., Paris, 1863). Henry Grandeau, son of Louis, translated agronomic works by P. E. Müller, by Julius Wolf, and by Shinkizi Nagai (on agriculture in Japan).

17 See Blaringhem, "Louis Grandeau," pp. 526–7. For a list of the publications on the *Expositions universelles*, see the catalog of the Bibliothèque nationale under Grandeau, Louis-Nicolas, pp. 427–42, especially pp. 431 and 442.

18 Felicitous phrases coined by Blaringhem, "Louis Grandeau," p. 526.

19 Figures are from Maurice Agulhon, Gabriel Désert, and Robert Specklin, *Apogée et crise de la civilisation paysanne* (Paris, 1976), vol. 3 of *Histoire de la France rurale*, sous la direction de Georges Duby et Armand Wallon.

20 Ruth Schwartz Cowan, "Gaston Bonnier" (1853–1922), *DSB*, II, 290–91.

21 Léon Dufour (directeur-adjoint), "Le Laboratoire de biologie végétale de Fontainebleau," *RIE*, 36 (1898), 127–30.

22 Bonnier's definition was simple enough. "La Biologie, c'est l'étude de la vie; la Biologie végétale, c'est l'étude de la vie des végétaux."

23 "Inauguration de la station de biologie végétale de Mauroc," *RS*, 5e série, 50e année (1912), 137–45.

24 S.B., "Grandeur et déchéance de la botanique," *RGSPA*, 46 (1935), 229–30.

25 Seymour H. Mauskopf, "Alexandre-Edouard Baudrimont" (1806–80), *DSB*, I, 517–19; and J. Ribéreau-Gayon (directeur de la Station agronomique et oenologique; directeur honoraire de l'Institut d'oenologie), *Les Sciences de la vigne et du vin à l'Université de Bordeaux* (Bordeaux, 1965).

26 U. Gayon, "L'Enseignement de la chimie appliquée à la faculté des sciences de Bordeaux," *RIE*, 36 (1898), 397–400.

27 Roesler edited the *Mittheilungen der k.k. chemischphysiologischen Versuchsstation für Wein-und Obsbau in Klosterneuburg bei Wien*.

28 See, e.g., J. Ribéreau-Gayon, *Contribution à l'étude des oxydations et réductions dans les vins* (thèse, Doctorat ès sciences physiques, Bordeaux, 1931); J.

Ribéreau et E. Peynaud, *Analyse et contrôle des vins* (1947; 2nd ed., 1958), and *Traité d'oenologie* (2 vols., 1960–1); E. Peynaud, *Contribution à l'étude biochimique de la maturation du raisin et de la composition du vin* (thèse d'Ingénieur-Docteur, Bordeaux, 1946). Professor Peynaud has acquired a popular following these days with his excellent coffee-table book, *Le Goût du vin ou le livre de la dégustation* (Paris, 1980); see also his *Connaissance et travail du vin* (new ed., Paris, 1981).

29 Louis Pasteur, *Etudes sur le vin* (Paris, 1886; second edition, 1875).

30 *Revue de viticulture*, 2 (1894), 136–40, review of U. Gayon, *Expériences sur la pasteurisation des vins de la Gironde*. Pasteurization is still one of the "manipulations permises" by the law. So is the standard practice of adding sulphur dioxide to the wine, as long as the product sold does not contain more than 350 mg (400 mg for naturally sweet wine) per liter. See Bernard Blanchet, *Code du vin* (Montpellier, 1970). On the male and female characteristics of wine, see Raymond Dumay, *Guide du vin* (Paris, 1967, p. 171: "Le Bourgogne, c'est *lui*; le Bordeaux, c'est *elle*, disait Charles Monselet. J'ajouterai que le Bordeaux, c'est aussi, souvent, *elle* et *lui*. Féminin par sa pudeur, sa réserve, sa suavité, ce grand vin est masculin par son corps, sa charpente et son étonnante santé." Bordeaux is agreeably androgynous. The great debate over the poor quality of burgundies may be followed in *Decanter* during August–October 1982; the excitement resulted from the publication of the highly critical book *Burgundy*, by Anthony Hanson. A most judicious commentary was made by Louis Latour in the issue for October, pp. 44–45.

31 J. Ribéreau-Gayon (professeur à la Faculté des sciences de Bordeaux, directeur de la Station agronomique et oenologique), *L'Oenologie d'hier et d'aujourd'hui* (Bordeaux, 1954).

32 See, for example, the manual by U. Gayon (directeur de la Station agronomique et oenologique de Bordeaux), *Vins: Vins ordinaires, vins mousseux, vins liquoreux et vins de liqueur* (Paris and Liège, 1912).

33 Georges Dupont, "La Fôret landaise . . . ," *Revue philomathique de Bordeaux et du Sud-Ouest*, 25e année (1922), 165–76.

34 Lovers of the detective story will remember one of the less tedious works of Freeman Wills Crofts, *The Pit-Prop Syndicate* (1922; New York, 1965, 1978).

35 Maurice Vèzes, "Une Fondation régionale: Le Laboratoire des résines à l'Université de Bordeaux," *RIE*, 41 (1901), 118–26.

36 The Laboratoire de physiologie végétale was soon following Brunel's advice. See Dr. G. Boyer, "Recherches sur les conditions de formation et de développement de la truffe mélanospore ou truffe de Périgord," *Procès verbaux de la Société des sciences physiques et naturelles de Bordeaux* (Dec. 19, 1907). Such studies had more than a regional interest; see Boyer, "Etudes sur la biologie de la truffe mélanospore," *Comptes rendus de l'Académie des sciences* (May 17, 1910), and Fernand Guéguerin, "La Truffe et le reboisement," *RS*, 49e année (1911).

37 In 1924, when the laboratory metamorphosized into an institute, a monthly, twenty-page *Bulletin de l'Institut du pin* was founded.

38 "Une Fondation régionale: L'Institut du pin de l'Université de Bordeaux," *RIE*, 88 (1934).

39 Georges Dupont, *La Forêt landaise vers l'exploitation méthodique de cette richesse nationale: Rôle de l'Institut du pin* (Bordeaux, 1923). First published in the *Revue philomathique de Bordeaux et du sud-ouest*, 1922–3. See M. Vèzes et G. Dupont, *Résines et térébenthines: Les Industries dérivées* (Paris, 1925), for all that you would ever want to know about resins. Discovery of synthetic resins has reduced the industrial use of the pine.

40 AN, F^{17} 13555 (Collège de France).

41 Berthelot, "Le Rôle de la science en agriculture," *RS*, 50 (1892), 65–7; P.-P. Dehérain, "Le Fumier de ferme," *RS*, 52 (1893), 289–95; see also E. Duclaux, "Le Rôle agricole des microbes," *RS*, 52 (1893), 834–8. The Muséum, in an experimental mood, gave Dehérain a chair of plant physiology in 1879. Duclaux went from Clermont to Lyon to the Institut agronomique, then to the Sorbonne as professor of biological chemistry. One of Pasteur's favorite students at the Normale, Duclaux became head of the Institut Pasteur in 1895. A Dreyfusard, Duclaux was one of the founders of the Ligue des droits de l'homme.

42 P. Dehérain, "Paul Thenard," *RS*, 34 (1884), 613–18. Paul was the son of Baron-Louis-Jacques Thenard, issued from poor peasants in the Aube; the father's *Traité de chimie élémentaire* (1813–16) gave four volumes of heavy enlightenment to a generation of students and teachers.

43 L. Blaringhem, *Action des traumatismes sur la variation et l'hérédité*, in *Bulletin scientifique de la France et de la Belgique*, 47 (1907). A monograph of 248 pages with plates. Blaringhem's later claim to have obtained new characteristics in maize was dismissed by Paul Becquerel, who saw them as atavistic or teratological characteristics long described by botanists in treatises on plant teratology. Becquerel's similar work on *Zinnia eleganas* also had no relevance for mutation theory but was interesting for the doctrine of evolution.

44 Gordon Wright, *Rural Revolution in France: The Peasantry in the Twentieth Century* (Stanford, 1964), pp. 12–34, 146.

45 Michel Boulet, "La Création de l'enseignement agricole en France (1848–1880)," a paper given at the annual colloquium on the history of education, Paris, 1981. I thank Linda Clark for bringing this paper to my attention. Boulet has written an analysis of the *Evolution de l'enseignement agricole (1789–1978)*, 2 volumes multigraphiés, Dijon, ENSSAA, 1979.

46 Roger Price, *An Economic History of Modern France, 1730–1914* (New York, 1981), pp. 71–84.

47 Emile Barbet (president of the Société des ingénieurs civils de France), "Les Grandes industries agricoles de l'alimentation en France," *La Technique moderne*, 1 (1908), 103–5.

48 It is odd that Barbet did not hold up as a model the industrial viticulture of the

Languedoc and the wine factory of the Domaine de Jouarres in the Aude. See Loubère, *The Red and the White*, pp. 191–206.

49 See Wright, *Rural Revolution in France*; Eugen Weber, "Comment la politique vint aux paysans: A Second Look at Peasant Politicization," *The American Historical Review*, 87 (1982); Price, *An Economic History of Modern France*, pp. 82, 228; and George W. Grantham, "The Persistence of Open-Field Farming in Nineteenth-Century France," *The Journal of Economic History*, 40 (1980), 515–31.

50 For French public health in early nineteenth-century France, see William Coleman, *Death Is a Social Disease: Public Health and Political Economy in Early Industrial France* (Madison, Wis., 1982). See also Dr. J.-P. Langlois, "Revue annuelle d'hygiène," *RGSPA*, 17 (1906), 829–30, and Gaston Trélat, "Santé publique et Paris de demain," *RGSPA*, 18 (1907), 267–76.

51 Henry Huet-Desaunay (avocat à la Cour d'appel de Paris), *Le Laboratoire municipal et les falsifications ou recueil des lois et circulaires concernant la vente des produits alimentaires et hygiène publique* (Paris, 1890).

52 P. Cazeneuve, *Les Colorants de la houille au point de vue toxiologique et hygiénique: Affaire de la succursale de la B. Anilin et Soda, Fabrik à Neuville-sur-Saône* (Lyon, 1887). Cazeneuve was professor of chemistry and toxicology in the faculty of medicine in Lyon, and Arloing was head of the veterinary school in Lyon.

53 Georges Bergeron, *Mémoire sur la fuchsine* (Toulouse, 1876), *Nouvelles recherches sur la fuchsine pure* (Rouen, 1876), and, with J. Cloüet, *Note sur l'innocuité absolue des mélanges colorants à base de fuchsine pure* (Rouen, 1876).

54 V. Feltz and E. Ritter, *Etude expérimentale de l'action de la fuchsine sur l'organisme* (Paris, 1877). Ritter wrote the much republished work *Des vins colorés par la fuchsine et des moyens employés pour les reconnaître*. Paul Cazeneuve, *La fuchsine au point de vue de la toxicologie et de l'hygiène*, reprint from *Lyon médical*, 1892. S. Arloing et P. Cazeneuve, *Sur les effets physiologiques de deux colorants rouges azoïques très employés pour colorer les substances alimentaires*, reprint from the *Archives de physiologie*, 1887. Georges Denigès, "Les Empoisonnements par l'arsenic," *RS*, 47e année (1909), 490–4. A report by G. Pouchet to the Académie de médecine, "Coloration artificielle des matières alimentaires," *RS*, 50e année (1912), 242–3, which dealt with the use of vegetable dyes. In 1913 a volume of the *Encyclopédie internationale d'assistance, prévoyance, hygiène et démographie* sounded the alarm: Paul Hubault, *Les Coulisses de la fraude: Comment on nous empoisonne* (Paris, 1913).

55 *Documents sur les falsifications des matières alimentaires et sur les travaux du laboratoire municipal: Rapport à Monsieur le préfet de police: Deuxième rapport* (Paris, 1885).

56 H. Pellet, *Examen de l'ouvrage de M. Charles Girard . . .* (Paris, 1883).

57 *Documents sur les falsifications . . . Deuxième rapport.*

58 J. Bruhat, *Le Laboratoire municipal de Paris . . .* (Paris, 1884).

59 F. Jarlaud, *Rapport sur le fonctionnement du laboratoire municipal présenté à la chambre de commerce de Paris le 21 février 1883 au nom de la commission des douanes, entrepôts et marchés* (Paris, 1883).

60 E. Jeannon (pharmacien de lre classe, etc.), *Conférence sur le laboratoire municipal de Paris, son fonctionnement et sa réorganisation* (Paris, 1890).

61 J. Bruhat, *La Loi Griffe et le laboratoire municipal de Paris: Lettres d'un chimiste (Jean de Metz)* (Paris, 1887).

62 Jarlaud, *Rapport sur le ... laboratoire municipal.* The "method of Paris" spread: J. Charles, "Les Vins de l'Allier," *Revue scientifique du Bourbonnais et du Centre de la France*, 2 (1889), 3–10, used the method of the Paris laboratory in analyzing the "vins de seconde cuvée" or "vins de sucre," made in great quantity in the Allier.

63 Chambre syndicale du commerce en gros des vins et spiriteux de Paris et du département de la Seine: *Rapport sur le fonctionnement du laboratoire municipal* (Paris, 1889).

64 *RGSPA*, 12 (1901), 113.

65 *RGSPA*, 17 (1906), 583–5.

66 Ville de Paris, Préfecture de police, *Le Laboratoire municipal de Paris: Son organisation: Son fonctionnement: Ses opérations* (Melun, 1915). Kling had been Chef adjoint de travaux à l'Ecole de physique et de chimie industrielles de la ville de Paris. See his study of "La Tautomérie," *RGSPA*, 18 (1907), 283–92, 311–18.

67 Ernst W. Stieb, *Drug Adulteration: Detection and Control in Nineteenth-Century Britain* (Madison, Wis., 1966). The *Revue internationale des falsifications* (organe officiel de la Commission internationale pour la répression des falsifications des denrées alimentaires) began in 1887 in Amsterdam with a roster of international collaborators, including Berthelot, Brouardel, Girard, Houzeau, Lajoux, Pabst, and G. Pouchet for France.

CHAPTER 6. SCIENCE IN THE CATHOLIC UNIVERSITIES

1 "Loi relative à la liberté de l'enseignement supérieur," *Journal officiel de la République française*, July 27, 185, p. 5921. The Archives of the *Archevêché de Paris*, série 5J, have some interesting material on the foundation of the university in Paris, but most of the archival material was transferred from the *Archevêché* to the *Institut catholique* in January, 1970. On the founding of the universities see Jacques Gadille, *La Pensée et l'action politique des évêques au début de la IIIe République, 1887–1883* (Paris, 1967), vol. 2, pp. 25–31.

2 Msgr. d'Hulst, "Discours de rentrée," Jan. 26, 1881, in *Mélanges oratoires* (Paris, 1901), II, 221 and 231.

3 An iconological note on part of the seal of the institute in Paris: "Le Bienheureux Albert le Grand, le patron de la Faculté des Sciences, montre du doigt l'aigle, l'un des sujets où se montre le mieux l'indépendance de son esprit dans ses études sur l'histoire naturelle."

4 *Rapport présenté à son Eminence le cardinal archevêque de Paris par M. l'abbé en la séance littéraire du 15 juillet 1857* (Ecole ecclésiastique des hautes études), p. 7, n. 1.

5   *Lettre pastorale* ... , Sept. 8, 1875. Archives, *Archevêché*, 5J, 1. On competition, the sociologist Joseph Ben-David agrees with the bishops: "Scientific Productivity and Academic Organization in Nineteenth-Century Medicine," *American Sociological Review*, 25 (1960), 828–43.

6   *Livret des facultés catholiques de Lille* (1892), 22–3. AN, F^{19} 4091.

7   Institut catholique, Lyon, *Procès-verbaux de la Commission exécutive de Fondation*, 1875–6.

8   *Bulletin de l'Université catholique de Lyon* (1879–80), 23.

9   See Paul Gerbod, *La Condition universitaire en France au XIXe siècle* (Paris, 1965), especially, 608–13. On anticlericalism, see Joseph N. Moody, *The Church as Enemy: Anticlericalism in Nineteenth-Century French Thought* (Washington, D.C., 1968), and "French Anticlericalism: Image and Reality," *Catholic Historical Review*, 56 (No. 4, 1971), 630–48.

10 See Harry W. Paul, "The Debate over the Bankruptcy of Science in 1895," *French Historical Studies*, 5 (1968), 299–327, and "The Crucifix and the Crucible: Catholic Scientists in the Third Republic," *Catholic Historical Review*, 58 (1972), 195–219.

11 *Journal officiel: Débats: Chambre.* Séance du 9 novembre 1891, p. 2099.

12 *Institut catholique de Paris: Cinquante ans d'enseignement libre: Mémorial, 1875–1925.* (Paris, 1925), pp. 3–4. See pp. 30–1 for a quotation by the Bishop of Châlons from *le père* Captier's plea, made before the founding of the universities, for an institutionalization of the defense against "une science incrédule." Captier effectively stated what would be the position of the Catholic intelligentsia on the Catholic universities, especially in view of the challenge of an impious science. "Aujourd'hui, en face d'une science incrédule qui n'accepte aucune direction, que doit-on faire, sinon s'emparer de la science par la science, en appeler de la nature mal étudiée? La science de la nature et la critique ne remplaceront jamais nos dogmes et ne feront rien directement pour le salut des âmes; mais elles rendront hommage à l'harmonie méconnue des oeuvres de Dieu ... Voici donc notre oeuvre, Messieurs; suivons la science partout où elle a pénétré: allons dans les laboratoires, dans les bibliothèques; observons les entrailles de la terre et tous les êtres qui la couvrent de leur vie ou de leurs ruines; examinons tout ce que nos adversaires examinent. Arrivés là, reprenons une à une les observations et les expériences, vérifions les calculs; complétons les découvertes; soyons surtout d'une franchise impitoyable: ne tolérons aucune feinte; n'hésitons pas à démasquer un seul mensonge, ni à soulever les doubles fonds des escamoteurs de la science; nous étonnerons les incrédules et nous les gagnerons peut-être en allant jusqu'au bout de la vérité avec cette énergique persévérance que donne l'esprit de la foi."

13 M. d'Hulst, *La mission chrétienne de la science* (Conférence faite à Evreux le 27 décembre 1883). This speech was given at a meeting of the *Union catholique de la Seine-Inférieure*, presided over by Paul Allard, a former magistrate.

14 M. d'Hulst, *Le Vrai terrain de la lutte entre croyants et incroyants* (Discours prononcé à la séance annuelle de l'Institut catholique de Paris, le 30 janvier 1884).

15 M. d'Hulst, *Le Rôle scientifique des facultés catholiques* (Paris, 1883).
16 M. d'Hulst, *L'Empoisonnement de la science*, pp. 186–7.
17 See circular letter of "les évêques protecteurs des facultés catholiques de l'ouest," *L'Univers* (22 mars 1896) and a pastoral letter of the Bishop of Angers (Freppel) in 1890. Both letters are in the *Archives nationales*, F^{19} 4091. The circular letter of the bishops also contained a good deal of fantasy and antirevolutionary conservatism. "Depuis un siècle, l'Eglise de France répare peu à peu les ruines faites par la Révolution. [The latest progress was the development of the Catholic universities.] Quelles pages glorieuses dans les annales de l'Eglise de France, que cette résurrection de son enseignement supérieur, il y a vingt ans!" Indulgences and special advantages were given by Pius IX and Leo XIII to the benefactors of the Institute at Lille. See *Bulletin de l'oeuvre des Facultés catholiques de Lille*, 10 (Aug. 1889, no. 119).
18 *Bulletin de l'université catholique de Lyon* (1879, no. 3).
19 Archives, Institut catholique, Lyon *Procès-verbaux... du Conseil Rectoral*, "Statuts provisoires."
20 Gustave Flaubert, *Madame Bovary: Moeurs de Province* (Paris, édition définitive, n.d.), pp. 84–85. See also the "Réquisitoire, plaidoirie et jugement du procès intenté à l'auteur devant le tribunal correctionnel de Paris" in 1857, pp. 389–461.
21 L. Baunard, *Rapport à l'assemblée annuelle de la Société générale d'éducation et d'enseignement* (May 13, 1892) and an appeal by Baunard printed in *L'Univers* (July 5, 1890). Both items in AN, F^{19} 4901. See also *Bulletin de l'université catholique de Lyon* (1879, no. 3), for similar arguments.
22 *La Vérité* (May 19, 1898) and *Livret des Facultés catholiques de Lille*, 24. AN, F^{19} 4091; AN, F^{17} 13396; *Revue de l'Institut catholique de Paris* (1906–7, nos. 11–12), 375.
23 Dr. Le Bec, "La Fondation d'une école de médecine à Paris," *Bulletin de l'Institut catholique de Paris*, 1891–2, pp. 171–6.
24 Robert Gilpin, *France in the Age of the Scientific State* (Princeton, N.J., 1968), especially pp. 111–2.
25 A. de Lapparent, "L'Institut catholique et l'enseignement de la géologie," *Revue de l'Institut catholique de Paris* (1906, no. 1), 1–21, and *Autobiographie manuscrite*, in the possession of the Abbé de Lapparent, professor of geology at the *Institut catholique* in Paris. Albert de Lapparent stipulated that it should never be published. He has a dossier in the Archives of the *Académie des Sciences*, and in AN, F^{14} 27307.
26 A large number of quasi-hagiographical French works exist on Branly. See, e.g., Jeanne Terrat-Branly, *Mon père Edouard Branly* (Paris, 1942) and G. Pelletier et J. Quinet, *Edouard Branly* (Paris, 1962). He has a dossier in the Archives of the Académie des sciences, as well as in AN, F^{17} 22769 and 40056. Freppel, Bishop of Angers, wrote the Archbishop of Paris in 1877 asking that Paris share Branly with the University at Angers for a year while their man got the doctorate required to satisfy state law. Paris and perhaps Branly refused to consider the idea seriously. Archives, Archevêché, Paris, 5j, 1.

27 The announcement of courses for 1888–9 lists the abbé d'Esclaibes for calculus and the abbé Fouet as *suppléant*, as well as the abbé Godefroy for chemistry. AN, F^{17} 13140. In 1903–4 three of the nine listed for the sciences are abbés: Hamonet, Nau (special mathematics), and Fouet. AN, F^{19} 4091. Information on Lemoine from AN, F^{14} 11575. Alfred Cardinal Baudrillart, *Vie de Mgr. Hulst* (2 vols., 2nd ed., Paris, 1912), vol. 1, pp. 344 ff., has some interesting details concerning the recruiting of Lapparent, Branly, and Lemoine. Abbé Paul de Foville, a graduate of the Polytechnique, played an important role in contacting his "anciens camarades." Advice was given by some of the great French scientists of the day. Naturally, those who gave advice, like Hermite and Puiseux, were Catholic.

28 *Bulletin de l'Institut catholique de Paris* (1892), ibid. (1894), 85–6. Rousselot later became a professor at the Collège de France. See *Les Modifications phonétiques du language étudiées dans le patois d'une famille de Cellefrouin* (thèse, Charente, 1891).

29 See A.-L. Donnadieu, *Université catholique de Lyon: Organisation du service de la zoologie à la Faculté des sciences* (Paris, 1879); extrait du *Contemporain* (March 1, 1879) and letter dated 26.V.91, from Donnadieu to the rector, in the Archives of the Institut catholique at Lyon.

30 AN, F^{19} 4091. Amédée de Margerie, professor of philosophy and dean of the faculty of letters at Lille, had also been a professor in the state faculty of letters at Nancy. Charles-Ernest Schmitt, listed for general chemistry at Lille, had been a *chargé de cours* at the Ecole supérieure de pharmacie in Strasbourg and in Nancy.

31 *Gazette des Tribunaux* (March 5 and 6, 1888), AN, F^{19} 4091.

32 AN, F^{17} 22715 and Abbé Girard, *Émile-Hilaire Amagat* (Sancerre, 1941).

33 Archives, Institut catholique, Lyon, *Procès-verbaux des différentes séances épiscopales et séances de la commission exécutive* (July 15, 1877).

34 Archives, Université catholique de l'Ouest, *Comité central, procès-verbal, 1892–1927*, and Archives départementales, Maine-et-Loire, 7T2. See A. de la Villerabel, "L'Ecole supérieure d'agriculture de l'Université catholique d'Angers," *Etudes*, 173 (1922), 641–61.

35 "Dossier des autographes" from Béchamp to Dumas, Archives, Académie des Sciences; and AN, F^{17} 20121.

36 The Rector's highest praise: "C'est de nos chimistes celui qui est le plus apte à donner à la science de la popularité en l'appliquant à l'industrie." AN, F^{17} 20121.

37 Ibid.

38 Archives, Académie des sciences. Letter is not dated, but it was probably written early in 1880. Dumas gave the reply to Taine's discourse when the latter was received into the French Academy on January 15, 1880. See J.-B. Dumas, *Discours et éloges académiques* (Paris, 1885), vol. II, pp. 115–150, "Réponse au discours prononcé par M. Taine pour sa réception à l'Académie française."

39 Béchamp to Dumas, Archives, Académie des sciences: "l'esprit français, naturellement . . . droit."

40 Cited in le Docteur P. David, "La Fondation de la faculté de médecine et de pharmacie de Lille," *Cahiers Féron-Vrau*, no. 1 (1954–5), p. 43.

41 He was a leader in the paternalistic Association catholique des Patrons du Nord. In 1901 Leo XIII turned *La Croix* over to him when the Assumptionist order was dissolved. See Msgr. Louis Baunard, *Les Deux frères: Cinquante années de l'action catholique à Lille: Philibert Vrau: Camille Féron-Vrau* (1829–1908) (Paris, 1910).

42 The lawyer for the institute described the "caractère difficile et ombrageux" of the ex-dean. See *Gazette des Tribunaux*, March 5 and 6, 1888; AN, F^{19} 4091.

43 Béchamp to Dumas, Archives, Académie des sciences. On Béchamp see Fr. Guermonprez, *Béchamp: Etudes et souvenirs* (Paris, 1927).

44 "Son oeuvre . . . étendue et fort diverse . . . incontestablement celle d'un botaniste . . . prend rang parmi les plus célèbres phytophysiologistes des temps modernes." René Souèges, *Notice sur la vie et les travaux de Henri Colin 1880–1943* (Institut de France – Académie des sciences – Séance du 27 décembre 1944). He had done research under the botanists Gaston Bonnier and Martin Molliard. Three times a laureate of the Institut de France, in 1912, 1922, and 1924 (Montagne, Lonchampt, and Vaillant prizes), he was a member of the Conseil de l'association des chimistes de sucrerie et de distillerie and a member of the Institut belge de la betterave.

45 The Comité de haut patronage of the Ecole des hautes études agricoles (Lille) included MM. le marquis de Dampierre, président de la Société des Agriculteurs de France, ancien député, président, le baron d'Aubigny, le comte de la Bouillerie, ancien ministre, président de l'Oeuvre des cercles catholiques d'ouvriers, le comte de Clésieux, le frère Eugène-Marie, directeur de l'Institut agricole de Beauvais, Fiévet, de Masny, ancien sénateur du Nord, le marquis de Gouvello, président de l'Oeuvre des orphelinats, ancien député, le comte de Luçay, membre de la Société d'agriculture, et secrétaire général adjoint de la Société des agriculteurs de France, le comte de Montalembert, député du Nord, le comte Yvert, secrétaire général de l'Union des oeuvres ouvrières, fondateur de l'Oeuvre des propriétaires chrétiens.

The Conseil de perfectionnement included MM. le Frère Antonis, directeur de l'enseignement agricole à l'Institut de Beauvais, Hellin, président du Comice agricole de Lille, le baron de Saint-Paul, T'serstevens, vice-président de la Société centrale d'agriculture de Belgique, membre fondateur de l'Institut agronomique de Louvain, Henri Lévêque de Vilmorin, membre de la Société nationale d'agriculture, et du Conseil de la Société des agriculteurs de France.

46 The Conseil de perfectionnement of the Ecole des hautes études industrielles included Mgr. Baunard, recteur des Facultés catholiques, président, MM. Descottes, inspecteur général des Mines, Dutilleul, Industriel à Armentières (Nord), Féron-Vrau, Industriel à Lille, Léon Harmel, Industriel au Val-des-Bois (Marne), Ch. Périn, membre correspondant de l'Institut, L. Cordonnier,

industriel à Roubaix (Nord), A. Prouvost, G. Motte, Ch. Flipo, industriel à Tourcoing, J. Motte-Barnard.

La Société anonyme de l'Université catholique de Lille included Henri Bernard, négociant, Lille, Paul Bernard, propriétaire, Lille, Louis-Charles, Anatole, comte de Caulaincourt, propriétaire, Lille, Louis-Jules-Elysée Cavrois, ancien auditeur au Conseil d'état, propriétaire, Arras, Louis Delcourt, filateur, Lille, Camille Féron-Vrau, filateur, Lille, Alexandre Jonglez de Ligne, ancien auditeur au Conseil d'état, propriétaire, Lille, Comte de Nicolaï, propriétaire, Boulogne-sur-mer, Jean-Baptiste-Auguste-Eustache-Marie Scalbert, banquier, Lille, Gustave Théry, avocat, Lille, Charles Verley, banquier, Lille, Philibert-Louis-Jules Vrau, filateur, Lille. See the *Livret des Facultés catholiques de Lille* (1892). In 1889–90 Angers added a course in Histoire des arts industriels, which for that year dealt with the organization of apprenticeship, and in 1894–5 added a course in Sciences agricoles dealing with "organisation de l'exploitation rurale; conditions diverses qui la déterminent, étude de diverses exploitations," taught by M. Nicolle, "ancien élève de l'Ecole polytechnique, directeur du Syndicat agricole." See Paul Fristot, "L'Enseignement supérieur des jeunes patrons de la grande industrie," *Etudes*, 97 (1903), 241–51.

In 1898 an Institut catholique d'arts et métiers began at Lille with 14 students and grew to 1,200 by 1923. Philibert Vrau and the Cercles catholiques d'ouvriers played important roles in its planning from 1876. The state had only three Arts et métiers with 300 openings in 1898. See J. Forbes, "Une Ecole catholique d'arts et métiers," *Etudes*, 73 (1897), 778–87, and Joseph Berteloot, "Instituts professionnels, collèges libres et collèges indépendants," *Etudes*, 160 (1919), 324–40, and "L'Institut catholique d'arts et métiers de Lille: A propos d'un anniversaire (1898–1923)," *Etudes*, 174 (1923), 666–83.

47 M. d'Hulst, *Le Rôle scientifique des facultés catholiques* (Paris, 1883), pp. 1–16; Léon Harmel, *Catéchisme du Patron* (Paris, 1881), pp. 163–75.
48 *Livret des facultés catholiques de Lille* (1892), pp. 1–31, 127–39.
49 Ibid., pp. 127–39.
50 Ibid.
51 Nathan Rosenberg, *Technology and American Economic Growth* (New York, 1972).
52 Information from *Le Figaro* (April 12, 1895), *L'Univers* (January 29, 1900), and E. Cantineau, *L'Université de Lille en 1896: Compte rendu fait par Cantineau* (Lille, 1897). All in AN, F^{19} 4091. See *Les Universités catholiques* (Lille, 1944). Lavisse's comments are in "La Question des Universités françaises," *RIE*, 32 (1886), 487–93. The *Livret des Facultés catholiques de Lille* proudly cited this secular republican recognition: "... L'un des maîtres les plus estimés de l'Université officielle ... Lavisse ... a vu Lille; ... 'cette université catholique est puissante et redoutable. Elle n'est pas seulement installée avec luxe et pourvue de tous les instruments d'enseignement et de travail; mais elle s'est incorporée à la Flandre, elle s'y est incarnée ... Je ne blâme, ni ne récrimine, je suis bien plus près d'admirer.'"

53 Archives, Institut catholique, Lyon, *Procès-verbaux . . . du Conseil Rectoral.*
54 The Ecole préparatoire Sainte-Geneviève (Normale [sciences], Polytechnique, les Mines, Saint-Cyr, Centrale, les Ponts, Institut agronomique, etc.) proudly advertised its great success in getting its students into the key institutions of science and technology. "Compte rendu annuel (juin 1908): 17 ont été admis à la Polytechnique, dans une promotion de 170 pour toute la France; – 6 à l'Ecole supérieure des Mines, dans une promotion de 30; – de plus, 5 qui y avaient été également admis, sont entrés à l'Ecole des Mines de Saint-Etienne; – 2 à l'Ecole nationale des Ponts et Chaussées, dans une promotion de 17; 1 à l'Ecole du Génie maritime sur 3 qu'on a pris; – 1 à l'Ecole des Mines de Saint-Etienne; – 35 à l'Ecole spéciale militaire de Saint-Cyr, dans une promotion de 225; – 22 à l'Ecole centrale des Arts et Manufactures sur une liste de 238; – 3 à l'Institut electrotechnique de Grenoble; – 2 au Polytechnicum de Zurich; – 1 à l'Ecole de physique et chimie industrielles; – 1 à l'Institut national agronomique; – 1 à l'Ecole centrale lyonnaise; – 3 sont entrés de plein saut dans les affaires; – 88 des plus jeunes ont été reçus aux divers baccalauréats en Sorbonne." *Etudes,* 116 (July–Sept. 1908), 864. The Jesuits were out of the school after 1901; and it was finally confiscated by the state in 1913. For a history of the school and its closing, see ibid., 136 (Aug. 5, 1913), 386–407.
55 Archives, Archevêché, 5 J, 1; Institut catholique de Paris, Ecole des sciences, *Procès-verbaux,* Jan. 6, 1898. The budget of the Institut catholique de Paris for 1889–90 was as follows:

Physics	20,100 francs
Chemistry	13,300
Geology	8,100
Garden	1,800
Math.	21,300
Letters	28,950
Law	137,000
Total	271,550

Additional expenses expanded the budget to 376,350 francs. The state budget grew from over 9.5 million to over 11.5 million in 1884 and to over 13 million in 1895. The Sorbonne faculty of sciences alone had slightly over 477,561 francs to spend in 1902.
56 See *Revue de l'Institut catholique* (Paris), 1905, 369–71, and 1906, 385–93; Raoul Allier, "Instituts," *Le Siècle* (July 23, 1906); and Auguste Cholat, "Du rôle actuel de l'enseignement supérieur libre en France," *Demain* (June 15 and 22, 1906). The issues had been raised in the *Observateur français* (Oct. 15, 1890 and Sept. 5, 1891) and in *L'Univers* as early as 1890. Some of this information is collected in AN, F¹⁹ 4091.

57 Cardinal Merry del Val to Cardinal Richard, Oct. 10, 1907, Archives, Archevêché, 5 J, 1; Archives, Faculté libre des sciences de Lille, *Procès-verbaux*, Feb. 12, 1925.
58 Archives, Archevêché, 5 J, 3.
59 See, e.g., Felix Ponteil, *Histoire de l'enseignement en France: Les Grandes étapes, 1789–1964* (Paris, 1966); Antoine Prost, *Histoire de l'enseignement en France, 1800–1967* (Paris, 1968); and Theodore Zeldin, "Higher Education in France, 1848–1940," *Journal of Contemporary History*, 2 (1967), 53–80.
60 AN, F¹⁷ 13663. *Conseil supérieur de l'instruction publique*, session of Dec. 27, 1907; le Docteur P. David, *Cahiers Féron-Vrau*, no. 1 (1954–5), pp. 33–46.
61 Copy of letter by Le Verrier, Aug. 28, 1875, in *Pièces annexes au procès-verbaux de la Faculté des sciences (Sorbonne)*, vol. 2, p. 102 former Archives de l'Académie de Paris, now in AN.
62 Georges Bertrin, "Pourquoi les instituts catholiques doivent vivre," *Revue du clergé français*, 9 (1896), 133–54.
63 Lavisse, *RIE*, 32 (1886), 493.

CHAPTER 7. SCIENTIFIC PUBLICATION

1 AN, F¹⁷ 3244.
2 *RQS*, 28 (1890), 290.
3 *RQS*, 31 (1892), 218–19.
4 *RQS*, 28 (1890), 290.
5 *RQS*, 37 (1895), 637–8.
6 Pierre Duhem, *The Aim and Structure of Physical Theory*, trans. Phillip Wiener (New York, 1962), p. 70.
7 *RQS*, 50 (1901), 50.
8 *RQS*, 17 (1885), 232.
9 *RQS*, 26 (1889), 26, 569–79.
10 C. C. Gillispie, *The Edge of Objectivity: An Essay in the History of Scientific Ideas* (Princeton, N.J., 1960), pp. 476, 493–4.
11 J. Bertrand, *Leçons sur la théorie mathématique de l'électricité* (Paris, 1889); *RQS*, 27 (1890), 58.
12 *RQS*, 31 (1892), 31, 245.
13 J. W. Herivel, "Aspects of French Theoretical Physics in the Nineteenth Century," *The British Journal for the History of Science*, 3 (1966), 109–10; T. S. Kuhn, *The Essential Tension: Selected Studies in Scientific Tradition and Change* (Chicago, 1977), p. 63.
14 *RQS*, 16 (1884), 219.
15 *RQS*, 23 (1888), 570–3.
16 *RQS*, 23 (1888), 570–3.
17 *RQS*, 31 (1892), 603–6.
18 *RQS*, 46 (1889), 46, 261.

19 *RQS*, 39 (1896), 231–43. E. G. Monod, *Stéréochimie: Exposé des théories de Le Bel et van 't Hoff, complétées par les travaux de MM. Fischer, Baeyer, Guye et Friedel* (Paris, 1895). See also Mary Jo Nye, "Berthelot's Anti-Atomism: A 'Matter of Taste'?," *Annals of Science*, 38 (1981), 585–90, and Terry Shinn, "Orthodoxy and Innovation in Science: The Atomist Controversy in French Cehmistry," *Minerva*, 18 (1980), 539–55.

20 *RQS*, 36 (1894), 618–19. In 1895 Baillère published a *Traité élémentaire de physique biologique* by Armand Imbert for physicians, physiologists, and naturalists.

21 *RQS*, 33 (1893), 257–65.

22 Pierre Duhem, "Thermochimie à propos d'un livre récent de M. Marcelin Berthelot," *RQS*, 42 (1897), 361–92, summarized in H. W. Paul, "The Role and Reception of the Monograph in Nineteenth-Century French Science," in A. J. Meadows, ed., *Development of Science Publishing in Europe* (Amsterdam, 1980), pp. 123–48. For a friendly review of the 1879 edition, see Edme Bourgoin, RS, 2e série, 17 (1879), 369–71; he thought Berthelot had provided in chemistry the model of modern physics for chemists to emulate. In 1902 a friendly review of Berthelot's *Les Carbures d'hydrogène, recherches expérimentales* (3 vols., 1901) appeared in the *RQS*, 51 (1902), 276–80, by H.D.G., S.J., who saluted Berthelot as the grand old man of chemistry.

23 Jacques Ellul, *Propaganda: The Formation of Men's Attitudes* (New York, 1973), pp. 39–41.

24 Robert Fox, "The *Savant* Confronts His Peers: Scientific Societies in France, 1815–1914," in Robert Fox and George Weisz, eds., *The Organization of Science and Technology in France, 1808–1914* (Cambridge, 1980), pp. 241–82.

25 *La Grande encyclopédie* (Paris, 1886–1902), vol. 30, pp. 147–8.

26 AN F[17] 13271

27 Art. 2: "Son objet est de concourir à l'avancement de la géologie en général, et particulièrement de faire connaître le sol de la France, tant en lui-même que dans ses rapports avec les arts industriels et l'agriculture." *Bulletin de la Société géologique de France*, (1830); séance du 17 mars 1830.

28 Among the French members of the society were Arago, Observatory, permanent secretary of the Academy of Sciences; Simon Bérard, director of Ponts et chaussées and a deputy; Caumont, secretary of the Société linnéenne; Delafosse, aide-naturaliste au Jardin-du-Roi; le baron de Férussac, editor of the *Bulletin universelle des sciences*; Marcel de Serres, faculty of sciences, Montpellier; le comte de Montlosier, president of the Académie royale des sciences et statistique universelle; Antoine Passy, préfet de l'Eure; Constant Prévost, professor of geology at the faculty of sciences, Paris; Ravergie, naturaliste-voyageur du Muséum; Théodore Virlet, member of the Commission scientifique de la Morée; Zuber-Karth, president de la Société industrielle de Mulhouse.

29 M. Godet, "Quelques observations sur la manière de travailler en histoire

naturelle, et en particulier sur les monographies," *Annales de la Société entomologique de France*, 1 (1832), 34–52.

30 *Annales scientifiques de l'École normale supérieure*, 1 (1864), and Craig Zwerling, "The Emergence of the Ecole normale supérieure as a Centre of Scientific Education in the Nineteenth Century," in Robert Fox and George Weisz, eds., *The Organization of Science and Technology in France, 1808–1914* (Cambridge, 1980), pp. 49–50. Pasteur secured a subsidy for the *Annales* from the Ministry of Education.

31 *Bulletin de la Société chimique de Paris*, 1 (1857); "Liste alphabétique des membres (1889)...," *Bulletin...*, 3e série, 1890, pp. 2–23.

32 Fox, "The *Savant* Confronts His Peers," p. 269.

33 *Journal de physique théorique et appliquée*, 1 (1873). The society published a *Bulletin des séances*. In 1907 secondary school teachers of the Association des professeurs de sciences physiques, chimiques et naturelles, des lycées et collèges de France (garçons et filles) started a *Bulletin de l'union des physiciens*. It was especially concerned with the problems arising from the reform of 1902, which made laboratory work compulsory.

34 By 1910 the Société française de physique had 1,558 members, compared with 1,124 for the Société chimique, but the chemical society had 61,822 francs of income, compared with 23,203 for physics. Fox, "The *Savant* Confronts His Peers," p. 275, table 2. See *Le Livre du cinquantenaire de la Société française de physique* (Paris, 1925).

35 Fox, "The *Savant* Confronts His Peers," p. 257. For some modern nostalgia concerning the lack of interest in understanding nature that is characteristic of modern biology, see Norman Macbeth, *Darwin Retried* (New York, 1971). The author, a good stylist, is a lawyer who would be quite at home in a provincial *société savante*.

36 *Bulletin de la Société d'histoire naturelle de Toulouse* (1966), tome 101, volume du centenaire. Table centennale des matières. Historique, catalogue de notices pour l'histoire des naturalistes.

37 Société des sciences physiques et naturelles de Bordeaux, "Historique de la Société...," III, 2, xxiib, "Table générale des matières des publications de la Société de 1850 à 1900." Another important publication series in Bordeaux was the *Annales des sciences naturelles de Bordeaux et du Sud-ouest* (physique du globe, géologie, botanique, zoologie, anthropologie). The first memoir (1883) was by Gaston Lespiault (professor at the faculty of sciences), "Des déboisements dans l'Amérique du Nord et de leur influence météorologique." Number two was by a professor of experimental medicine in the faculty of medicine, F. Jolyet, "Recherches sur la torpille électrique," a piece of work done in the laboratory of the Société scientifique in Arcachon. From 1882 to 1885 twelve memoirs were published.

38 *Bulletin de la Société d'histoire naturelle de Colmar* (1865–6), 6e et 7e années, 343–83, "rapport général" by M. le Dr Faudel. By the year 1868 income was up to 5,629 francs, of which 3,330 francs came from dues. From 1861 to

1866 the departmental subsidy was 4,500 francs, the ministerial subvention 1,350 francs, and that of the city of Colmar over 4,150 francs.

39 After 1871 the old Société des sciences naturelles de Strasbourg (1828) transferred with three-quarters of its members to Nancy; its minutes appeared in the *Revue médicale de l'Est*. It had only sixty-two members in 1899, but it included Brillouin, Poincaré, and Willm among corresponding members. The name changed as well. See *Bulletin de la Société des sciences de Nancy*, series II, 1 (1873) for details on the transfer. The Ministry of Education, like some scientists, felt morally obligated to support the society of *Francia irredentia*.

40 Fox, "The *Savant* Confronts His Peers," pp. 249, 252, limits the importance of the society. Cf. Pasteur, "Dans l'une de mes dernières communications à la Société . . ." *Mémoires de la Société impériale des sciences, de l'agriculture et des arts de Lille*, 2nd series, 5 (1858), 13–26.

41 *Bulletin de la Société des sciences historiques et naturelles de l'Yonne*, tables analytiques, 2nd series, 1867–78.

42 Boreau (1803–75) was a pharmacist who became a director of the Jardin des plantes d'Angers and professor of botany at the Ecole supérieure. His *Flore du centre de la France* (1840) was in its third edition by 1857. *Bulletin de la Société d'études scientifiques*, 1re année, 1871.

43 AN, F^{17} 13271.

44 *Revue des sciences naturelles*, 1, no. 1 (1872). Yvette Conry, *L'Introduction du Darwinisme en France au XIXe siècle* (Paris, 1974), pp. 229–55, "Lille et Montpellier."

45 *Revue des sciences naturelles de l'Ouest (et de leurs applications à l'agriculture, la pisciculture, l'ostréiculture et aux pêches maritimes)*, 1 (1891).

46 *Bulletin de la Société zoologique d'acclimatation*, 1 (1854); *Revue des sciences naturelles appliquées* (Bulletin bimensuel de la Société nationale d'acclimatation de France), 4e série, 6 (1889), 36e année.

47 Fox, "The *Savant* Confronts His Peers," p. 275. See *Bulletin de la Société botanique de France*, 1854–1910.

48 See Conry, *L'Introduction du Darwinisme*, pp. 305–17.

49 Memoirs published in the *Annales du Musée d'histoire naturelle de Marseille* during 1882–3 were: A.-F. Marion, *Esquisse d'une topographie zoologique du golfe de Marseille*; Louis Roule, *Recherches sur les ascidies simples des côtes de Provence (Phallusiadées)*; A.-F. Marion, *Considérations sur les faunes profondes de la Méditerranée d'après les dragages opérés au larges des côtes méridionales de France*; René Koehler (préparateur de zoologie à la Faculté des sciences de Nancy), *Recherches sur les échinides des côtes de Provence*; MM. les professeurs A. Kowalevsky et A.-F. Marion, *Documents pour l'histoire embryogénique*; M. A. Kowalevsky, *Embryogénie du chiton polii (philippi)*; M. A. Kowalevsky, *Etude sur l'embryogénie du dentale* (dedicated to Lacaze-Duthiers).

50 The *Annales*, 1 (1891) carried the following articles: J. Macé de Lépinay, "Sur la double réfraction du quartz," P. Appell, V. Jamet, Ch. Fabry, "Théorie de la visibilité et de l'orientation des franges d'interférence," and A. Perot, "Etude de l'oscillateur Blondlot."

51 "Avertissement," *Annales de l'Université de Grenoble*, 1 (1889).
52 Robert Fox, "La Société zoologique de France: Ses origines et ses premières années," *Bulletin de la société zoologique de France*, 101 (1976), 799–812. Volume 1 (1876) stated that the aim of the society was to encourage zoological studies and contribute to the progress of this science. Edmond Perrier's *Traité de zoologie* (10 fascicules in 4 volumes) started appearing in 1893, finishing up in 1932, with an index in 1933. Advertising the avoidance of theories, to let the facts speak, the work was hailed for its style and for providing an alternative to translations of foreign works.
53 Yves Delage, *La Structure du protoplasma et les théories sur l'hérédité et les grands problèmes de la biologie générale* (Paris, 1895). *L'Année biologique* ("comptes rendus annuels des travaux de biologie générale"), edited by Delage, also began in 1895.
54 Derek de Solla Price, *Science since Babylon* (New Haven, Conn., 1973), ch. 8, "Diseases of Science," and *Little Science, Big Science* (New York, 1963).
55 Zwerling, "The Emergence of the Ecole normale supérieure," pp. 38–41.
56 George Weisz, *The Emergence of Modern Universities in France, 1863–1914* (Princeton, N.J., 1983), pp. 203–5.
57 Camille Limoges, "The Development of the Muséum d'histoire naturelle of Paris, c. 1800–1914," in Fox and Weisz, eds., *The Organization of Science and Technology*, pp. 219–21.
58 Weisz, *The Emergence of Modern Universities*, p. 205.
59 Terry Shinn, "The French Science Faculty System, 1808–1914: Institutional Change and Research Potential in Mathematics and the Physical Sciences," *Historical Studies in the Physical Sciences*, 10 (1979), pp. 302, 315, 328–9.
60 Weisz, *The Emergence of Modern Universities*, pp. 205–7.

CHAPTER 8. SCIENCE FUNDING IN THE TWENTIETH CENTURY

1 For the dark side, see Albert Ranc, *Le Budget du personnel des recherches scientifiques en France* (Paris, 1926), one of "les enquêtes de la société de chimie industrielle," with a preface by G. Urbain. Counterbalancing historical views are in Terry Shinn, "The French Science Faculty System, 1808–1914: Institutional Change and Research Potential in Mathematics and the Physical Sciences," *Historical Studies in the Physical Sciences*, 10 (1979); Elisabeth Crawford, "The Prize System of the Academy of Sciences, 1850–1914," in Robert Fox and George Weisz, eds., *The Organization of Science and Technology in France 1808–1914* (Cambridge, 1980); and Paul Forman, John L. Heilbron, and Spencer Weart, "Physics *circa* 1900: Personnel, Funding, and Productivity of the Academic Establishments," *Historical Studies in the Physical Sciences*, 5 (1975), 1–185.
2 Crawford, "The Prize System of the Academy of Sciences." Archival material on legacies and gifts to science is in AN, F^{17} 13004-5-6-7-8 and 13017-18-19-20.

3   Charles Richet, "Les Études sur la fermentation au laboratoire de Carlsberg," *RS*, 25 (1880), 851–5. Richet marvels at Jacobsen's gift of the equivalent of nearly 1.5 million francs with an annual income of 28,000 francs up to his and his wife's deaths. B.M., "La Fondation Carlsberg," *RGSPA*, 41 (1930), 230, and Léon Bertin, "Le Marinbiologisk Laboratorium de Copenhague," *RGSPA*, 47 (1936), 623–30. When the oceanographer Johannes Schmidt died in 1933, oceanography got its own laboratory while genetics kept the quarters both had been occupying.

4   E. Duclaux, "Les Instituts bactériologiques en France et à l'étranger," *RS*, 48 (1891), 481–3.

5   An article favorable to the Caisse, "Pour la science," in *Eclair* (Oct. 15, 1900), also attacked prizes – "une mauvaise méthode."

6   AN, F^{17} 13004. Becquerel left 100,000 francs to the academy (1910).

7   Crawford, "The Prize System of the Academy of Sciences," pp. 292, 294, and 297 ff.

8   Ibid., p. 305

9   Ranc, *Le Budget du personnel*, ch. 2.

10 AN, F^{17} 17431, 17462, 17486, 17487, 17491; also *Rapports scientifiques sur les travaux entrepris . . . au moyen des subventions de la caisse des recherches scientifiques* (Melun, 1905, etc.)

11 Frédéric Blancpain, "La Création du CNRS: Histoire d'une décision 1901–39," *Bulletin de l'institut international d'administration publique*, no. 32 (Oct.–Dec. 1974), p. 99 (757).

12 "Revue des sciences," *Journal des débats politiques et littéraires*, Feb. 8, 1912. Audiffred himself collected 60,000 francs for the Caisse in 1901. Gifts included 4,000 francs from the Crédit foncier in 1904 and 5,000 francs from the municipal council of Paris in 1907.

13 AN, F^{17} 17437-38-39 and 17450.

14 Charles Richet, "La Physiologie et la médecine," *RS*, 42 (1888), 810; see *RS*, 58e année (1920), 225 ff. But in 1913 Paris had one of the highest cancer rates of industrial cities, followed by Berlin and Boston.

15 AN, F^{17} 17434.

16 "Publications savantes" accounted for 993,650 francs of the credits for 1930. Blancpain, "La Création du CNRS," p. 139 (797).

17 Blancpain, "La Création du CNRS," p. 98 (756).

18 Georges Lemoine, "Le Développement de la recherche scientifique," *RS*, 60e année (1922), pp. 106–7.

19 Report on the Congrès de Lille of 1874 in *RS*, 14 (1874), 141, 169, 194.

20 This is argued by Joseph de Joannis in "Berthelot chimiste," *Etudes*, 3 (1907), 317–41, 454–66. See Mary Jo Nye, "Berthelot's Anti-Atomism: A 'Matter of Taste'?," *Annals of Science*, 38 (1981), 585–90.

21 *Etude de quelques laboratoires industriels et des écoles supérieures techniques en Allemagne: Rapport présenté à M. le Ministre du commerce par M. Albert Granger, docteur ès sciences* (Paris, 1901).

22 *Bulletin de l'enseignement technique*, 4e année, no. 1 (Jan. 5, 1901), report on speech by Léon Bourgeois to the Chambre des députés, Dec. 24, 1900, on the Conservatoire.

23 See "Les laboratoires nationaux de recherches scientifiques," Compte rendu des séances de l'Académie des sciences, 163 (Nov. 13, 1916), 581, report of the Commission d'action extérieure de l'académie (Jordan, d'Arsonval, Lippmann, Picard, Haller, Lacroix, Tisserand, Le Chatelier). A report by Le Chatelier on the creation of the new section of the academy, "Applications de la science à l'industrie," is in *La Technique moderne*, 10 (1918). The journal *Science et industrie* began in April 1917 as an organ of industry covering metallurgy, electricity, automobiles, and aviation, with the aim of promoting research sections in French industry.

24 On the school, see Terry Shinn, "Des sciences industrielles aux sciences fondamentales: La Mutation de l'Ecole supérieure de physique et de chimie (1882–1970)," *Revue française de sociologie*, 22 (1981), 167–82, and "Orthodoxy and Innovation in Science: The Atomist Controversy in French Chemistry," *Minerva*, 18 (1980), 539–55.

25 E. Tassilly, "Charles Lauth" (1836–1914), *RGSPA*, 25 (1914), 217–18.

26 The first issue of the *Revue générale des matières colorantes et des industries qui s'y rattachent*, 1 (1897), 1–2, has a "Présentation" by Schützenberger dealing with the aim of the journal: an intimate collaboration between theory and practice directed toward industrial progress. This would enable the French to catch up on developments in organic chemistry and make research science an integral part of the French chemical industry. According to *Science et industrie*, no. 1, April 1917, 87 percent of dyestuffs used in France were bought from German companies.

27 The advertisement pages for the Syndicat des fabricants de jumelles et parties connexes in the *Revue d'optique*, 2 (1923), listed thirty-six manufacturers.

28 Charles Fabry (1867–1945), "L'Institut d'optique théorique et appliquée," *RIE*, 82 (1928), 257–65. Fabry, professor at Marseille (1894–1920), then the Sorbonne (1920–37) and also the Ecole polytechnique after 1928, the Directeur général de l'Institut d'optique. The Conseil de l'Institut d'optique included the mathematician Emile Picard. The editorial committee of the *Revue d'optique* in 1924 was listed as Fabry, Maurice de Broglie, and Arnaud de Gramont de Guiche, but Gramont had died in 1923. Gramont's procedures of spectroscopic analysis were adopted by the U.S. Bureau of Standards. For the comment on German optics and some details of Prussian subsidies to the Jena laboratories, see Frank Pfetsch, "Scientific Organisation and Science Policy in Imperial Germany, 1871–1914: The Foundation of the Imperial Institute of Physics and Technology," *Minerva*, 8 (1970), 557–80.

29 L. Lossier, "La Fabrication des montres et l'enseignement de l'horlogerie à Besançon," *RS*, 47 (1891), 196–203.

30 M. J. Roux (Directeur de l'Ecole nationale professionnelle), *Les Laboratoires industriels d'essais en Allemagne* (Paris, 1913). Prefaced by d'Arsonval, this work

is a report done for the chamber of commerce of Limoges. See also Georges Claude, "La Recherche scientifique, ses applications à l'industrie et la synthèse industrielle de l'ammoniaque," *RGSPA*, 32 (1921), 534–43, 570–81.

31 *Travaux du laboratoire central d'électricité*, 1 (1884–1905), 2 (1904–11), published in 1910 and 1912, respectively, for the Société internationale des électriciens by P. Janet, director of the laboratory and the school, with a preface by E. Bouty, president of the Commission administrative du laboratoire.

32 S. A. R. Monseigneur le comte de Paris, "premier pilote de France," also wrote an article in the *Revue universelle* (June 15, 1933), extracted in the *Almanach*, pp. 40–43, deploring the state of French aviation, especially in the military. Being rightist did not necessarily mean being wrong.

33 See L. Marchis, *L'Epopée aérienne*, (Paris, 1910), which is full of photographs and brief biographies. More serious fare is *L'Aviation*, by Paul Painlevé, Emile Borel, and Charles Maurain (Paris, 1923). See also Emile Berrubé, *Flottes aériennes en France et en Allemagne* (Paris and Nancy, 1910), which deals with German superiority in numbers of dirigibles and balloons. For a list of journals and works, see Georges de Bothézat, "Etude de la stabilité de l'aéroplane" (Paris, 1911), a doctoral thesis with a preface by Painlevé. *La Technique aéronautique* (*Revue internationale des sciences appliquées à la locomotion aérienne*) began in 1910. The editor was Lt. Col. G. Espitallier; the editorial committee included Appell, A. Blondel, Lecornu, Marchis, Painlevé, H. Poincaré, and Witz.

34 The Ecole supérieure d'aéronautique et de construction mécanique gave a boost to the ailing refrigeration industry because its graduates could easily specialize in a similar technology. See *La Revue générale du froid*, 1 (1909), 104–6, the official organ of the Association française du froid, founded in 1909.

35 The Ecole nationale supérieure de l'aéronautique became the Ecole nationale supérieure de l'aéronautique et de l'espace in 1968 and moved to Toulouse in 1972. The Ecole nationale d'ingénieurs de constructions (Ministère des armées), created in 1946, had moved to Toulouse in 1963; it trains engineers for the aeronautics industry as well as military engineers for the Ministry of Air.

36 A. Caquot (directeur général technique du Ministère de l'air), "L'Orientation actuelle des industries aéronautiques," *RS*, 68e année (1930), 198–204.

37 A. Toussaint (Ingénieur de l'Institut aérotechnique de Saint-Cyr), "Les Laboratoires aérodynamiques," *L'Air*, no. 1 (1919), 17–19.

38 A. Lepresle (directeur du Laboratoire Eiffel), "Résultats des travaux effectués au Laboratoire aérodynamique de Göttingen," *Revue générale de l'aéronautique*, no. 1 (1922), 110–51.

39 A. Caquot (directeur général technique du Ministère de l'air), "L'Orientation actuelle des industries aéronautiques," *RS*, 68e année (1930), pp. 198–204.

40 Patrick Fridenson et Jean Lecuir, *La France et la Grande-Bretagne face aux problèmes aériens, 1935–mai 1940* (Paris, 1976). Further information on aviation in the 1930s may be found in a series of articles by Gustave Delage, "Les Forces aériennes dans le monde en 1936," *RS*, 74e année (1936); two books by Pierre Barjot: *L'Aviation militaire française* (3rd ed., Paris, 1937), and

Les Forces aériennes mondiales (Paris, 1938); Jean Romeyer, *L'Aviation civile française* (1936); and E. Vellay, "L'Expérimentation technique en vol des avions," *RS*, 76e année (1938), 474–85. R. J. Overy, "German Air Strength 1933 to 1939," *The Historical Journal*, 27 (1984), 465–71, points out that "until 1938 Germany's air power was a paper tiger," while the Luftwaffe waited for a new generation of aircraft and the training of enough crews.

41 Saint-Exupéry, *Pilote de guerre* (Paris, 1942), p. 83.

42 AN, F[17] 17481.

43 Maurice Crosland, "Marcelin Berthelot," *DSB*, II, and "Science and the Franco-Prussian War," *Social Studies of Science*, 6 (1976), 185–214.

44 Ministère de la guerre, Archives de l'artillerie, 4.d.2.

45 Weiss, advised by Painlevé, went back to his professorship at the Polytechnique in Zurich in 1916 to maintain the French presence.

46 Charles Moureu, *La Chimie et la guerre: Science et avenir* (Paris, 1920).

47 A. Béhal, A. Haller, Charles Moureu, "La Crise de notre industrie chimique organique et la défense nationale," *RS*, 60e année (1922), p. 683. By the time of the armistice the factory in the Fort de la Double-Couronne (Saint-Denis) employed 200 people to turn out daily sixteen tons of *pâtes fumigènes*.

48 Comité et section technique du Génie, *Archives, Génie: Gaz*. Long study on Z companies by Commandant Winkler.

49 See Moureu, *La Chimie et la guerre*, passim.

50 Blancpain, "La Création du CNRS," p. 101: "la guerre ... provoque la coordination des services, la promotion des techniciens et le renforcement du dirigisme."

51 Ibid., p. 102.

52 A list of the chief propositions is given in the report by Louis Marin for the budget commission in 1918, reprinted in the long article on the Office national des inventions in *Recherches et inventions*, no. 276 (July–Oct. 1938), pp. 103–202. Historians of technology will be intrigued by references to instruments for altitude measurement, variable surface airplanes, secret infrared communication, aerial photography, etc. See also R. Legendre, "L'Office national des recherches scientifiques et industrielles et des inventions," *RS*, 61e année (1923), 165–72.

53 Henry Le Chatelier, "Les Laboratoires de recherche scientifique dans l'industrie," *Revue universelle des Mines*, 6th series, V, no. 5 (1920).

54 The members of the subcommittee were Painlevé, (president) Charles Dumont, Pottevin, (vice presidents) Appell, Emile Borel, Hippolyte Bouchayer, de Courville, Guillet, Lacroix, Maurin, colonel Mercier, Lucien Poincaré, Renault, and Sagourin.

55 These two aims were clearly spelled out in the law: "Il est créé un Office national ... Cet Office a pour objet:

"1. De provoquer, de coordonner et d'encourager les recherches scientifiques de tout ordre qui se poursuivent dans les établissements scientifiques ou que peuvent entreprendre des savants en dehors des ces organisations;

"2. De développer et de coordonner spécialement les recherches scientifiques appliquées au progrès de l'agriculture et de l'industrie nationales, ainsi que d'assurer les études demandées par les services publics et d'aider les inventeurs." *Recherches et inventions*, p. 112. See also Daniel Berthelot, "Les Carburants nationaux et leur étude scientifique." *RS*, 60e année (1922), 789–96.

56 Blancpain, "La Création du CNRS," Tableau II, p. 140.

57 Aimé Cotton, "La Production des champs magnétiques et le projet du gros électro-aimant de l'université de Paris," *RGSPA*, 25 (1914), 622, 639, 665–9. Initial figures for the magnet were 43,500 gauss or, modified, 46,400 gauss. See also Cotton, "Pourquoi faut-il construire un très gros électro-aimant?", *RS* 52e année (1914), 513–19.

58 "Pour les laboratoires de France [la journée de Pasteur]," *RIE*, 76 (1922), 368–9.

59 Maurice Hamy, note on Lorentz, who was an "associé étranger" of the Association scientifique after 1910, *RGSPA*, 39 (1928), 97–98.

60 J. van den Handel, "Heike Kamerlingh Onnes," *DSB*, VII, 220–2.

61 Spencer R. Weart, *Scientists in Power* (Cambridge, Mass., 1979), pp. 117–18.

62 Blancpain, "La Création du CNRS," p. 764.

63 *Centre national de la recherche scientifique: Groupe des laboratoires de Bellevue* (Paris, 1958).

64 Blancpain, "La Création du CNRS," pp. 107–10.

CHAPTER 9. THE DENOUEMENT OF THE 1930S

1 Charles Moureu, presidential speech given in 1922 to a Confédération banquet for interested members of parliament. *RS*, 60e année (1922), pp. 274–5.

2 Spencer R. Weart, *Scientists in Power* (Cambridge, Mass., 1979), p. 14.

3 W. Kilian, "Sur le recrutement du personnel des laboratoires scientifiques," *RS*, 60e année (1922), pp. 717–19. This article points out the mediocrity of many people working in laboratories. See also Charles Richet *fils*, "La Situation matérielle des savants," *RS*, 63e année (1925), pp. 97–100, for the same dreary theme: "C'est la misère," but Richet also makes the unreasonable recommendation that the amount of prize money for scientists be increased, giving scientists the social status of cyclists and jockeys. Barrès's superb propaganda is collected in *Pour la haute intelligence française* (Paris, 1925), with a preface by Moreu.

4 Albert Ranc, "La Cité des sciences," *RS*, 73e année (1935), pp. 6–10.

5 *Jubilé scientifique de M. Emile Borel* (Paris, 1940); *Annales de l'Institut Henri Poincaré* (*Recueil de conférences et mémoires de calcul des probabilités et physique théorique*), 1 (1930); 3 (1932); Edward M. MacKinnon, *Scientific Explanation and Atomic Physics* (Chicago, 1982), p. 191.

6 For MacKinnon's critical view of the influence of Louis de Broglie on French physics see his article "De Broglie's Thesis: A Critical Retrospective," *American Journal of Physics*, 44 (1976), 1047–55, a view echoed by Pierre Papon, *Pour*

une prospective de la science. Recherche et technologie: les enjeux de l'avenir (Paris, 1983), p. 369, note 12. But Bruce Wheaton, *The Tiger and the Shark: Empirical roots of Wave-Particle Dualism* (Cambridge, 1983), p. 293, n. 111, finds MacKinnon "ahistorical and physically misleading." Wheaton's book definitively shows the fertile role of the French, especially Maurice and Louis de Broglie, in twentieth-century physics. Like Spencer Weart's *Scientists in Power* and Dominique Pestre's *Physique et physiciens en France, 1918–1940* (Paris, 1984), it is a key book on modern physics in France.

7   Albert Ranc, "L'Organisation d'un service national de la recherche scientifique en France," *RS*, 68e année (1930), pp. 688–93.

8   J. Delsarte, "De l'enseignement supérieur en France et spécialement des facultés des sciences," *RS*, 77e année, (1939), pp. 140–3; "Sur un projet de réforme de l'enseignement supérieur," ibid., pp. 299–303; and R. Anthony, "A propos de l'état actuel de l'enseignement supérieur en France," ibid., pp. 455–60.

9   J. Delsarte, "De l'organisation de la recherche scientifique," *RS*, 77e année (1939), pp. 1–3.

10   Emmanuel Dubois, "Sur l'administration de la recherche scientifique en France," *RS*, 77e année (1939), pp. 68–9.

BIBLIOGRAPHY

Several recent publications contain excellent bibliographies. George Weisz, *The Emergence of Modern Universities in France, 1863–1914* (Princeton, N.J., 1983), has a basic bibliographical note supplemented by a source index. An excellent bibliography is provided by Barnett Singer in his fine synthesis of scholarship on "The Ascendancy of the Sorbonne: The Relations between Centre and Periphery in the Academic Order of the Third Republic," *Minerva*, 20 (1982), 269–300. For the organizational aspects of science and technology, see Robert Fox and George Weisz, editors, *The Organization of Science and Technology in France 1808–1914* (Cambridge, 1980). The annotated bibliography in John H. Weiss, *The Making of Technological Man: The Social Origins of French Engineering Education* (Cambridge, Mass., 1982), is very useful on science as well as technology; see also Terry Shinn, *Savoir scientifique et pouvoir social: L'Ecole polytechnique, 1794–1914* (Paris, 1980). Some guidance to the hard sciences may be found in Mary Jo Nye, *Molecular Reality: A Perspective on the Scientific Work of Jean Perrin* (London, 1972), and Bruce R. Wheaton, *The Tiger and the Shark: Empirical Roots of Wave-Particle Dualism* (Cambridge, 1983). For biology, see Yvette Conry, *L'Introduction du Darwinisme en France au XIXe siècle* (Paris, 1974); Robert E. Stebbins, "France," in Thomas F. Glick, ed., *The Comparative Reception of Darwinism* (Austin, Tex., 1974), pp. 117–63; and Linda L. Clark, *Social Darwinism in France* (Tuscaloosa, Ala., 1984). Some guidance to the literature on science, religion, and ideology may be found in Harry W. Paul, *The Edge of Contingency: French Catholic Reaction to Scientific Change from Darwin to Duhem* (Gainesville, Fla., 1979), in Clark, *Social Darwinism*, and in an issue of the *Revue de Synthèse*, III s., nos. 106–8 (1982), devoted to Littré.

The following works appeared too late for me to profit from them: Gérard Brun, *Techniciens et technocratie en France, 1918–1945* (Paris, 1985); Georges Canguilhem et al., *Du développement à l'évolution au XIXᵉ siècle* (Paris, 1985); Elisabeth Crawford, *The Beginnings of the Nobel Institution: The Science Prizes, 1901–1915* (Cambridge, 1984); Stanley L. Jaki, *Uneasy Genius: The Life and Words of Pierre Duhem* (The Hague, 1984). F. W. J. McCosh, *Boussingault: Chemist and Agriculturist* (Dordrecht, 1984); and Dominique Pestre, *Physique et physiciens en France, 1918–1940* (Paris, 1984).

INDEX

geology, 19, 28, 29, 33, 34, 51, 143, 166, 182, 183, 229, 231, 236, 246, 258, 268, 269, 273, 277, 297, 301; paleontological, 30; publications, 266, 267; stratigraphic, 29
Gerhardt, Charles, 33, 140, 310
Germany: agriculture, 188, 200, 207–9, 219; Catholics, 246; education, 24, 55, 157, 158, 168, 178, 207, 308–9; industry, 16, 151, 165, 168, 169, 176–8, 200, 208, 209–10 (sugar), 212, 218 (wine), 308, 309, 311, 321 (chemical), 324, 325, 346; science, 55, 284, 307, 310, 347; surplus laboratory equipment, 335–6; universities, 43, 44, 46, 56, 228, 250
Giard, Alfred, 71, 97, 101–2, 105, 108, 109, 112–14, 116–33, 206, 266, 278, 279, 283, 290; career at Lille, 118–19; quarrel with Delage, 119–29
Gilbert, Philippe, 256–8
Gillispie, Charles C., 3
Girard, Charles, 215–17, 219, 310
Girardin, Jean, 20, 142, 185
Gley, Eugène, 63, 65, 79, 298
Gohau, Gabriel, 88, 130
Gossart, E., 199, 201
Gosselet, Jules, 138, 143, 172, 173, 176
Goy, Joseph, 176
Gramont, Armont de, duc de Guiche, 314, 317
Grandeau, Louis, 9, 181, 183, 187–97, 211
La Grande encyclopédie, 252
grandes écoles (see also under individual écoles), 8, 16, 17, 36, 58–9, 138, 150, 153, 158, 160, 162, 172–6, 184, 204, 250, 261, 338
grants (see also funding; funds) 289–91, 293, 295–304, 352
Gratiolet, Pierre, 24, 277
Grignard, Victor, 170, 301, 321
Grimaux, Edouard, 310
Grmek, M. D., 61, 79
Guinchant, J., 166, 301
Guizot, François, 21, 43

Hadamard, Jacques, 263, 275, 320, 336
Haeckel, Ernst, 113, 119, 266

Haller, Albin, 34, 49, 134, 154, 163, 170, 309, 324, 325
Hallez, Paul, 113, 114, 116, 119
Hamonet, abbé, 229, 230, 250
Harmel, Léon, 239, 240, 243
Heckel, Edouard, 278, 282
Henneguy, L., F., 71, 127
heredity, 124, 130, 133
Hermite, Charles, 32, 49, 262, 270, 283
Herriot, Edouard, 343, 346, 348
Hertz, Heinrich, 53, 257
histology, 29, 77, 93, 101, 112, 123, 124, 127, 284
Hofmann, A. W. von, 212–13
Holmes, Frederick, 79
Houel, Charles, 61, 62
Hoüel, J., 261, 262, 275
Hulst, Msgr. d', 223, 239, 245, 250
hygiene, 65, 66, 183, 212, 213, 237, 242, 306

industry, 11, 13, 15, 16, 58, 66, 134–6, 138–40, 142, 143, 145, 148–56, 163–73, 175–7, 186, 187, 191, 194, 196, 199, 203–5, 207, 209–10, 213, 235–6, 241–3, 249, 268, 271, 272, 275, 307–11, 313, 316, 322–4, 326, 329, 333, 335, 338, 341, 345, 348, 349; agricultural, 137, 148; aviation, 315–17, 319; beer, 142, 152–4, 163, 172, 209, 290 (see also brewing); chemical, 8, 15, 140, 148, 150, 153, 164, 165, 167, 169, 175, 177, 199, 200, 203, 213, 321 (see also chemistry: industrial); dyeing and textile, 142, 143, 148, 150, 154, 166, 167–9, 236, 309; electrical, 8, 15, 157, 158, 165, 175, 313; expansion of, 143, 153, 158, 187, 244; fish, 106–7; hydroelectric, 140, 168, 175; machines for, 143, 156; mechanical, 200; metallurgical, 8, 15, 34, 141, 148, 153, 154, 168, 169, 175, 177; mining, 34, 177 (see also mining); optical, 311; paper, 166, 200; physics of, 142; precision tool, 319; research for, 2, 141, 170, 177, 178, 325, 327, 339; science of, 140, 142, 171; societies for, 135; steel, 135, 140, 153, 175, 319; sugar, 32, 51, 137, 208, 210, 211, 215; technology of, 153, 175, 269

Printed in the United States
By Bookmasters